Numerical Analysis in Pascal ABC
Studies in Applied Mathematics
Leonid Zhavoronkov

The book is an adopted translation of the monograph "Numerical analysis and modelling in Pascal ABC", in russian, 366 pages.

The monograph is translated and edited by the author in 2019.

Categories

Applied mathematics. Numerical analysis

Key words:

Transcendental equation, numerical differentiation, integration, interpolation, extrapolation, digital filter, cryptography.

Preface

The aim of the book is maintenance of task setting, mathematic description and computer solving of the investigated problem. There are 79 code examples in the book that explain how to solve practical problems in mathematics, mechanics, economics, experimental data processing, numeric row prediction, automatic control, dynamic modelling and information security. These items are associated by numerical analysis methods and using of programming language Pascal ABC as the main research instrument.

The monograph is destined to broad sections of readers who apply information technologies in mathematics and mathematic methods in different practical activities. They are students of professional middle and high technical schools, educational and economic universities as well as specialists that are not professional programmers but are to use numerical methods and computer modelling in practice. Software package is applied.

Book review

The monograph is intended to persons who wants to master programming languages for their professional activities. Firstly, the book may be useful for students of colleges and high schools – technical, pedagogical, economical. It may be interesting also for working specialists that are not professional programmers.

The monograph continues edition of series in Russian devoted to solution of scientific and practical problems by means of numerical methods. These methods became necessary part of mathematical provision of modern computers with huge productivity and memory resources. But these opportunities can be realized only under condition that investigator is able to use corresponding methods of modelling complex processes and systems in different spheres of activity.

Obvious dignity of the monograph is multi-discipline, wide scope of actual science problems. It is an excellent manual for many disciplines such as Informatic, Programming technology, Information technology, Mathematic modelling, Econometric, Automatic control, Information security.

The book has 11 chapters. They are heterogeneous as to complexity of problems and solution methods. Some of them demand serious previous mathematical training of readers. It is explainable quite enough – universal programming language may be illustrated using diverse tasks only. But it is the reason to appear different mathematical methods of approach – from elementary algebra to higher algebra and number theory. For instance, digit filters and solution of differential equations are based on z-transform theory. However, this subject is still out of education programs in many high schools in Russia.

These theoretical materials may be interested to aspirants who study profoundly control methods in economic, engineering, information security. Despite of these restrictions, the complex theoretical materials are expedient in the book and do not prevent from studying by less trained readers. They can omit the theoretical computations and use concluding formulae. The last are enough sufficient to develop own programming codes.

The author suggests multiple examples of codes in Pascal ABC, but in the same time calls the reader to seek new solution variants and optimize his codes.

The book examples and illustrations are worked out in details. They are adjusted in many years of teaching information technologies in professional middle and high schools. All examples are based on own methodical books, scientific publications and inventions of the author.

In conclusion one may affirm that the book "Numerical analysis and modelling in Pascal ABC" will be interesting both to beginners and more experienced in programming and mathematical analysis.

Reviewer:

V.M. Troyanovsky, candidate of economic sciences,

Federal Research Center "Information and Management",

Russian Academy of Sciences.

Moscow, 17.05.2017

CONTENT

11.12. Digital signature

INTRODUCTION

Interest for methods of approximate solution of applied problems had arisen long before Christ, but only in the 17th century significant incentives had occurred owing to development of european science, growth of industry, astronomy, navigation. Share of numerical methods in mathematical analysis was arising continuous during the next centuries. No serious scientific research is possible now without above-mentioned methods.

Achievements in numerical analysis are owing to works of notable mathematicians such as I. Newton, P. Fermat, G. Leibniz, L. Euler, G. Lagrange, A. Legendre, S. Poisson, A. Cauchy, K. Gauss, F. Bessel, P. Chebyshev. Nowadays progress in discrete mathematic keeps step with invention of the first electro-mechanical and then electronical computers in the middle of the 20th century. It should be rendered in that cause oustanding mathematicians – C. Shannon, N. Wiener, A. Lyapunov, L. Pontryagin, A. Kholmogorov, A. Vitushkin.

High-capacity computers are available now at the working place of every specialist due to fantastic success in microelectronic technology. That is why educational standards involve using a computer in any professional activity. However new educatoinal problem occurred in that field of knowledge. The matter is how to teach student to automated approximate solution of applied problems and in the same time to keep in mind methods of fundamenal mathematic based on the principles of infinitesimal quantity. The situation is complicated by the next circumstances:

- limited duration of learning in educational institutions;

- a lot of programming languages;

- a lot of mathematical applications for personal computer.

The education standard for information technologies in secondary school (at least in Russian Federation) provides being able to use basics even if one of high-level programming languages - Visual Basic, Visual Basic for Applications, Pascal, Visual C++.

History of the language Pascal is very notable. It was developed by Niklaus Wirth in 1970 for educational need. The aim has been achieved completely. Pascal turned out the best language for beginners as well for professional programmers for a period of 30 years. Several successive basic versions were issued then - Turbo, Extended, Object. Nowaday these are not more at the market of software products, but a great number of enthusiastics are making attempts to promote succesfull educational language. Compilators Free Pascal, Pascal Lazarus and Pascal ABC have been developed as a result.

Pascal ABC is destined to beginners but the compilator is also in fact the excellent instrument for solving the very complicated professional problems. One ought only to study vast potential possibilities of Pascal ABC and to set worthy practical problems.

Some manuals concerning basics of Pascal ABC for secondary school in Russia can be found in World Wide Web [1], but there is acute dificiency of practical textbooks for professional training. Three categories of potential Pascal users can be seen distinctly:

- pupils of secondary schools and students of professional colleges;

- students of high school in economic, pedagogic and technology;

- working professional specialists.

Hence, three levels of programming skill are aimed in educational process. In the secondary school pupils study methods of algebraical description and algorithmical solution of problems, basic elements of the language including variables, constants, conditions, cycles, arrays, vector and raster graphics.

Students of the high school study methods of mathematical analysis in continuous and discrete descriptions, extend their knowledge of programming language and begin solution of problems in accordance with future profession under guidance of teachers. It may be electric circuit analysis, developing of an economic model, automatical design of machinery, information security and so further.

Working specilist is to set a problem independenly, fulfil formal specification, develop an algorithm for computer solution and realize it in proper code. The very important stage of programming involves right setting of the problem, an adequate mathematical description and developing algorithm for a computer. Wide mental outlook, knowledge of classic mathematical analysis and methods of numerical analysis are needed at this stage.

We ought to remember sparsness in mathematical competentions of potential readers. That is why you will not find out strict proofs in this monograph. Mathematical abstracts concerning any problem are given only in such volume that is necessary for understanding origin of conceptions, definitions and expressions. Readers with secondary education can omit some formula deducing (for example, interpolation and extrapolation, differential equations solution, digital filter theory, matrix algebra) without damage to programming skill. They can simply develop codes using prepared and proved formulae. In addition to that one ought to keep in mind that Laplace transform and z-transform simetimes are not required course in high-school nowaday. Very likely the more advanced readers will need such computations.

In our opinion, numerical analysis is worthy more attention in educational process than it has now in mathematical cycle of disciplines. There are two reasons for that:

- an eternal problem of limited hour fund in educational process;

- insufficient experience in teaching numerical methods.

This experience can not be aquired from textbooks and manuals only. The teacher ought to take part in real scientific research where problems arise and can be solved by numeric methods.

The monograph is written to demonstrate huge possibilities of Pascal ABC in many activities in the 21st century. It is destined to specialists who have already known Pascal bases and principles of mathematical analysis, but are not still conceive solution methods of problems arising in their profession. They are the largest user group consisting of students and working specialists for which computer has became the main instrument for experimental data treatment, technological processes of supervision and working-out arguments for a decision.

The first version of the book was published in russian in 2017 [2]. This book in your hands is the second edition in author's translation.

The monograph [2] which partly repeats the structure of author's prototype [3], but contains new items to which an interest of specialists has arisen at the last years. Author's experience shows that perfection in programming can be achived best of all using proved codes and of course reference books. That is why every solved problem in the monograph is illustrated with according variant of code and resulting screenshot.

The book is written on the base of many years experience in teaching disciplines "Principles of control devices and automatic regulation", "Programming technologies", "Informatic", "Information control technologies", "Information technologies in professional activities". And at last many considered problems were taken from author's practice of scientific research and developing automatic control systems.

Chapter 1. DRAWING AND ANALYSIS OF FUNCTIONS

1.1. Function graphics

Any line (straight or curve) on a screen is approximated by a broken line drawn with straight segments that join two neighbour points indicated by coordinates (x1, y1) and (x2, y2). In the Pascal ABC the straight segment is drawed on the screen by the library function Line(x1, y1, x2, y2). If the distance between points is small enough then a smooth curve of functional dependence can be seen on the screen.

There are two ways to compute all Cartesian coordinates of points on the curve:

- to accumulate arrays of all point coordinates;

- to calculate two coordinates of the current point (x1, y1) and then save them by means of appropriation operations x2:=x1, y2:=y1; thus the current values became the previous for the next step of calculation.

The second mode saves memory considerably and often proves to be preferable. Sometimes the first mode can be acceptable also if quantity of points on a curve is not too large.

Let us consider a problem from the school physic course, namely from the cinematics. A solid body is thrown with start velocity vector V directed under angle of slope A and further the body is in free flight. It is necessary to build flight trajectory under the next restrictions that are acceptable if horizontal distance is less than 100 meters. These are:

1. Earth surface is supposed plane; vector of Earth attraction does not change its direction. The flight takes place in plane-parallel gravity field.

2. Air resistance is neglected as well as air movement (wind).

3. The flight is stopped in the point where hight becames zero. Elastic rebound is excluded.

Under such conditions the trajectory (named ballistic) is described in vertical flatness by the expression:

$$y(x) = x *tgA - g*x^2 / (2V^{2}*\cos^{2}A),$$

where

    x - horizontal distance;

    y – hight;

    $g = 9.81$ m/sec^2 – acceleration of gravity.

Example 1.1.

Write a program for research trajectories of a ball that gets start velocity 25 m/sec with angle of slope 30, 45 and 60 grades. Calculate trajectory $y(x)$ with step $\Delta x = 1$ m. Build ballistic trajectories.

Solution.

Code Prog_1_1 (Listing 1.1) contains:

- a subroutine "CoordSys" that draws Cartesian coordinate system and marks up axes;

- a subroutine "Parabola" that receives value of the parameter A, computes trajectory points using the cycle "do … while" and draws the function graphic; the subroutine is run three times (fig.1).

The main program calls the subroutine "Parabola", transfers value of parameter A into the subroutine and prints its value on display when trajectory drawing has been over.

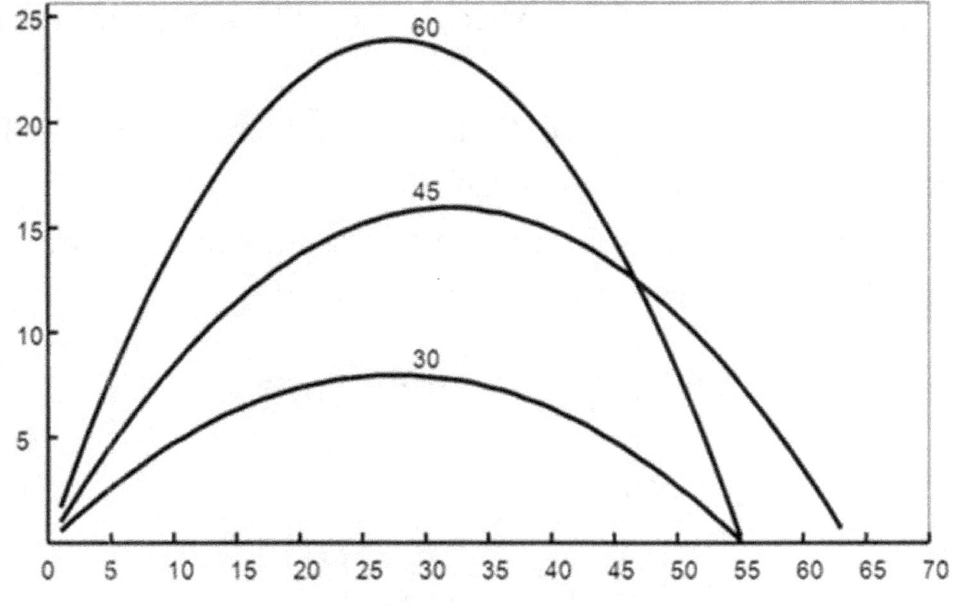

Figure 1. Ballistic trajectories

```
Program Prog_1_1;
Uses GraphABC, ABCobjects;

Procedure CoordSys;
  var   j, x, num: integer;
  var   d: string;
begin
  writeln('    Vo = 25 m/sec.    Angle of slope (grades):
                       A1 = 30,   A2 = 45,    A3 = 60');
  SetWindowSize(850, 500);
  SetPenStyle(psClear);      SetPenColor(clBlack);
  SetPenWidth(1);            Line(100, 30, 800, 30);
  Line(800, 30, 800, 440);   SetPenWidth(3);
  Line(100, 440, 800, 440); Line(100, 30, 100, 440);
  for j := 1 to 5 do
  begin
    SetPenWidth(3);               SetPenColor(clBlack);
    Line(100, 440 - 80 * j, 108, 440 - 80 * j);
  end;
  for j := 1 to 14 do
  begin
    SetPenWidth(3);   SetPenColor(clBlack);
    Line(100 + 50 * j, 440, 100 + 50 * j, 432);
```

```pascal
      end;
   for j := 0 to 14 do
     begin
       x := 5 * j;
       d := IntToStr(x);
       var t := new TextABC(95 + 50 * j, 450, 14, d);
     end;
    for j := 1 to 5 do
     begin
       x := 5 * j;
       d := IntToStr(x);
      var t := new TextABC(75, 432 - 80 * j, 14, d);
    end;
  end;


Procedure Parabola(ang: real);
   const g = 9.81; v=25;
   const x0=100; y0=440; mx=10; my=16;
   var  k, x1, x2, y1, y2: integer;
   var  x, y, a: real;
   begin
     SetWindowSize(850, 500); SetPenStyle(psClear);
     SetPenColor(clBlack);  SetPenWidth(4);
     a := ang * pi / 180;
     x := 0;
     for k := 1 to 70 do
       begin
         x := x + 1;
         y := x * tan(a) - g * x * x / (2 * v * v * cos(a) * cos(a));
         x1:=x0+Round(x*mx);       y1:=y0-Round(y*my);
         SetPenWidth(4);           SetPenColor(clBlack);
         if (k>1) and (y>0) then Line(x1,y1,x2,y2);
         x2 := x1; y2 := y1;
       end;
   end;
//========Main========
var ang: real;
var alp1, alp2, alp3: string;
begin
  CoordSys;
  ang:= 30;  Parabola(ang);
```

```
ang:= 45;  Parabola(ang);
ang:= 60;  Parabola(ang);
alp1:=IntToStr(30);
var t1:=new TextABC(390,290,14,alp1);
alp2:=IntToStr(45);
var t2:=new TextABC(390,162,14,alp2);
alp3:=IntToStr(60);
var t3:=new TextABC(390,38,14,alp3);
end.
```

Listing 1.1

The function family (fig.1) shows:

- the ballistic trajectory has a parabola form;

- the trajectory becames optimum with parameter value 45^0; it means that the maximum distance is achieved with given start energy pulse as well minimum energy is necessary for achievement of given distance.

1.2. High order curves

There are remarkable curves on the plane that were known long ago since Archimedes epoch. Now we talk about high order curves. Above-mentioned parabola is one-valued function $y(x)$: for every value x exists the single value y. Cartesian coordinates are convenient for drawing the graphic of such function.

High-order curve proves to be many-valued function in Cartesian system. The polar coordinates are more convenient to descript such curve as a trajectory of the end of the vector rotating around the origin on a supporting axis. The vector module is dependent on the rotation angle. A man is in habit of observing this dependence in Cartesian coordinates however. Hence, it is necessary to calculate rectangle coordinates of every point using parametric equations of the curve. Let us consider some examples of high order curves.

Example 1.2.

Develop the program that draws the left spiral of Archimedes. Spiral equation in the polar coordinates is:

$$R(\varphi) = a\varphi,$$

where: R - radius-vector;

    a - constant;

    φ – rotation angle.

Spiral parametric equations in Cartesian coordinates (x, y) are:

$$x = R(\varphi){*}\cos(\varphi); \quad y = R(\varphi){*}\sin(\varphi);$$

Solution.

The code Prog_1_2 (listing 1.2) makes the next actions:

- runs the subroutine "CoordSys" that draws and marks up rectangle coordinate axes;

- uses the subroutine "Spiral" to calculate 1080 function values with step $\Delta\varphi = 2^0$, a = 0.5 and to draw spiral graphic (fig.2).

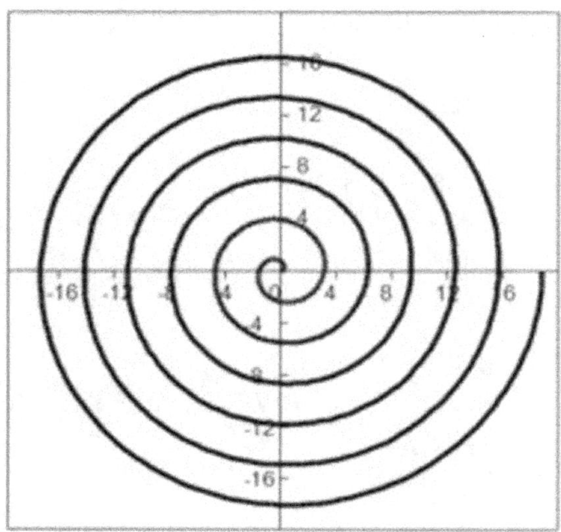

Figure 2. Archimedes spiral

```pascal
Program Prog_1_2;
Uses GraphABC, ABCobjects;

Procedure CoordSys;
 var x1, x2, y1, y2, z: integer;
 var k, xc, yc: integer;   var s:string;
 begin
   SetWindowSize(580,480); SetPenStyle(psClear);
   SetPenWidth(1);          SetPenColor(clBlack);
   Line(154,240,550,240);   Line(350, 40,350,440);
   Line(154,40,550, 40);    Line(154,440,550, 440);
   Line(154,40,154, 440);   Line(550,40,550, 440);
  for k:=1 to 10 do
   begin
    Line(150+40*k, 240, 150+40*k, 246);
   end;
   for k:=1 to 10 do
    begin
     Line(350, 40+40*k, 356, 40+40*k);
    end;
  for k:=1 to 9 do
   begin
    xc:=k*4-20;  s:=IntToStr(xc);
    var t:= new TextABC(142+40*k,248,12,s);
   end;
  for k:=1 to 4 do
   begin
    yc:=k*4;      s:=IntToStr(yc);
    var t:= new TextABC(362,230-40*k,12,s);
   end;
  for k:=1 to 4 do
   begin
     yc:=-k*4;    s:=IntToStr(yc);
     var t:= new TextABC(324,232+40*k,12,s);
   end;
 end;

Procedure Spiral;
const p=3.14158; mx=10; my=10; x0=350; y0=240;
var j, k, x1, x2, y1, y2: integer;
var ag, ar, R, x, y: real;
begin
 R:=0;  ag:=0;  x:=0;  y:=0; j:=0;
 var t:=new TextABC(280,10,11,'Archimedes spiral');
```

```pascal
  for k:=1 to 1080 do
    begin
      ag:=ag+2;  ar:=ag*pi/180;   R:=0.5*ar;
      x:=R*cos(ar);    y:=R*sin(ar);
      x1:=x0+Round(x*mx);
      y1:=y0-Round(y*my);
      SetPenWidth(2);    SetPenColor(clBlue);
      if k>1 then  Line(x1, y1, x2, y2);
      x2:=x1;  y2:=y1;
      if (x>0) and  (Abs(y)<0.0003) then
        begin   j:=j+1;   end;
        Sleep(5);
    end;
    writeln(' ');    writeln(' Argument step:');
    writeln('      2 grades');
end;
//==== Main ====
begin
  CoordSys; Spiral;
end.
```

Listing 1.2

Remarks.

1. The computer processes the task during some milleseconds only. It is useful to enlarge this time artificially for visual observation how the spiral is drawing. For that purpose every computation cycle is made slow down by means of built-in function Sleep(5).

2. An angle coordinate is to be involved in both spiral description in the radians. Hence, when giving angle value in grades, one ought to translate it in radian using the constant π with accuracy even if up to the 5th decimal digit, for instance $\pi \approx 3.14158$.

The second remark is actual also in the rest parts of the monograph where you can meet trigonometrical functions.

Example 1.3.

Develop the program that draws a curve named as "Pascal snail". It is described in the parametrical form by the expressions:

$$x = A*\cos^2\varphi + L*\cos\varphi;$$
$$y = A*\cos\varphi*\sin(\varphi) + L*\sin\varphi.$$

Compute and draw the snail with constant values $A = 4.8$ and $L = 2.1$ using an argument step $\Delta\varphi = 2^0$.

Solution.

The code Prog_1_3 (listing 1.3) makes the next actions:

- runs the subroutine "CoordSys" that draws and marks up rectangle coordinate axes;

- uses the subroutine "Snail" which computes 200 function values.

The snail with given parameter values A and L is shown at fig.3.

```
Program Prog_1_3;
Uses GraphABC, ABCobjects;

Procedure CoordSys;
  var x1, x2, y1, y2, z: integer;
  var k, xc, yc: integer;   var s:string;
  begin
   SetWindowSize(550,480); SetPenStyle(psClear);
   SetPenWidth(1);                SetPenColor(clBlack);
   Line(150,240,510,240);    Line(210, 40,210,440);
   Line(150,40,510, 40);      Line(150,440,510, 440);
   Line(150,40,150, 440);    Line(510,40,510, 440);
   for k:=1 to 10 do
     begin
        Line(90+60*k, 240, 90+60*k, 232);
     end;
   for k:=1 to 10 do
     begin
       Line(204, 40*k, 210, 40*k);
```

```pascal
   end;
  s:=IntToStr(-2);
  var tt:= new TextABC(130,245,12,s);
  for k:=2 to 5 do
      begin
          xc:=(k-1)*2; s:=IntToStr(xc);
          var t:= new TextABC(146+60*k,245,12,s);
      end;
  for k:=1 to 4 do
      begin
          yc:=k*1; s:=IntToStr(yc);
          var t:= new TextABC(190,230-40*k,12,s);
      end;
  for k:=1 to 4 do
      begin
          yc:=-k*1;   s:=IntToStr(yc);
          var t:= new TextABC(185,230+40*k,12,s);
      end;
end;

Procedure Snail;
const p=3.14158; mx=40; my=40; x0=210; y0=240;
var j, k, x1, x2, y1, y2: integer;
var ph, phr, A, L, x, y: real;
begin
  A:=4.8; L:=2.1; ph:=0; x:=0; y:=0; ;j:=0;
  var t:=new TextABC(280,10,11,'Pascal snail');
  for k:=1 to 200 do
   begin
    ph:=ph+2;        phr:=ph*pi/180;
    x:=A*cos(phr)*cos(phr)+L*cos(phr);
    y:=A*cos(phr)*sin(phr)+L*sin(phr);
    x1:=x0+Round(x*mx);   y1:=y0-Round(y*my);
    SetPenWidth(2);         SetPenColor(clBlue);
    if k>1 then  Line(x1, y1, x2, y2); x2:=x1; y2:=y1;
   end;
  writeln(' '); writeln('Parameters:');
  writeln('    A = ',A:2:1); writeln('    L = ',L:2:1);
  writeln(' '); writeln('Argument step0:');
  writeln('    2 grades'); writeln(' ');
  writeln('Step quant.:'); writeln('      200');
 end;
```

```
//===== Main ====
begin
  CoordSys; Snail;
end.
```

Listing 1.3

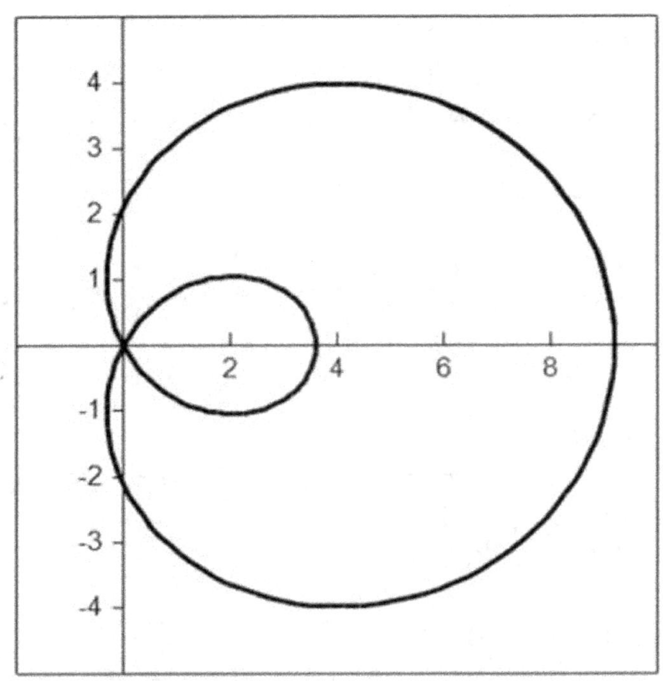

Figure 3. Pascal snail

Remark.

It is recommended to choose some pairs of parameters without assistance and see how the internal snail loop is dependent on the parameters.

Figure "epicycloid" is a trajectory of the point on the cycle which rolls outside another immovable cycle without slip. "Hypocycloid" is a trajectory of the point on the cycle which rolls inside another immovable cycle.

Example 1.4.

Develop the program that draws an epicycloid decribed by parametric equations.

$$x = A*\cos(\varphi) - B*\cos(A*\varphi);$$
$$y = A*\sin(\varphi) - B*\sin(A*\varphi).$$

Provide the next actions:

- changing of values A and B;

- uniform rotating of the epicycloid at some revolutions anti-clockwise and then clockwise at the same quantity.

Remark.

The rotating of a plane figure is put in practice by means of turning Cartesian axes at angle $\beta(t)$, where variable t is the current time. In that case the coordinate transform takes place. Initial coordinates (x, y) in immovable system are reduced into coordinates (u, v) of the movable system in accordance with the formulae:

$$u = x*\cos\beta + y*\sin\beta;$$
$$v = -x*\sin\beta + y*\cos\beta.$$

Solution.

The code Prog_1_4 (listing 1.4) makes the next actions:

- runs the subroutine "Square" that prepares square field to place a figure and values of parameters;

- uses the subroutine "Epicycle" which receives the parameter values as well as the rotation increment value and then calculates 250 values of the curve for every new position;

- prints out values A and B under the graphic of the epicycloid (fig.4 – fig.9).

```pascal
Program Prog_1_4;

Uses GraphABC, ABCobjects;

Procedure Square;
begin
  SetWindowSize(600,500);  SetPenColor(clBlack);
  SetPenWidth(1);            Rectangle(145,40,550,450);
  var t4:=new TextABC(20,86,11,'Parameters:');
  var t5:=new TextABC(40,110,11,'A      B');
  var t6:=new TextABC(30,118,11,'-------------');
  var t7:=new TextABC(35,136,11,'3.0    2.0');
  var t8:=new TextABC(35,154,11,'5.0    2.5');
  var t9:=new TextABC(35,172,11,'4.0    2.5');
  var t10:=new TextABC(35,190,11,'2.5    1.0');
  var t11:=new TextABC(35,208,11,'3.5    2.0');
  var t12:=new TextABC(35,226,11,'5.5    2.0');
  var t13:=new TextABC(35,244,11,'2.5    2.0');
  var t14:=new TextABC(35,262,11,'4.0    2.0');
  var t15:=new TextABC(35,280,11,'5.0    2.0');
  var t16:=new TextABC(35,298,11,'4.0    3.0');
  var t17:=new TextABC(35,316,11,'2.0    3.0');
  var t18:=new TextABC(35,334,11,'3.0    3.0');
  var t19:=new TextABC(35,352,11,'2.5    2.2');
  var t20:=new TextABC(35,370,11,'3.0    3.0');
  var t21:=new TextABC(35,388,11,'2.5    2.2');
end;


Procedure Epicycle(A,B,db: real);
const p=3.14158; mx=20; my=20; x0=350; y0=240;
var j, k, x1, x2, y1, y2: integer;
var ph, pr, x, y, be, u, v: real;
begin
   ph:=0; x:=0; y:=0;
   for j:=2 to 50 do
     begin
        SetPenColor(clWhite);
        Rectangle(150,45,540,430); be:=db*j*pi/180;
        for k:=1 to 250 do
           begin
              ph:=ph+4;  pr:=ph*pi/180;
              x:=A*cos(pr)-B*cos(A*pr);
              y:=A*sin(pr)-B*sin(A*pr);
              u:=x*cos(be)+y*sin(be);    v:=-x*sin(be)+y*cos(be);
              x1:=x0+Round(u*mx);        y1:=y0-Round(v*my);
```

```
            SetPenWidth(2);  SetPenColor(clBlue);
            if k>1 then Line(x1, y1, x2, y2);
            x2:=x1; y2:=y1;
        end;
    end;
//==================== Main ==================
var A, B, db: real;
var ta, tb: string;
begin
  Square;
  A:=5.5; B:= 2.0;  //---- Figure parameters
  db:=9;   Epicycle(A,B,db);
  Sleep(10);
  db:=-9;  Epicycle(A,B,db);
  Str(A:2:1,ta);
  Str(B:2:1,tb);
  var tc:=new TextABC(300,460,12,'A = '+ta+'   B = '+tb);
end.
```

Listing 1.4

Example 1.5.

Develop the program that draws a hypocycloid decribed by parametric equations.

$$x = (A - B)*\cos(\varphi) - B*C*\cos((1 - A/B)*\varphi);$$

$$y = (A - B)*\sin(\varphi) - B*C*\sin((1 - A/B)*\varphi).$$

Provide the next actions:

- changing of values A and B;

- uniform rotating of the epicycloid at some revolutions anti-clockwise and then clockwise at the same revoluton quantity.

Solution.

The code Prog_1_5 (listing 1.7) makes the next actions:

- runs the subroutine "Square" that prepares square field to place a figure and values of parameters;

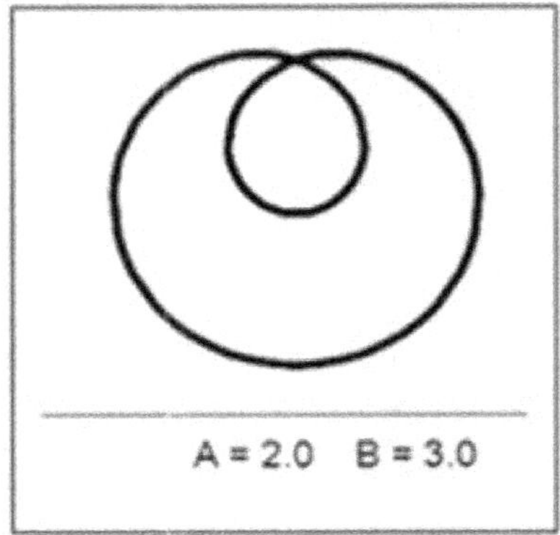

Figure 4. Epicycloid-1

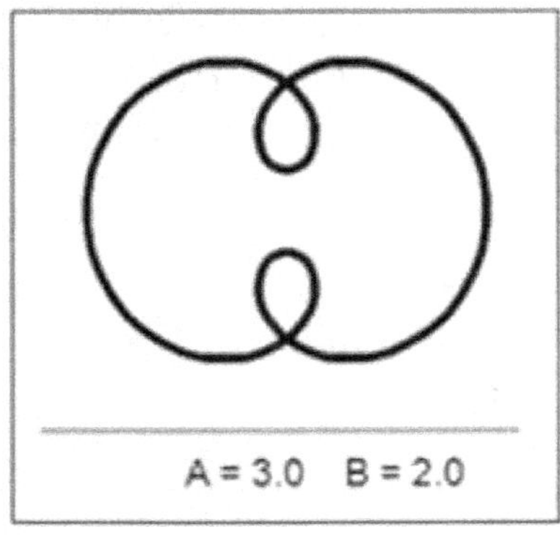

Figure 5. Epicycloid-2

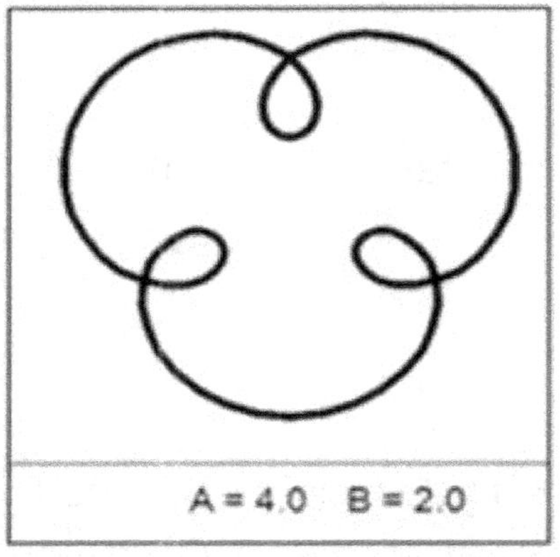

Figure 6. Epicycloid-3

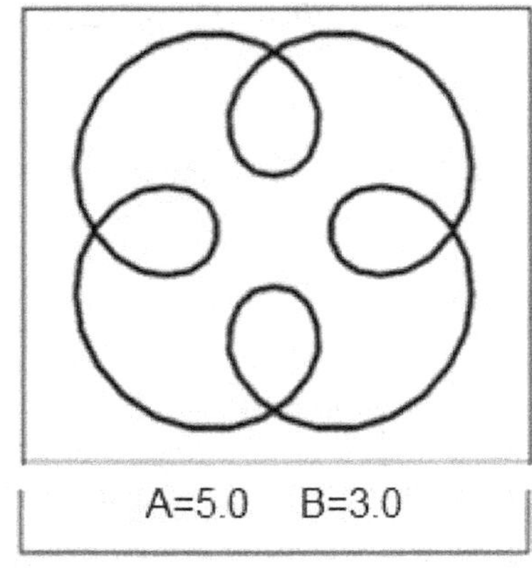

Figure 7. Epicycloid-4

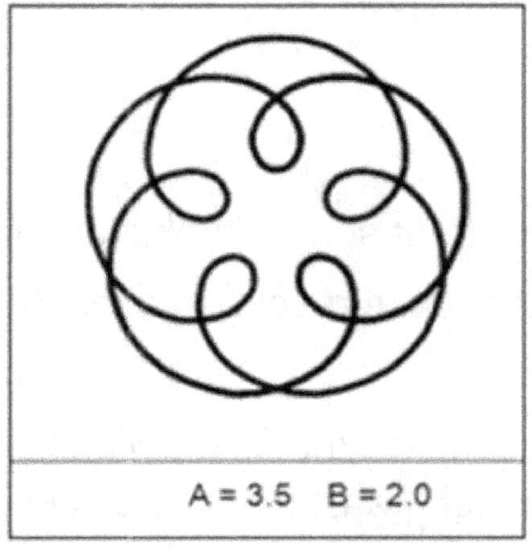

Figure 8. Epicycloid-5

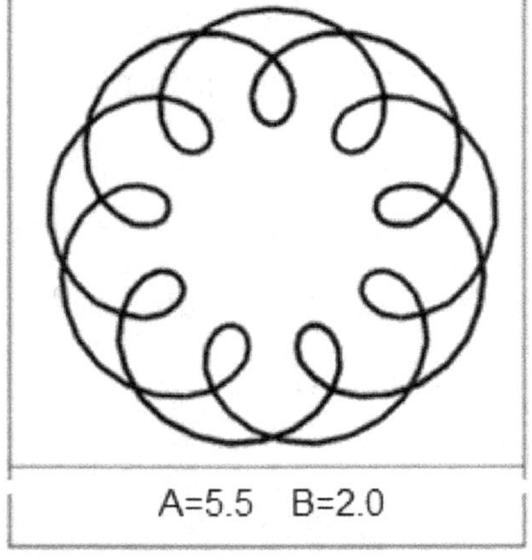

Figure 9. Epicycloid -9

- uses the subroutine "Hypo" which receives the parameter values as well as the rotation increment value and then calculates 350 values of the figure for every new position;

- prints out values A and B under the of the hypocycloid (fig. 10 – fig. 15).

```
Program Prog_1_5;
Uses GraphABC, ABCobjects;

Procedure Square;
begin
    SetWindowSize(590,500);  SetPenColor(clBlack);
    SetPenWidth(1);               Rectangle(145,40,570,460);
    var t1:=new TextABC(300,10,12,'Hypocycloid');
    var t2:=new TextABC(20,50,11,'Try values');
    var t3:=new TextABC(20,68,11,'of' three);
    var t4:=new TextABC(20,86,11,'parameters');
    var t5:=new TextABC(30,110,11,'A     B     C');
    var t6:=new TextABC(15,118,11,'--------------------');
    var t7:=new TextABC(20,136,11,'2.5   1.0   0.8');
    var t8:=new TextABC(20,154,11,'4.0   0.8   1.5');
    var t9:=new TextABC(20,172,11,'4.0   1.0   1.4');
    var t10:=new TextABC(20,190,11,'3.5   1.0   1.0');
    var t11:=new TextABC(20,208,11,'3.5   1.0   1.4');
    var t12:=new TextABC(20,226,11,'3.0   1.0   1.4');
 end;

Procedure Hypo(A, B, C, db: real);
const p=3.14158; mx=35; my=35 ; x0=350; y0=250;
var j, k, x1, x2, y1, y2: integer;
var ph, pr, x, y, be, u, v: real;
begin
  ph:=0;  x:=0;  y:=0;
  for j:=2 to 50 do
    begin
     SetPenColor(clWhite);
     Rectangle(150,45,550,450);
     be:=db*j*pi/180;
     for k:=1 to 350 do
       begin
          ph:=ph+4;   r:=ph*pi/180;
```

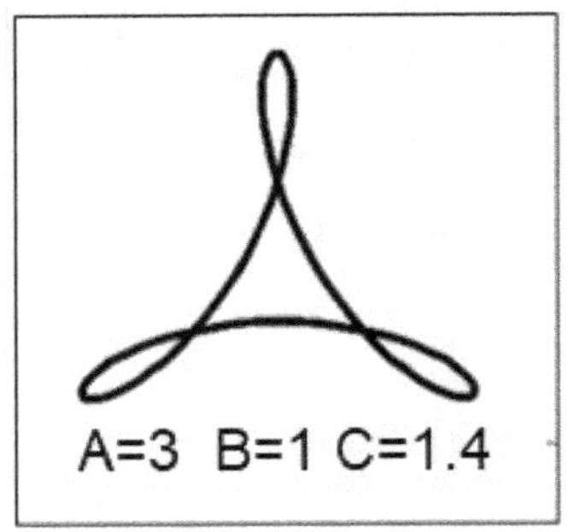

A=3 B=1 C=1.4

Figure 10. Hypocycloid-3

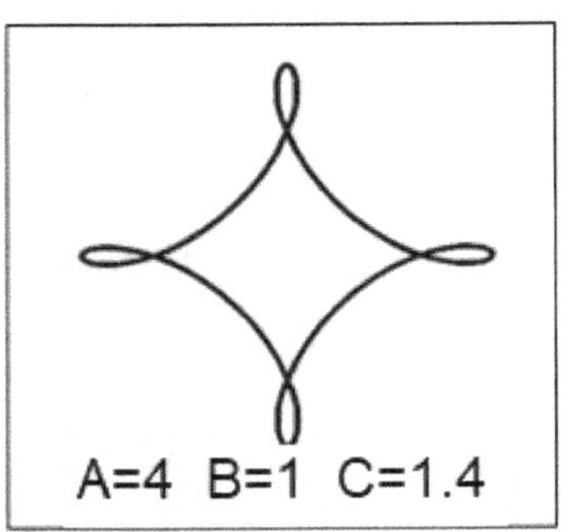

A=4 B=1 C=1.4

Figure 11. Hypocycloid-4

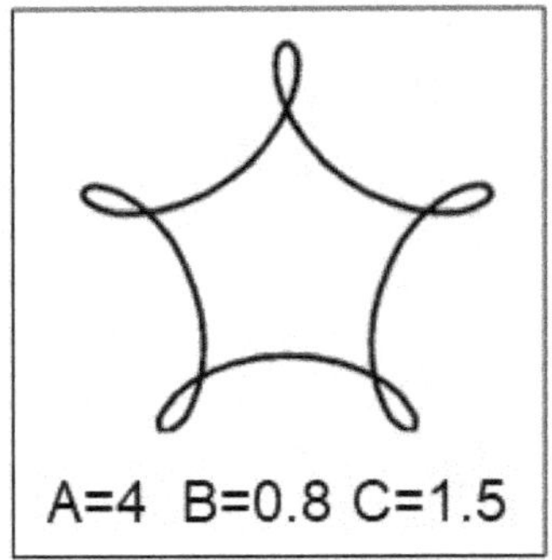

A=4 B=0.8 C=1.5

Figure 12. Hypocycloid-5.1

A=2.5 B=1 C=0.8

Figure 13. Hypocycloid-5.2

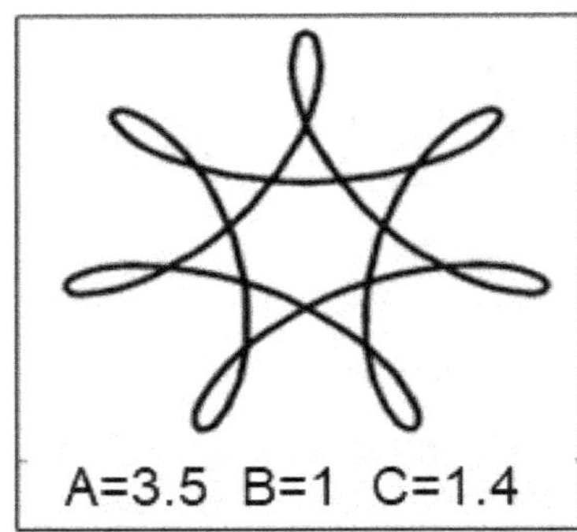

A=3.5 B=1 C=1.4

Figure 14. Hypocycloid-7.1

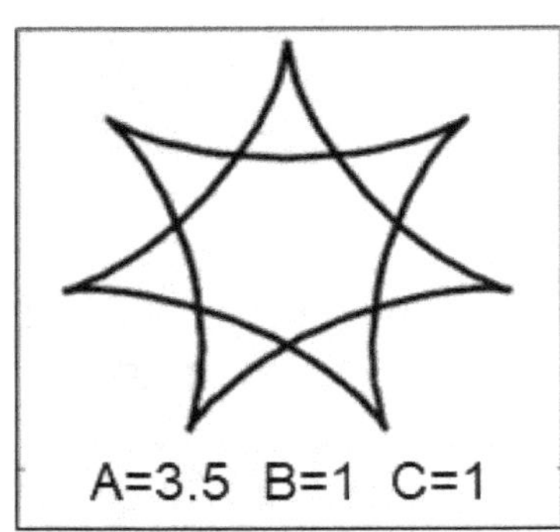

A=3.5 B=1 C=1

Figure 15. Hypocycloid-7.2

```
x:=(A-B)*cos(pr)+B*C*cos((1-A/B)*pr);
y:=(A-B)*sin(pr)+B*C*sin((1-A/B)*pr);
u:=x*cos(be)+y*sin(be);  v:=-x*sin(be)+y*cos(be);
x1:=x0+Round(u*mx);       y1:=y0-Round(v*my);
SetPenWidth(2);           SetPenColor(clBlue);
```

```
      if k>1 then  Line(x1, y1, x2, y2);
      x2:=x1;  y2:=y1;
    end;
    end;
    Sleep(5);
  end;
end;//================ Main =====================
var A, B, C, db: real;
var ta, tb, tc: string;
begin
  Square;          A:=3.5; B:= 1.0; C:=1.0;
  db:=9;           Hypo(A,B,C,db);
  Sleep(5);        db:=-9;Hypo(A,B,C,db);
  Str(A:2:1,ta);   Str(B:2:1,tb);    Str(C:2:1,tc);
  var tt:=new TextABC(260,470,12,
      'A = '+ta+' B = '+tb +'  C = '+tc);
end.
```

Listing 1.5

1.3. Root points

An equation with a single unknown x is an equality of two functions of the same variable:

$$U(x) = V(x), \qquad (1.1)$$

which is true by some definite values of x.

Variable values x_1, x_2, ..., x_n, which make true equality (1.1), are named as roots of the equation. The equation (1.1) is algebraical if both functions are algebraical. Any algebraical equation can be reduced by means of transformations to the canonical form:

$$P(x) = a_0x^n + a_1x^{n-1} + \ldots + a_n = 0; \qquad (1.2)$$

$$P(x) = \sum_{k=0}^{n} a_k x^{n-k} = 0$$

The equation (1.1) is named transcendental if though one of the functions is not algebraical. In that case it may be exponential, logarithmical, trigonometrical or hyperbolical, depending on the function form.

There are two groups of solution methods. The first group is joined by methods which due to some transformations bring to formulae for roots computing. Such transformations are quite simple for algebraical equations if degree is not higher than 5. It is the reason that such methods are not of great practical value. The second group includes numerical methods which allow to find roots of any equation either transcendental or high degree algebraical one despite of complexity.

Numerical solution methods are two-stage as a rule. Intervals of roots existing are determined at the first stage. Root values are defined more exactly at the second stage.

There are many methods of approximate calculus of roots with given solution accuracy [5]. The choice of method is dependent on equation form and on some demands claimed to a process and to result of a solution.

Approximate methods of transcendent equation solution are described in the book [3] where some examples are shown. The main of them are a Newton (or "tangent") method and a chord method. Both provide the solution accuracy at least to the 14th decimal digit. Similar tasks are solved effectively in the application Microsoft Excel even without programming procedures.

Algebraical and transcendent equations are solved by means of computer in two ways when accounting formulae do not exist:

The manner 1.

A graphic provides a rough but quick solution. It is necessary to compute values of the functions at the left and right parts of the equation (1.1), to draw graphics of these functioins and find coordinates of cross points. This way is often used in the secondary school algebra.

The manner 2.

An equation is reduced to the form (1.2). Values of the function in the left part are calculated. A function graphic is drawed and the points are localized, where the function crosses the argument axis and the changes of value sign take place. Then the axis X is scanned in every localized region using a small step which is enough for root values selection with a given accuracy.

At the beginning let us solve a task from the school algebra course.

Example 1.6.

Find roots of the equation system:

$$\begin{cases} x = 3y\,/\,4; \\ xy + x + y = 62; \end{cases}$$

Remark.

The system can be reduced to a quadratic equation with well known formula for exact computation of two roots. We shall not use it however but shall develop a program that will calculate and draw three graphics – two for origin equations and the third one for a reduced parabola.

Solution.

The program Prog_1_6 (listing 1.6) makes the next actions:

- runs the subroutine "CoordSys" that draws and marks up rectangle coordinate axes;

- runs the subroutine "XYZ" that calculates and draws graphics of three functions (fig.16).

```
Program Prog_1_6;
uses ABCobjects, GraphABC;
Procedure CoordSys;
 var k, z: integer;
 var xc, yc: integer;
 var s: string;
 begin
  SetWindowSize(800,480);  SetPenWidth(2);
  SetPenStyle(psClear);        Line(100, 240, 760, 240);
  Line(430, 20, 430, 452);     SetPenWidth(1);
  Line(100, 22, 760, 20);      Line(100, 452, 760, 452);
  Line(100, 20, 100, 452);     Line(760, 20, 760, 452);
  for k:=1 to 23 do
    begin
      Line(70+30*k, 240, 70+30*k, 232);
    end;
  for k:=1 to 5 do
    begin
      xc:=2*(k-1)-10;  Str(xc,s);
      var t:= new TextABC(62+60*k,210,12,s);
   end;
   for k:=2 to 6 do
     begin
       xc:=2*(k-1); Str(xc,s);
       var t:= new TextABC(366+60*k,248,12,s);
     end;
  for k:=1 to 13 do
     begin
       Line(430, 36*k-14, 438, 36*k-14);
     end;
  for k:=1 to 5 do
```

```pascal
    begin
        yc:=30-5*k; Str(yc,s);
        var t:= new TextABC(405,12+36*k,12,s);
    end;
  for k:=1 to 5 do
      begin
        yc:=-5*k;  Str(yc,s);
        var t:= new TextABC(400,228+36*k,12,s);
      end;
end;


Procedure XYZ;
  const x0=430; y0=240; mx=30; my=8; mu=0.7;
  var k, x1, x2, y1, y2, z1, z2, u1, u2: integer;
  var a, b, c, d, e, x, dx, yp, yq,  zp, zq: real;
  var r1, r2, yc, zc, up, uq: real;
  var r1t, r2t, yct, zct: string;
  var m1, m2, r1d, r2d, ycd, zcd: integer;
  begin
    SetWindowSize(800,500); SetPenStyle(psClear);
    x:=-10; dx:=0.002; yp:=4*x/3; m1:=1; m2:=1;
    for k:=1 to 5000 do
        begin
          x:=x+dx;     yp:=4*x/3;   zp:=(62-x)/(1+x);
          up:=4*x*x+7*x-186;      x:=x+dx;     yq:= 4*x/3;
          zq:=(62-x)/(1+x);          d:=abs(yq-zq);
          x1:=x0+Round(x*mx);    y1:=y0-Round(yp*my);
          z1:=y0-Round(zp*my);   u1:=y0-Round(up*mu);
          if (x>-8) and (x<-7) and (d<0.007) then
            begin
              r1:=x;  yc:=yq;  m1:=0;
            end;
          if (x>5) and (x<7) and (d<0.007) then
            begin
              r2:=x;  zc:=zq;  m2:=0;
            end;
          SetPenWidth(4);  SetPenColor(clBlack);
          if k>1 then Line(x1, y1, x2, y2);
          SetPenColor(clBlack);
          if (k>1) and (x<-3.6) then Line(x1, z1, x2, z2);
          if x>1.4 then Line(x1, z1, x2, z2);
```

```pascal
      SetPenColor(clBlack);
      if k>1 then Line(x1, u1, x2, u2);
      x2:=x1; y2:=y1;  z2:=z1; u2:=u1;
    end;
  var t1:= new TextABC(170,60,12,'The 1st root');
  Str(r1:3:2,r1t);
  var t2:= new TextABC(190,90,12,'x1 = '+r1t);
  Str(yc:3:2,yct);
  var t3:= new TextABC(190,110,12,'y1 = '+yct);
  var t4:= new TextABC(600,280,12,'The 2nd root');
  Str(r2:3:2,r2t);
  var t5:= new TextABC(620,310,12,'x2 = '+r2t);
  Str(zc:3:2,zct);
  var t6:= new TextABC(620,330,12,'y2 = '+zct);
  r1d:= x0+Round(r1*mx);    r2d:= x0+Round(r2*mx);
  ycd:=y0-Round(yc*my);     zcd:=y0-Round(zc*my);
  SetPenStyle(psSolid);     SetPenWidth(3);
  SetPenColor(clBlack);
  Circle(r1d, y0,6);        Circle(r1d,ycd,6);
  Circle(x0,ycd,6);         Circle(r2d, y0,6);
  Circle(r2d,zcd,6);        Circle(x0,zcd,6);
  SetPenStyle(psDot);
  Line(r1d, y0,r1d,ycd);  Line(r1d,ycd,x0,ycd);
  Line(r2d, y0,r2d,zcd);  Line(r2d,zcd,x0,zcd);
end;
//==============Main============================
begin
CoordSys; XYZ;
  var t7:= new TextABC(100,0,12, 'Origin equation system');
  var t8:= new TextABC(350,0,12,  'x = 3y / 4;    xy + x + y =
62;');
  var t9:=new TextABC(480, 370,12,  'The system is reduced');
  var t10:=new TextABC(480, 390,12, 'to the single equation:');
  var t11:=new TextABC(480, 420,12
          '4x^2 + 7x - 186 = 0   (parabola)');
end.
```

Listing 1.6

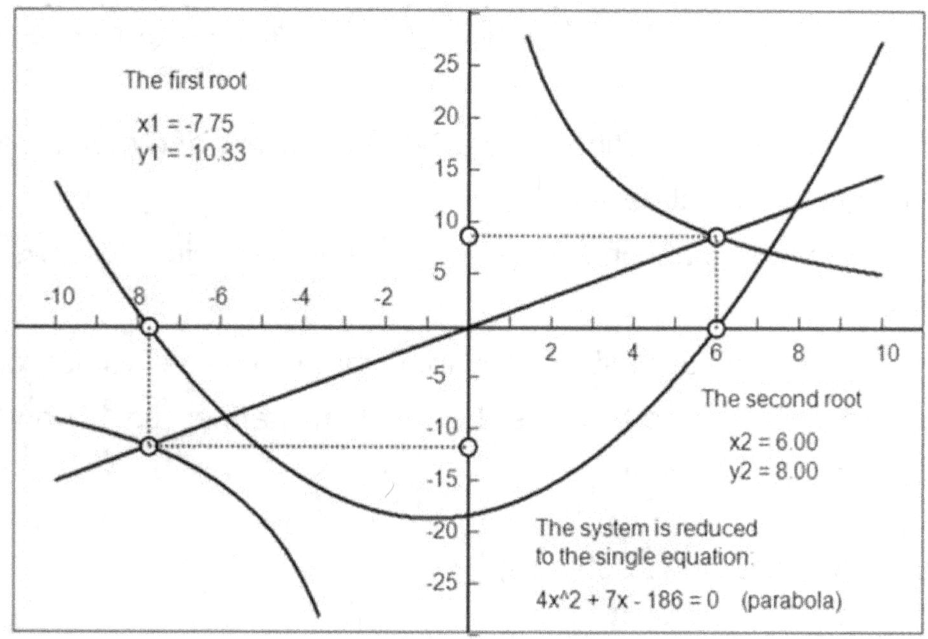

Figure 16. Solution of the algebraic equation system

Conclusion.

As it is seen from fig.16, it would be simpler to reduce the equation system to a quadratic equation and then to find roots using standard formulae even without a computer or a calculator. But the methodical value of a graphic solution of the equation system in the secondary school is evident too. It gives an obvious demonstration that pupil will remember for a long time. The next example confirms a practical value of algebraical equation solution using root selection.

Example 1.7.

Find roots of the high degree equation:

$$ax^6 + bx^5 + cx^2 + dx + e = 0$$

in the interval $0 < x < 1.6$ with an accuracy 0.001 if the coefficient values are:

$$a = -3.8; \; b = 7.2; \; c = -7.3; \; d = 3.5; \; e = -0.1.$$

Solution.

The program Prog_1_7 (listing 1.7) realizes the next actions:

- runs the subroutine "CoordSys" that draws and marks up rectangle coordinate axes;

- uses the function XPN(x, n) that rises variable x to power n;

- runs the subroutine "Polynom" that calculates values of the equa-tion left part, draws the function graphic (fig.17) and finds the equation roots.

```
Program Prog_1_7;
Uses GraphABC, ABCobjects;

Function xpn(x: real; n: integer): real;
   var j: integer;   var acc: real;
   begin
     acc:=1.0;
     for j:=1 to n do   acc:=acc*x;
     xpn:=acc;
   end;
Procedure CoordSys;
  const x0=100;
  var k, z: integer;
  var xc, yc: real;   var s: string;
  begin
    SetWindowSize(800,500);  SetPenStyle(psClear);
    for k:=1 to 13 do
       begin
          if k=6 then SetPenWidth(3) else SetPenWidth(1);
          SetPenColor(clBlacl);  Line(100, 40*k, 780, 40*k);
       end;
     for k:=1 to 19 do
       begin
          if k=2 then SetPenWidth(3) else SetPenWidth(1);
          SetPenColor(clBlack);
        Line(60+40*k, 40, 60+40*k, 440);
     end;
  for k:=2 to 16 do
```

```pascal
      begin
        xc:=0.1*(k-1);    Str(xc:3:1,s);
        var t:= new TextABC(92+40*k,245,12,s);
      end;
  for k:=1 to 5 do
    begin
      yc:=0.2*(k-1);    Str(yc:2:1,s);
      var t:= new TextABC(112,260-40*k,12,s);
    end;
  for k:=1 to 5 do
    begin
      yc:=-0.2*k;    Str(yc:2:1,s);
      var t:= new TextABC(108,222+40*k,12,s);
    end;
  end;

Procedure Polynom;
    const x0=140; y0=240; mx=400; my=200;
    var k, x1, x2, y1, y2, sp, sq, cs, m1, m2, m3, m4: integer;
    var a, b, c, d, e, x, yp, yq, dx, r1, r2, r3, r4: real;
    begin
      SetWindowSize(800,500);    SetPenStyle(psClear);
      a:=-3.8;  b:=7.2; c:=-7.3; d:=3.5;
      e:=-0.1; x:=-0.18;
      yp:=a*xpn(x,6)+b*xpn(x,5)+c*x*x+d*x+e;
      dx:=0.001; m1:=1; m2:=1;  m3:=1; m4:=1; x:=-0.05;
      for k:=1 to 810 do
        begin
          x:=x+dx;
          yp:=a*xpn(x,6)+b*xpn(x,5)+c*x*x+d*x+e;
          if yp>=0 then  sp:=1 else sp:=-1;    x:=x+dx;
          yq:=a*xpn(x,6)+b*xpn(x,5)+c*x*x+d*x+e;
          if yq>=0 then sq:=1 else sq:=-1; cs:=sp*sq;
          if (m1=1) and (x<0.2) and (sq=1)  then
            begin
              r1:=x;  m1:=0;
            end;
          if (m2=1) and (x>0.3) and (cs=-1) then
            begin
              r2:=x; m2:=0
            end;
          if (m3=1) and (x>0.7) and (cs=-1) then
```

```
      begin
        r3:=x; m3:=0
      end;
    if (m4=1) and (x>1.3) and (cs=-1) then
      begin
        r4:=x; m3:=0
      end;
    x1:=x0+Round(x*mx);   y1:=y0-Round(yp*my);
    SetPenWidth(3);        SetPenColor(clBlack);
    if k>1 then Line(x1, y1, x2, y2); x2:=x1; y2:=y1;
  end;
  writeln(' ax^6 + bx^5 + cx2 + dx + e = 0');
  writeln('Roots Coefficients: a = -3.8  b = 7.2
                    c = -7.3  d = 3.5  e = -0.1');
  writeln(' x1 = ',r1:4:3);  writeln(' x2 = ',r2:4:3);
  writeln(' x3 = ',r3:4:3);  writeln(' x4 = ',r4:4:3);
end;
```

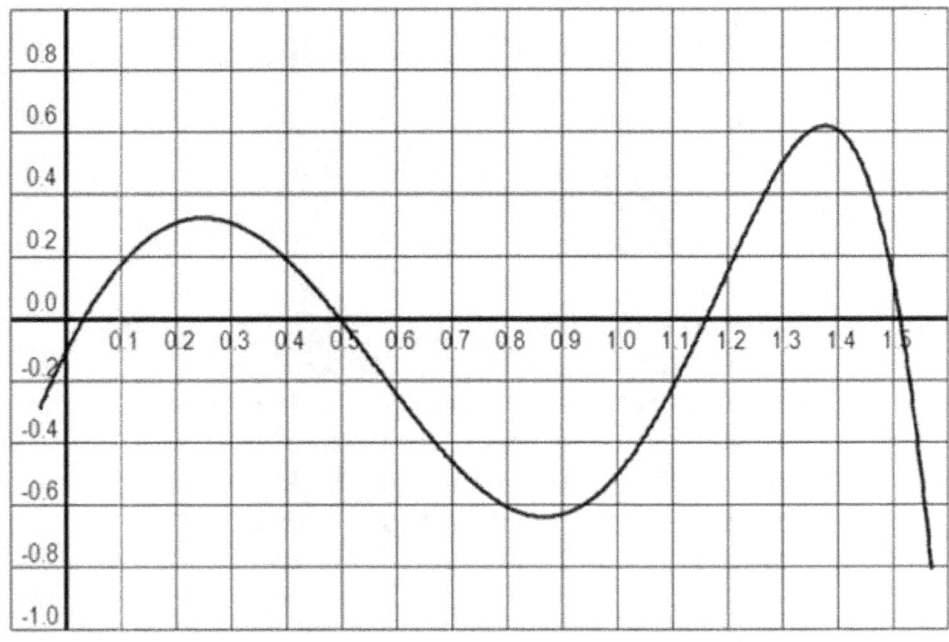

Figure 17. Solution of the algebraic equation:
Ax^6+Bx^5+Cx^2+Dx+E=0;
A=-3.8; B=7.2; C=-7.3; D=3.5; E=-0.1;
Roots: x1=0.032; x2=0.496; x3=1.16; x4=1.512

```
//======Main======
begin
  CoordSys; Polynom;
end.
```

Listing 1.7

An example of a transcendental equation follows further.

Example 1.8.

Find roots of the equation:

$$0.2x^2 + 10\cos^2(x/3) - 8 = 0$$

in interval -6.4 < x < 6.4 with an accuracy 0.001.

Solution.

The program Prog_1_8 (listing 1.8) contains two subroutines:

- "CoordSys" that draws and marks up rectangle coordinate axes;

- "Equat" that calculates values of the equation left part, draws the function graphic (fig.18), detects the equation roots and checks the solution accuracy.

```
Program Prog_1_8;
Uses GraphABC, ABCobjects;

Procedure CoordSys;
  const x0=100;
  var k, z, xd, yd: integer;
  var s: string;
  begin
    SetWindowSize(800,490); SetPenStyle(psClear);
    SetPenColor(clBlack);      SetPenWidth(1);
    Line(100, 40, 750, 40);      Line(100, 418, 750, 418);
    Line(100, 40, 100, 420);     Line(750, 40, 750, 420);
    Line(100, 230, 750, 230);  Line(425, 40, 425, 420);
    for k:=1 to 26 do
      begin
```

```pascal
        Line(99+25*k, 230, 99+25*k, 238);
      end;
    for k:=1 to 6 do
      begin
        xd:=1*k-7;    s:=IntToStr(xd);
        var t:= new TextABC(66+50*k,242,12,s);
      end;
    for k:=1 to 6 do
      begin
        xd:=1*k;        s:=IntToStr(xd);
        var t:= new TextABC(422+50*k,242,12,s);
      end;
     for k:=1 to 22 do
        begin
          Line(425,434-17*k, 432, 434-17*k);
        end;
    for k:=1 to 5 do
        begin
          yd:=1*k;   s:=IntToStr(yd);
          var t:= new TextABC(440,220-34*k,12,s);
        end;
     for k:=1 to 5 do
        begin
          yd:=1*k-6;   s:=IntToStr(yd);
          var t:= new TextABC(440,424-34*k,12,s);
        end;
  end;

Procedure Equat;
  const x0=425; y0=230; mx=50; my=35;
  var k, u, v, u1, u2, v1, v2: integer;
  var x, dx, y1, y2, dy, r1, r2, r3, r4: real;
  var s1, s2, s3, s4, p1, p2, p3, p4: string;
  begin
    SetPenWidth(3);  SetPenColor(clBlack);
    x:=-6.4; dx:=0.0001;
    for k:=1 to 64000 do
      begin
        x:=x+dx;   y1:=x*x/5+10*cos(x/3)*cos(x/3)-8;
        x:=x+dx;   y2:=x*x/5+10*cos(x/3)*cos(x/3)-8;
        u1:=x0+Round(x*mx); v1:=y0-Round(y1*my);
        if k>1 then Line(u1, v1, u2, v2);
```

```pascal
      u2:=u1; v2:=v1;
      if (x>5.5) and (x<6) and (y1*y2<0) then r1:=x;
      if (x>1.3) and (x<2) and (y1*y2<0) then r2:=x;
    end;
  r3:=-r2;  r4:=-r1;
  var t0:=new TextABC(160,10,12,'Root points:');
  Str(r1:4:3,s1);
  var t1:=new TextABC(310,10,12,'x1 = '+s1);
  Str(r2:4:3,s2);
  var t2:=new TextABC(410,10,12,'x2 = '+s2);
  Str(r3:4:3,s3);
  var t3:=new TextABC(510,10,12,'x3 = '+s3);
  Str(r4:4:3,s4);
  var t4:=new TextABC(610,10,12,'x4 = '+s4);
  var t5:=new TextABC(120,440,12,'Checking:');
  y1:=r1*r1/5+10*cos(r1/3)*cos(r1/3)-8;
  Str(y1:5:4,p1);
  var t6:=new TextABC(220,440,12, 'y(x1) = '+p1);
  y1:=r2*r2/5+10*cos(r2/3)*cos(r2/3)-8;
  Str(y1:5:4,p2);
  var t7:=new TextABC(350,440,12, 'y(x2) = '+p2);
  y1:=r3*r3/5+10*cos(r3/3)*cos(r3/3)-8;
  Str(y1:5:4,p3);
  var t8:=new TextABC(480,440,12, 'y(x3) = '+p3);
  y1:=r4*r4/5+10*cos(r4/3)*cos(r4/3)-8;
  Str(y1:5:4,p4);
  var t9:=new TextABC(610,440,12, 'y(x4) = '+p4);
  v:=y0;   SetPenStyle(psSolid);
  SetPenWidth(2);  SetPenColor(clBlack);
  u:=x0+Round(r1*mx);   Circle(u,v,8);
  u:=x0+Round(r2*mx);   Circle(u,v,8);
  u:=x0+Round(r3*mx);   Circle(u,v,8);
  u:=x0+Round(r4*mx);   Circle(u,v,8);
 end;
//======Main=====
begin
 CoordSys; Equat;
end.
```

Listing 1.8

Remark 1.

The function y(x) at the left part of the equation is even. It means that its roots are symmetrical relatively the origin. The subroutine "Equat" calculates the positive roots only and forms the negative roots by changing signs of the calculated roots.

Remark 2.

The two-stage search of roots is not necessary here due to a high-performance processor. Scanning of the axis X is being done at single stage with a small step $\Delta x = 0.00001$. Thus 64000 function values are calculated and the found roots are rounded to 0.001.

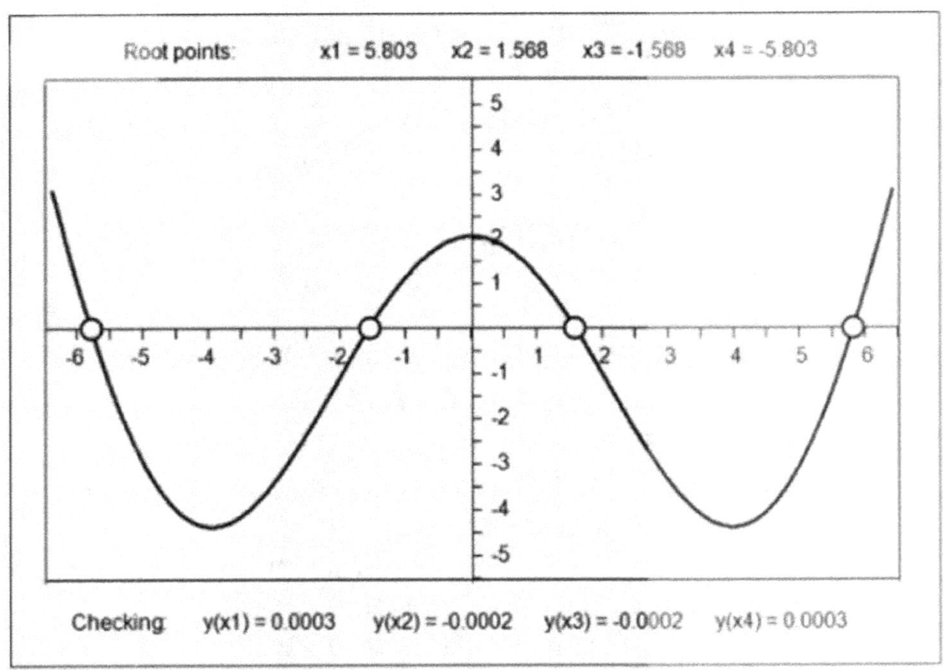

Figure 18. Solution of the transcendental equation

Conclusion.

One ought to substitute the found root value at the left part of the equation to check the accuracy of the root selection. In this case the result is less then 0.0003 though the value 0 was expected.

It is easy to decrease the error essentialy by diminishing of the step Δx for instance 10 times.

1.4. Extreme points

The smooth functions have a continuous first derivative in the interval of the extremums research. In accordance with the Fermat theorem a local extremum is detected at point x_0 where the first derivative is equal to zero. There are two adequate conditions of the extremum existence:

- the first derivative changes its sign when it is crossing the point x_0; a direction of sign change (plus to minus or minus to plus) is an indication of the extremum type (maximum or minimum);

- the second derivative is not equal to zero at the point x_0; a functiom maximum differs from a minimum by a sign of the second derivative.

There are known the best "pre-computer epoch" methods of the extremums search that might be found in the literature:

- the method of the consecutive bisection of an interval;

- the golden section method;

- the method of a parabolical approximation of an aim function.

The method of a continuous interval scanning was considered always as the least effective one. The modified procedure was suggested to improve the scanning method by means of a variable step. At the first a large step was used for a quick interval scanning and further it was decreased when an extremum was suspected. It was necessary to achieve a given search accuracy. By analogy with roots selection the one-stage scanning procedure can be used now due to the modern high-capacity computers. A convincing example can be seen by the analysis of the function y(x) from the previous task.

Examlpe 1.9.

Find the extreme points of the function

$$y(x) = 0.2x^2 + 10\cos^2(x/3) - 8$$

in the interval -6.4 < x < 6.4 with an accuracy 0.001.

Solution.

The program Prog_1_9 (listing 1.9) contains two subroutines:

- "CoordSys" that draws and marks up rectangle axes by analogy with the example 1.8;

- "Extrem" that calculates values of the function, draws the graphic (fig.19), detects the extreme points and displays its coordinates.

```
Program Prog_1_9;
Uses GraphABC, ABCobjects;

Procedure CoordSys;
 const x0=100;
 var k, z, xd, yd: integer;
 var s: string;
 begin
  SetWindowSize(800,490); SetPenStyle(psClear);
  SetPenColor(clBlack);      SetPenWidth(1);
  Line(100, 40, 750, 40);    Line(100, 418, 750, 418)
  Line(100, 40, 100, 420);   Line(750, 40, 750, 420);
  Line(100, 230, 750, 230);  Line(425, 40, 425, 420);
  for k:=1 to 26 do
    begin
     Line(99+25*k, 230, 99+25*k, 238);
    end;
  for k:=1 to 6 do
    begin
      xd:=1*k-7; s:=IntToStr(xd);
      var t:= new TextABC(66+50*k,242,12,s);
```

```pascal
  end;
 for k:=1 to 6 do
   begin
     xd:=1*k;  s:=IntToStr(xd);
     var t:= new TextABC(422+50*k,242,12,s);
   end;
 for k:=1 to 22 do
   begin
     Line(425,434-17*k, 432, 434-17*k);
   end;
 for k:=1 to 5 do
   begin
     yd:=1*k;    s:=IntToStr(yd);
     var t:= new TextABC(440,220-34*k,12,s);
   end;
  for k:=1 to 5 do
   begin
     yd:=1*k-6;    s:=IntToStr(yd);
     var t:= new TextABC(440,424-34*k,12,s);
   end;
end;

Procedure Extrem;
  const x0=425; y0=230; mx=50; my=35;
  var k, u, v, u1, u2, v1, v2, m1, m2, m3, m4: integer;
  var x, dx, y1, y2, dy, r1, r2, r3, r4: real;
  var xe1, ye1, xe2, ye2, xe3, ye3: real;
  var s1, s2, s3: string;
  begin
   SetPenWidth(3);     SetPenColor(clBlack);
   x:=-6.4;  dx:=0.0001;  m1:=1;  m2:=1;  m3:=1;
   for k:=1 to 64000 do
     begin
       x:=x+dx;  y1:=x*x/5+10*cos(x/3)*cos(x/3)-8;
       x:=x+dx;  y2:=x*x/5+10*cos(x/3)*cos(x/3)-8;
       dy:=y2-y1;
       u1:=x0+Round(x*mx);
       v1:=y0-Round(y1*my);
       if k>1 then Line(u1, v1, u2, v2);
       u2:=u1; v2:=v1;
       if (m1=1) and (x>-5) and (x<-0.3) and (dy>0) then
         begin
```

```pascal
          xe1:=x; ye1:=y2;   m1:=0
        end;
      if (m2=1) and (x>-1) and (x<1) and (dy<0) then
        begin
          xe2:=x; ye2:=y2; m2:=0;
        end;
      if (m3=1) and (x>3) and (dy>0) then
        begin
          xe3:=x; ye3:=y2;    m3:=0;
        end;
      end;
    var t0:=new TextABC(140,10,12,'The extreme points:');
    Str(xe1:4:3,s1);
    var t1:=new TextABC(330,10,12,'Xmin1 = '+s1);
    Str(xe2:4:3,s2);
    var t2:=new TextABC(460,10,12,'Xmax = '+s2);
    Str(xe3:4:3,s3);
    var t3:=new TextABC(590,10,12,'Xmin2 = '+s3);
    var t4:=new TextABC(140,440,12,'The extreme points:');
    Str(ye1:4:3,s1);
    var t5:=new TextABC(330,440,12,'Ymin1 = '+s1);
    Str(ye2:4:3,s2);
    var t6:=new TextABC(460,440,12,'Ymax = '+s2);
    Str(ye3:4:3,s3);
    var t7:=new TextABC(590,440,12,'Ymin2 = '+s3);
    v:=y0;                    SetPenStyle(psSolid);
    SetPenWidth(2);           SetPenColor(clBlack);
    u:=x0+Round(xe1*mx);  Circle(u,v,8);
    u:=x0+Round(xe2*mx);  Circle(u,v,8);
    u:=x0+Round(xe3*mx);  Circle(u,v,8);
    SetPenWidth(2);           u:=x0+Round(xe1*mx);
    v:=y0-Round(ye1*my);  Circle(u,v,8);
    SetPenWidth(1);           Line(u,y0,u,v);
    SetPenWidth(3);           u:=x0+Round(xe2*mx);
    v:=y0-Round(ye2*my);   Circle(u,v,8);
    u:=x0+Round(xe3*mx);   v:=y0-Round(ye3*my); Cir-
cle(u,v,8);
    SetPenWidth(1);           Line(u,y0,u,v);
end;
```

```
//======Main====
```
begin
  CoordSys; Extrem;
end.

Listing 1.9

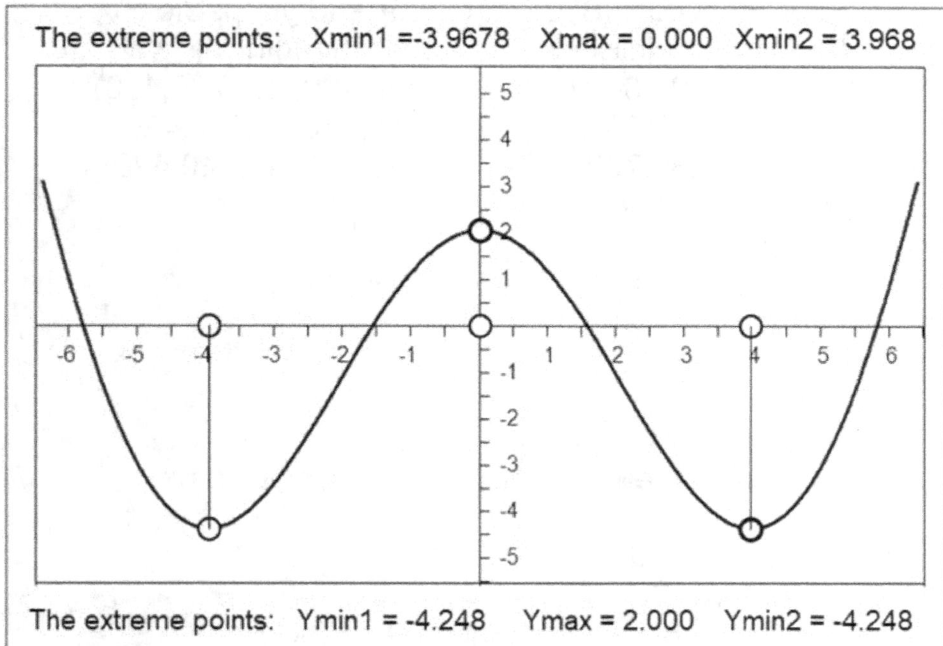

Fig.19. Three extremum points

Example 1.10.

Find the extreme points of the function

$$y(x) = -3.8x^6 + 7.2x^5 - 7.3x^2 + 3.5x - 0.17$$

in the interval $-0.1 < x < 16.0$ with the accuracy 0.001.

Solution.

The program Prog_1_10 (listing 1.10) contains two sub-routines:

   - "CoordSys" that draws and marks up rectangle axes;

   - "Polynom" that calculates values of the function, draws the graphic (fig.20), detects the extremal points and displays its coordinates.

```pascal
Program Prog_1_10;
Uses GraphABC, ABCobjects;

Procedure CoordSys;
  var k, z: integer; var xc, yc: real;
  var s: string;
  begin
   SetWindowSize(850,500);  SetPenStyle(psClear);
   SetPenColor(clBlack);        SetPenWidth(1);
   Line(100,40,780,40);         Line(100,420,780,420);
   Line(100,40,100,420);        Line(780,40,780,420);
   Line(100,230,780,230);       Line(140, 40,140,420);
  for k:=1 to 19 do
    begin
       Line(20+40*k, 230, 20+40*k, 238);
    end;
   for k:=2 to 16 do
    begin
       xc:=0.1*(k-1);Str(xc:3:1,s);
       var t:= new TextABC(92+40*k,240,11,s);
   end;
   for k:=1 to 9 do
      begin
          Line(132, 40*k+30, 140, 40*k+30);
      end;
  for k:=1 to 4 do
     begin
       yc:=1-0.2*k;    Str(yc:2:1,s);
       var t:= new TextABC(108,20+40*k,12,s);
     end;
  for k:=1 to 4 do
     begin
       yc:=-0.2*k;    Str(yc:2:1,s);
       var t:= new TextABC(146,220+40*k,12,s);
     end;
end;
Function xpn(x: real; n: integer): real;
 var j: integer;
 var  acc: real;
 begin
  acc:=1.0;
  for j:=1 to n do  acc:=acc*x;
```

```pascal
    xpn:=acc;
  end;

Procedure Polynom;
 const x0=140; y0=230; mx=400; my=200;
 var k, x1, x2, y1, y2, m1, m2, m3, u, v : integer;
 var a, b, c, d, e, x, yp, yq, dx, dy: real;
 var xe1, ye1, xe2, ye2, xe3, ye3: real;
 var s1, s2, s3: string;

  begin
    SetWindowSize(800,500);  SetPenStyle(psClear);
    a:=-3.8;  b:=7.2;  c:=-7.3;  d:=3.5;  e:=-0.1;  x:=-0.18;
    yp:=a*xpn(x,6)+b*xpn(x,5)+c*x*x+d*x+e;
    x:=-0.1;  dx:=0.001;  m1:=1; m2:=1; m3:=1;
   for k:=1 to 835 do
     begin
       x:=x+dx;  yp:=a*xpn(x,6)+b*xpn(x,5)+c*x*x+d*x+e;
       x:=x+dx;  yq:=a*xpn(x,6)+b*xpn(x,5)+c*x*x+d*x+e;
       dy:=yq-yp;
       if (m1=1) and (x<0.4) and (dy<0) then
         begin
           xe1:=x; ye1:=yq; m1:=0;
         end;
       if (m2=1) and (x>0.7) and (dy>0) then
         begin
           xe2:=x; ye2:=yq; m2:=0;
         end;
       if (m3=1) and (x>1.2) and (dy<0) then
         begin
           xe3:=x; ye3:=yq; m3:=0;
         end;
       x1:=x0+Round(x*mx);      y1:=y0-Round(yp*my);
       SetPenWidth(3);          SetPenColor(clBlack);
       if k>1 then Line(x1, y1, x2, y2);
       x2:=x1;   y2:=y1;
end;
var t0:=new TextABC(140,10,12,  'The extreme points:');
Str(xe1:4:3,s1);
var t1:=new TextABC(330,10,12,'Xmax1  = '+s1);
Str(xe2:4:3,s2);
```

```
  var t2:=new TextABC(460,10,12,'Xmin = '+s2);
  Str(xe3:4:3,s3);
  var t3:=new TextABC(590,10,12,'Xmax2 = '+s3);
  var t4:=new TextABC(140,430,12, 'The extreme points:');
  Str(ye1:4:3,s1);
  var t5:=new TextABC(330,430,12,'Ymax1 = '+s1);
  Str(ye2:4:3,s2);
  var t6:=new TextABC(460,430,12,'Ymin = '+s2);
  Str(ye3:4:3,s3);
  var t7:=new TextABC(590,430,12,'Ymin2 = 3);
  v:=y0;  SetPenStyle(psSolid);   SetPenWidth(2);
  SetPenColor(clBlack);
  u:=x0+Round(xe1*mx);    Circle(u,v,8);
  u:=x0+Round(xe2*mx);    Circle(u,v,8);
  u:=x0+Round(xe3*mx);    Circle(u,v,8);
  u:=x0+Round(xe1*mx);    v:=y0-Round(ye1*my);
  Circle(u,v,8);          SetPenWidth(1);
  Line(u,y0,u,v);         SetPenWidth(2);
  u:=x0+Round(xe2*mx);    v:=y0-Round(ye2*my);
  Circle(u,v,8);          SetPenWidth(1);
  Line(u,y0,u,v);         u:=x0+Round(xe3*mx);
  v:=y0-Round(ye3*my);    Circle(u,v,8);
  SetPenWidth(1);         Line(u,y0,u,v);
end;
//=======Main=====
begin
  CoordSys; Polynom;
end.
```

Listing 1.10

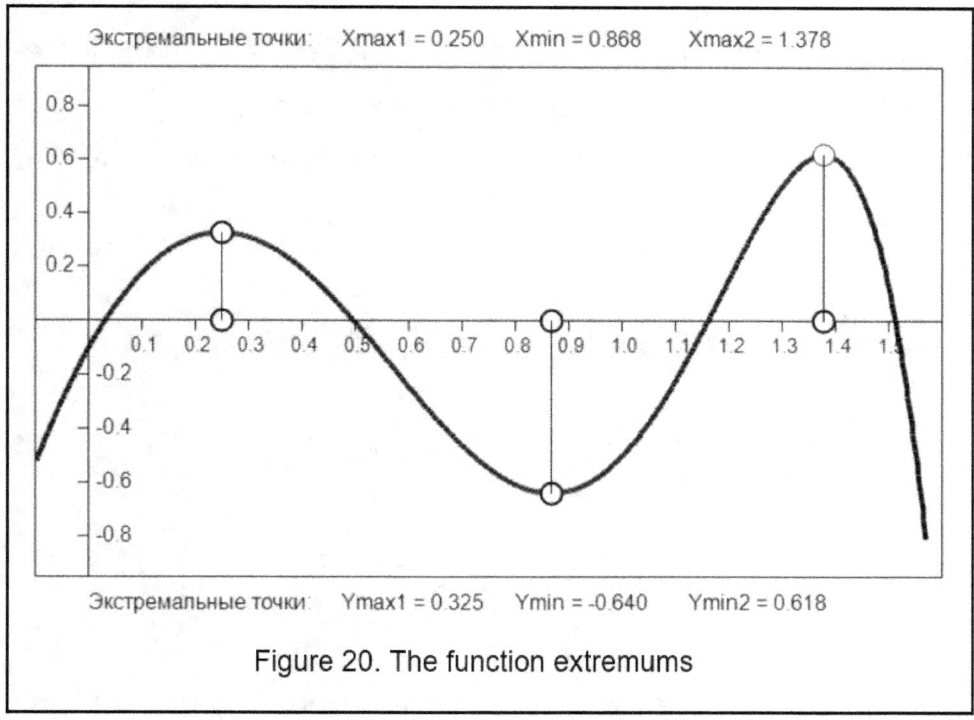

Figure 20. The function extremums

Chapter 2. NUMERICAL DIFFERENTIATION

2.1. Methods of derivative approximation

An advantage of the numerical differentiation is in the fact that the result can be achieved even if an analytical expression of a function is absent. It is quite enough to have a numerical row of function values and a fixed argument step is not obligatory.

A numerical differentiation is done in accordance with a derivative definition but with reserve that an argument increment Δx remains finite though it can be very small.

There are three modes to substitute the first derivative for a ratio:

$$\left.\frac{dy(x)}{dx}\right|_{right} \approx \frac{y_{k+1}-y_k}{\Delta x}, \qquad 0 \le k \le n-1; \qquad (2.1)$$

$$\left.\frac{dy(x)}{dx}\right|_{left} \approx \frac{y_k-y_{k-1}}{\Delta x}, \qquad 1 \le k \le n; \qquad (2.2)$$

$$\left.\frac{dy(x)}{dx}\right|_0 \approx \frac{y_{k+1}-y_{k-1}}{2\Delta x}, \qquad 1 \le k \le n-1. \qquad (2.3)$$

The expressions (2.1) – (2.3) are called the right, left and central approximations of the derivative accordingly.

The right and the left ones provide aproximations to the first derivative with the first order accuracy and the central approximation ensures the second order accuracy. The left approximation is recommended for a differentiation of dynamic processes and the central one – for an analysis of net functions.

The second derivative can be got as relation of finite differences defined for the first derivative.

2.2. Solution of applied problems

Differentiation of a function has a great many applications. There are the most expanded problems:

- seeking of local extremums in optimisation tasks;

- calculations a velocity and an acceleration of the object in auto-matic control systems;

- suppression of constant and low frequence spectral har-monics in the mixture of useful signal plus noise.

Example 2.1.

Calculate the first derivative of the function

$$y(x) = 90*\sin(x/3)/(x + 3)$$

in the interval $0 < x < 10$ using two modes:

- analytical using exact differential rules of an infinite mathematic;

- approximately with numerical methods.

It is necessary to draw graphics of the functions, derivative and errors of the numerical differentiation.

Solution.

The program Prog_2_1 (listing 2.1) contains two subrou-tines:

- "CoordSys" that draws and marks up rectangle axes;

- "Diff" that calculates values of the function, derivative and errors and then draws three graphics (fig.21).

The analytical differentiation provides the exact pattern of the first derivative:

$$dy/dx = 90[(x + 3)*\cos(x/3)/3 - \sin(x/3)]/(x^2 + 6x + 9)$$

```
Program Prog_2_1;
Uses GraphABC, ABCobjects;
Procedure CoordSys;
```

```pascal
var  k, z, xc, yc: integer;
var  s: string;
begin
  SetWindowSize(850, 500);   SetPenStyle(psClear);
  SetPenColor(clBlack);         SetPenWidth(1);
  Line(100, 40, 784, 40);       Line(100, 440, 784, 440);
  Line(100, 40, 100, 440);      Line(784, 40, 784, 440);
  Line(100, 320, 784, 320);

  for k := 1 to 20 do
    begin
      Line(66 + 34 * k, 320, 66 + 34 * k, 328);
    end;
  for k := 2 to 10 do
      begin
        xc := 1 * (k - 1);        s:= IntToStr(xc);
        var t := new TextABC(28 + 68 * k, 330, 11, s);
      end;
  for k := 1 to 10 do
    begin
      Line(100, 10 + 40 * k + 30, 108, 10 + 40 * k + 30);
    end;
    for k := 1 to 11 do
    begin
      yc := 14 - 2 * (k - 1);
      s := IntToStr(yc);
      var t := new TextABC(76, 40 * k - 10, 12, s);
    end;
end;

Procedure Diff;
const  x0 = 100;  y0 = 320;  mx = 68;  my = 20;
var k, x1, x2, y1, y2, z1, z2, w1, w2, m, u, v: integer;
var x, xm, ym, yp, yq, dx, dy, z, w, dz: real;
var  s1, s2: string;
begin
  SetWindowSize(850, 500);  SetPenStyle(psClear);
  x := 0;dx := 0.001;M := 1;
  SetPenStyle(psClear);
  for k := 1 to 5000 do
  begin
    x := x + dx;  yp := 90 * sin(x / 3) / (x + 3);
```

```pascal
    x := x + dx;  yq := 90 * sin(x / 3) / (x + 3);
    w := 90 * ((x + 3) * cos(x / 3) / 3 - sin(x / 3)) / (x * x
    + 6 * x + 9);
    x1 := x0 + Round(x * mx);     y1 := y0 - Round(yp * my);
    dy := yq - yp;
    if dy > 0 then
      begin
        xm := x; ym := yp;  m := 0;
      end;
    z := dy / dx; dz := (z - w) / 10;
    if k < 6 then writeln(dz * 100:5:4, ' %');
    z1 := y0 - Round(z * my);
    w1 := y0 - Round(dz * my * 40000);
    SetPenWidth(3);
    if k > 1 then
    begin
      SetPenColor(clBlack);  Line(x1, y1, x2, y2);
      SetPenColor(clBlack);  Line(x1, z1, x2, z2);
    end;
  if k > 1 then
    begin
      SetPenColor(clBlack);   Line(x1, y1, x2, y2);
      SetPenColor(clBlack);   Line(x1, z1, x2, z2);
      SetPenWidth(3);         SetPenColor(clBlack);
      Line(x1, w1, x2, w2);
    end;
    x2 := x1;y2 := y1;z2 := z1;w2 := w1;
  end;
  var t0 := new TextABC(120, 10, 12,
            'Point of the maximum:');
  Str(xm:4:3, s1);
  var t1 := new TextABC(300, 10, 12, 'x = ' + s1);
  Str(ym:4:3, s2);
  var t2 := new TextABC(400, 10, 12, 'Ymax = ' + s2);
  SetPenStyle(psSolid);     SetPenWidth(1);
  SetPenColor(clBlack);     u := x0 + Round(xm * mx);
  Circle(u, y0, 8);         yq := 90 * sin(xm / 3) / (xm + 3);
  v := y0 - Round(yq * my); Circle(u, v, 8);
  SetPenWidth(1);           Line(u, y0, u, v);
end;
//===================Main===========
```

```
begin
  CoordSys; Diff;
  var t1 := new TextABC(460, 80, 12, 'y(x)');
  var t2 := new TextABC(180, 270, 12, 'dy / dx');
  var t3 := new TextABC(360, 210, 11, 'Error');
  var t4 := new TextABC(360, 230, 11,
          'of the differentiation');
  var t5 := new TextABC(360,250,11,
          'Emax = 0.033%  at  x = 0');
end.
```

Listing 2.1

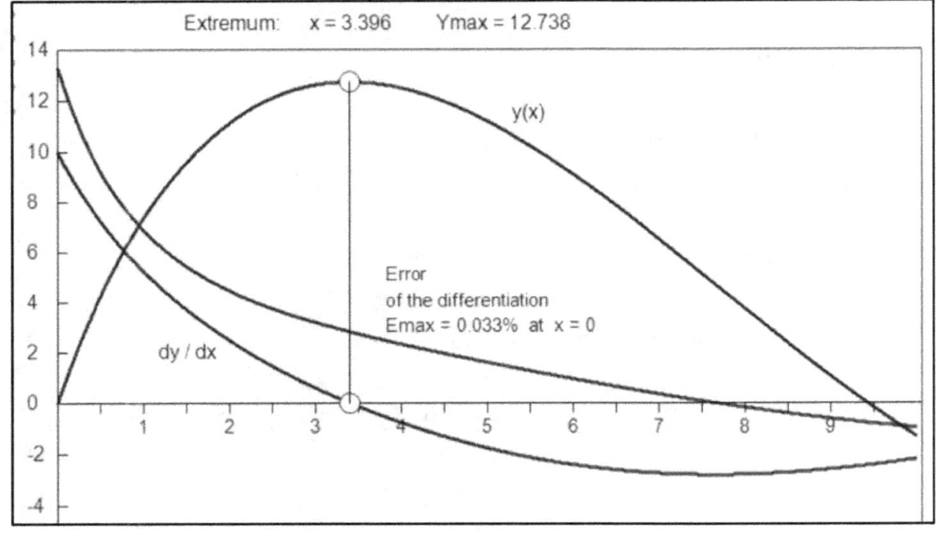

Figure 21. The first derivative of the function

The high frequence pulse with rectangle envelope is used in the range finders (radar, laser, sonar). Reflected from some heterogeneity, the pulse returns at the radiation point being very smoothed. Due to a loss of the energy its form becomes like a bell in vertical section. Such a form can be well approximated by the "cosine-square" function. The procedure of the week signsl selecrtion from the mixture with a noise as a rule includes the differentiation of the smoothed pulse.

Example 2.2.

Fulfil double differentiation of the function:

$$y(x) = 5*[\cos (x) - 1]^2$$

in the interval $0 < x < 2\pi$ by the classical method using the differentiation rules of the exponential and trigonometric function.

Solution.

The program Prog_2_2 (listing 2.2) contains two subroutines:

- "CoordSys" that draws and marks up rectangle axes;

- "ExactDif" that calculates values of the function and derivatives and errors.

These are the formulae of the first and the second derivatives.

$$dy/dx = -5*(\cos (x) - 1)*\sin(x);$$

$$d^2y/dx^2 = -5*[\cos^2(x) - \cos (x) - \sin^2(x)].$$

The result of the differentiation is at fig.22.

```
Program Prog_2_2;
Uses GraphABC, ABCobjects;

Procedure CoordSys;
var  k, z, xc, yc: integer;   var d: real;   var  s: string;
begin
  SetWindowSize(850, 480); SetPenStyle(psClear);
  SetPenColor(clBlack);       SetPenWidth(1);
  Line(80, 40, 800, 40);      Line(80, 360, 800, 360);
  Line(80, 40, 80, 360);      Line(800, 40, 800, 360);
  Line(80, 240, 800, 240);    Line(440, 80, 440, 320);
  for k := 1 to 25 do
  begin
   Line(50 + 30 * k, 240, 50 + 30 * k, 248);
  end;
for k := 2 to 12 do
   begin
```

```pascal
      xc := 30 * (k - 1);
      s := IntToStr(xc);
      var t := new TextABC(10+60 * k, 250, 10, s);
    end;
for k := 1 to 8 do
    begin
      Line(80, 10 + 40 * k + 30, 90, 10 + 40 * k + 30);
    end;
for k := 1 to 9 do
    begin
      yc := 30 - 5 * k;
      s:=IntToStr(yc);
      var t := new TextABC(60, 40 * k -8, 10, s);
    end;
 for k := 1 to 8 do
    begin
      Line(790, 10 + 40 * k + 30, 800, 10 + 40 * k + 30);
    end;
 for k := 1 to 9 do
     begin
       yc := 30 - 5 * k; s:=IntToStr(yc);
      var t := new TextABC(808, 40 * k -8, 10, s);
    end;
end;

Procedure ExactDif;
const  pi=3.14158; x0 = 80;  y0 = 240;
mx = 115; my = 8; me1=1000; me2=1000;
var k, xx, x1, x2, y1, y2, z1, z2, w1, w2: integer;
var m, u1, u2, v1, v2, ew1, ew2, ev1, ev2: integer;
var x, p, dp ,xm, ym, yp, yq, yr,dx, dy, v, va, w, wa, z, dz: real;
var a1, a2, a3, b1, b2, c1, c2: real;
var wp, wq, zp, zq, up, uq, vp, vq, ew, ev: real
begin
  SetPenStyle(psClear);
  x := 0;   dx:=0.001;
  for k := 1 to 2080 do
  begin
  x:=x+dx;   a1 := 5 * (cos(x)-1)*(cos(x)-1);
  x:=x+dx;   a2 := 5 * (cos(x)-1)*(cos(x)-1);
  x:=x+dx;   a3 := 5 * (cos(x)-1)*(cos(x)-1);
  v:=-5*(cos(x)-1)*sin(x);
```

```pascal
  w:=-5*(cos(x)*cos(x)-cos(x)-sin(x)*sin(x));
  dy:=a3-a1;    va:=dy/(4*dx);
  wa:=(a3-2*a2+a1)/(2*dx*dx);
  x1 := x0 + Round(x * mx);    y1 := y0 - Round(a3 * my);
  v1 := y0 - Round(va * my);   w1 := y0 - Round(w * my);
  SetPenWidth(2);
  if k > 1 then
  begin
    SetPenColor(clBlack);   SetPenWidth(3);
    Line(x1, y1, x2, y2);       Line(x1, v1, x2, v2);
    Line(x1, w1, x2, w2);       SetPenColor(clBlue);
    SetPenWidth(3);
  end;
  x2 := x1; y2 := y1; v2 := v1; w2:=w1;
 end;
end;
//================ Main =================
begin
 CoordSys; ExactDif;
 var t1 := new TextABC(410, 60, 12, 'y(x)');
 var t2 := new TextABC(360, 174, 12, 'dy / dx');
 var t3 := new TextABC(420, 324, 11, 'd2y / dx2');
end.
```

Listing 2.2

The program Prog_2_3 (listing 2.3) solves the same problem by means of the numeric method and estimates the solution accuracy. The function and both derivatives are drawn with thin curves on the fig.23. The approximation errors are drawn with the bold curves.

```pascal
Program Prog_2_3;
Uses GraphABC, ABCobjects;

procedure CoordSys;
var  k, z, xc, yc: integer;
var d: real;
var  s: string;
```

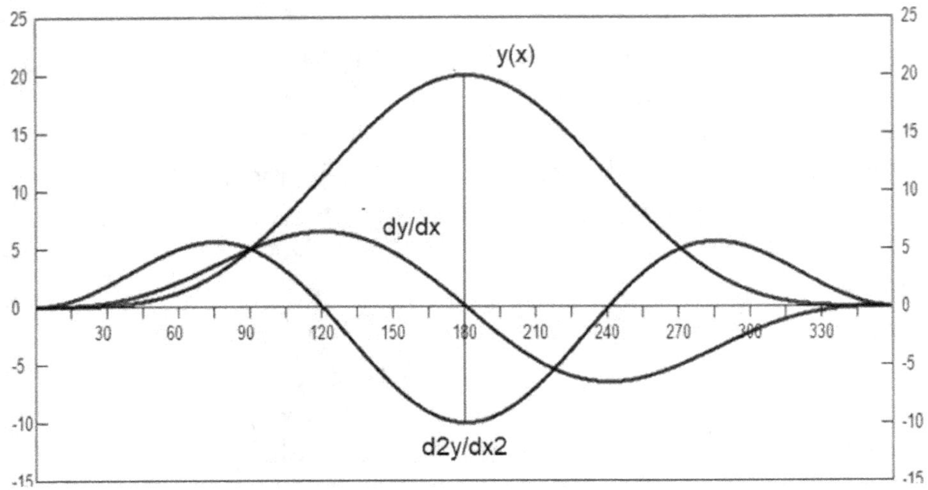

Figure 22. Double differentiation

```
begin
  SetWindowSize(850, 480);    SetPenStyle(psClear);
  SetPenColor(clBlack);       SetPenWidth(1);
  Line(80, 40, 800, 40);      Line(80, 440, 800, 440);
  Line(80, 40, 80, 440);      Line(800, 40, 800, 440);
  Line(80, 240, 800, 240);
  for k := 1 to 25 do
     begin
       Line(50 + 30 * k, 240, 50 + 30 * k, 248);
     end;
  for k := 2 to 12 do
     begin
       xc := 30 * (k - 1);  s := IntToStr(xc);
       var t := new TextABC(10+60 * k, 250, 10, s);
     end;

  for k := 1 to 10 do
     begin
       Line(80, 10 + 40 * k + 30, 88, 10 + 40 * k + 30);
     end;
  for k := 1 to 11 do
     begin
       d := (24 - 4 * k)/100;  Str(d:3:2,s);
       var t := new TextABC(48, 40 * k -8, 10, s);
     end;
  for k := 1 to 10 do
    begin
```

```pascal
      Line(794, 10 + 40 * k + 30, 800, 10 + 40 * k + 30);
    end;
  for k := 1 to 11 do
    begin
      yc := 30 - 5 * k; s:=IntToStr(yc);
      var t := new TextABC(808, 40 * k -8, 10, s);
    end;
end;

Procedure NumericDif;
const  pi=3.14158; x0 = 80;  y0 = 240;
       mx = 115;  my = 8; me1=1000; me2=1000;
var k, xx, x1, x2, y1, y2, z1, z2, w1, w2: integer;
var m, u1, u2, v1, v2, ew1, ew2, ev1, ev2: integer;
var x, p, dp, xm, ym, yp, yq, yr, dx, dy: real;
var a1, a2, a3, b1, b2, c1, c2, v, va, w, wa, real;
var z, dz, wp, wq, zp, zq, up, uq, vp, vq, ew, ev: real;
var s1, s2: string;
begin
  SetPenStyle(psClear);
  x := 0;   dx:=0.001;
  for k := 1 to 2080 do
  begin
   x:=x+dx;    a1 := 5 * (cos(x)-1)*(cos(x)-1);
   x:=x+dx;    a2 := 5 * (cos(x)-1)*(cos(x)-1);
   x:=x+dx;    a3 := 5 * (cos(x)-1)*(cos(x)-1);
   v:=-5*(cos(x)-1)*sin(x);
   w:=-5*(cos(x)*cos(x)-cos(x)-sin(x)*sin(x));
   dy:=a3-a1; va:=dy/(4*dx);
   ev:=100*(v-va)/6.99;
   wa:=(a3-2*a2+a1)/(2*dx*dx);
   ew:=10*(w-wa);
   x1 := x0 + Round(x * mx);
   y1 := y0 - Round(a3 * my);
   v1 := y0 - Round(va * my);
   w1 := y0 - Round(w * my);
   ev1 := y0 - Round(ev * me1);
   ew1 := y0 - Round(ew * me2);
   SetPenWidth(2);
   if k > 1 then
   begin
     SetPenColor(clBlack);  SetPenWidth(2);
```

```
      Line(x1, y1, x2, y2);      Line(x1, v1, x2, v2);
      Line(x1, w1, x2, w2);   SetPenColor(clBlue);
      SetPenWidth(4);            Line(x1, ev1, x2, ev2);
      Line(x1, ew1, x2, ew2);
    end;
    x2 := x1;        y2 := y1;
    v2 := v1;        w2:=w1;
    ew2:=ew1;    ev2:=ev1;
  end;
end;
```

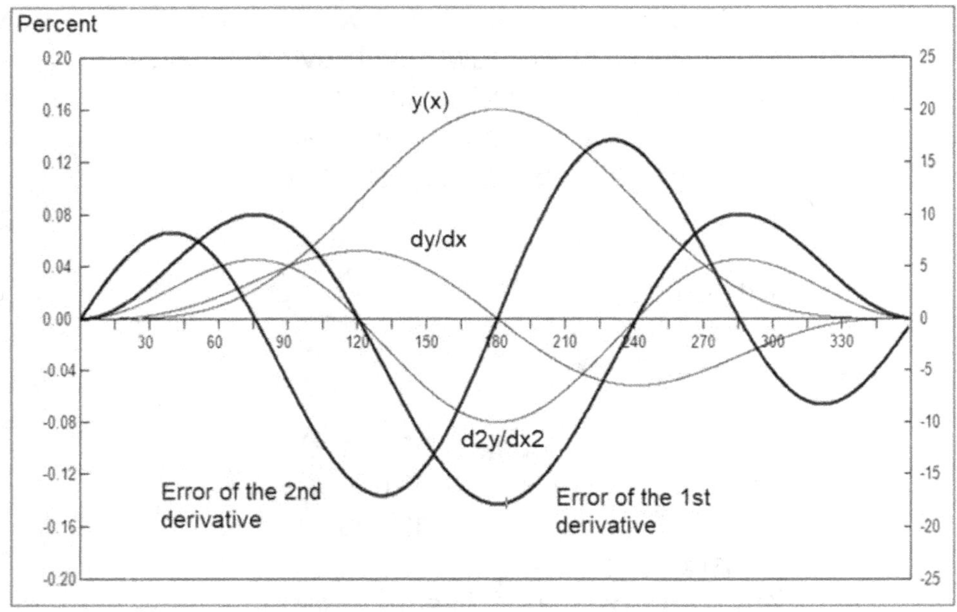

Figure 23. Errors of the numerical differentiation

```
//========================Main=====================
begin
  CoordSys; NumericDif;
  var t0 := new TextABC(30, 10, 11,  'Percent');
  var t1 := new TextABC(300, 10, 12, 'Numeric differentiation');
  var t2 := new TextABC(410, 60, 12, 'y(x)');
  var t3 := new TextABC(360, 174, 12, 'dy / dx');
  var t4 := new TextABC(420, 324, 11, 'd2y / dx2');
  var t5 := new TextABC(490, 370, 11, 'Errors');
  var t6 := new TextABC(490, 390, 11, 'of the 1st derivative');
  var t7 := new TextABC(210, 370, 11, 'Errors');
```

```
  var t8 := new TextABC(210, 390, 11, 'of the 2nd derivative');
end.
```

Listing 2.3

Chapter 3. NUMERICAL INTEGRATION

3.1. Integral approximation methods

There are problems in engineering and economic where integration of one-variable function is required. If the integrable function f(x) can be decomposed to some elementary functions in the interval [a, b] then Newton-Leibniz formula is applied:

$$\int_a^b f(x)dx = F(b) - F(a),$$

where F(x) – an antiderivative function of the f(x).

Numerous reference books deal with integration methods. The situations arise quite often however when f(x) or F(x) are not expressed by elementary functions or an analytical record of the f(x) does not exist. A quadrature formula for an approximate integration is used in this case:

$$\int_a^b f(x)dx = \sum_k w_k f(x_k) + R_n,$$

where
w_k – a weight coefficient of the function value with the index k;

R_n – a residual (or an error) of integration.

Weight coefficients are selected in such manner that an error becomes not more then a given value. An integration accuracy is raised by means of an enlarging of the addends quantity or by a rational choice of the quadrature formula that is by an approximation method.

Among these methods in the literature about the numerical analysis one can find the formulae of a rectangle rule, a trapezoid rule, methods of Simpson, Gauss, Chebyshev. It is advantageous to apply the simplest methods that provide a quick preparaton for a problem solution using a computer. The necessary approximation accuracy can be achieved by decrease of the integration step that is by increase of calculaton volume for a digital computer.

Rectangle rule

The integration algorithm consists of the next actions:

1. The argument interval between the point $a = x_0$ and the point $b = x_k$ is broken up into n equal parts by the length $h = (b - a)/n$.

2. The integrable function $y(x)$ inside every interval is substituted by the horizontal straight segment at the left or at the right side of the current point $[x_k; y(x_k)]$.

3. The area of the elementary rectangle is calculated:

$$S_k = h^* \, y(x_k).$$

4. The definite integral is an area under the curve $y(x)$. This area is approximated by the sum of values S_k:

$$\int_a^b y(x)dx \approx h^*(y_0 + y_1 + ... + y_{n-1}) = h\sum_{k=0}^{n-1} y(x_k)$$

It is evident that the approximation accuracy will arise with deminishing of the value h. If the exact integral value is failed to compute then the approximation error R is estimated by means of the double integrations $S_1(h_1)$, $S_2(h_1)$, made with step values h_1 and $h_2 = 0.5h_1$:

$$R \le \frac{\left|S_1(h_1) - S_2(h_2)\right|}{3}$$

Trapezoid rule

The integrable function y(x) in the interval [a; b] is substituted by a broken line z(x) formed with straight pieces that join points belonging to the function y(x). These points have tabulated values $y(x_k)$ that are called a grid sampling. Let us consider that the grid is formed with the constant argument step h = (a − b)/h. The linear segments are described by the equation:

$$z_k(x) = y_{k-1} + (y_k - y_{k-1})*(x - x_{k-1})/h, \qquad k = 1, 2, \dots, n$$

The integral of the approximating function is a sum of integrals:

$$\int_a^b z(x)dx \approx \sum_{k=0}^{n-1} \int_{x_{k-1}}^{x_k} z(x)dx =$$

$$= \frac{b-a}{n}[0.5y(a) + y(x_1) + y(x_2) + \dots + 0.5y(b)]$$

The trapezoid rule in accordance with its definition naturally provides the higher approximation accuracy in every piece of the broken line in comparison with the rectangle rule. It is clear of course - the second order approximation is better than the first order method in a separate interval. But one ought to remember that the effect of a mutual error compensation may occure inside the lengths where the integrable function has segments of growth and slump. In such case the rectangle rule seems not too bad as to an accuracy.

Simpson method

The main idea of this method is the piece-nonlinear approximation of the integrable function y(x) between the knot points with tabulated ordinates. A quadratic parabola is chosen as an approximating segment. Such line is the Lagrange interpolation polynom of the second degree. It provides the forth order accuracy of the approximation (for the comparison – the second order accuracy is achieved only with the trapezoid rule). The integration interval is broken at even quantity of pieces. Every pair of pieces has three knot points x_0, x_1, x_2 and three ordinates y_0, y_1, y_2. The coefficients $L_k(x)$ and the polynom values $P_k(x)$ are calculated at the doubled segment:

$$P(x) = y_0 * L_0(x) + y_1 * L_1(x) + y_2 * L_2(x),$$

where:

$$L_0(x) = \frac{(x - x_1)(x - x_2)}{(x_0 - x_1)(x_0 - x_2)};$$

$$L_1(x) = \frac{(x - x_0)(x - x_2)}{(x_1 - x_0)(x_1 - x_2)};$$

$$L_2(x) = \frac{(x - x_0)(x - x_1)}{(x_2 - x_0)(x_2 - x_1)}.$$

It is necessary to say once more that the rectangle rule is the very preferable for the approximated integration nowadays due to the minimum waste of time for the mathematical preparation to the problem solution. Any accuracy can be achieved at the same time by means of the simple decrease of the integration step that is by enlarging quantity of computer operations. It means that the processor working time is to be increased at some milliseconds only. Such time expenditure cannot be compared with waste of working time of the programmer if he would solve the integration problem using the Lagrange method for instance. No proof is needy that the programmer labour is estimated higher then the computer's work.

3.2. Solution of applied tasks

At first it is useful to take a problem that has an exact solution. It means we need a function that is convenient to integrate using known methods of the mathematic analysis, the Newton-Leibniz formula and the reference books. The exact integration allows to estimate the error of approximation.

Example 3.1.

Calculate the energy of the pulse that is modelled by a function

$$p(t) = A*[\cos(t) - 1]^2, \quad 0 \le t \le T,$$

where $\quad$ $A = 10$ – the pulse amplitude;

$\quad\quad\quad$ $T = 10$ – the pulse duration.

The function $p(t)$ is easy integrated "by hand".

Solution.

Pprogram Prog_3_1 (listing 3.1) contains two subroutines:

- "CoordSys" that draws and marks up rectangle axes;

- "Energy" that computes the values of the pulse, the integral and the approximation errors (fig.24).

```pascal
Program Prog_3_1;
Uses GraphABC, ABCobjects;

Procedure CoordSys;
var k, xd, yd: integer;
var x: real;   var d: string;
begin
  SetWindowSize(850, 520);  SetPenStyle(psClear);
  SetPenColor(clBlack);        SetPenWidth(1);
  Line(100, 10, 800, 10);        Line(100, 440, 800, 440);
  Line(100, 10, 100, 440);       Line(800, 10, 800, 440);
  for k := 1 to 20 do
    begin
      Line(100+35*k, 440, 100+35*k, 448);
    end;
  for k := 1 to 9 do
    begin
      xd := 1 * k; d:=IntToStr(xd);
      var t := new TextABC(166 + 70 * (k-1), 450, 11, d);
    end;
for k := 1 to 9 do
    begin
      Line(100, 48 * k+8, 108, 48*k+8);
    end;
  for k := 0 to 9 do
    begin
      yd := 5 * k;  d := IntToStr(yd);
      var t := new TextABC(75, 432 - 48 * k, 11, d);
    end;
  for k := 0 to 8 do
    begin
      Line(792, 50 * k-2, 800, 50*k-2);
    end;
    for k := 0 to 8 do
    begin
      yd := 20 * k; d := IntToStr(yd);
      var t := new TextABC(808, 440 - 50 * k, 11, d);
    end;
end;
```

```pascal
Procedure Energy;
const pi=3.14158; x0 = 100;  y0 = 440;
        mx = 107.5;  my = 9.6; mz=0.04;
var k, x1, x2, y1, y2, z1, z2: integer;
var x, dx, yp, yq, dy, z, w, dz, ye, yd, zp, zq, pd, ed: real;
var s: string;
begin
  dx:= 0.01; x:=0; pd:=10; zp:=0;
for k := 1 to 650 do
    begin
    yp := 10 * (cos(x)-1)*(cos(x)-1);  zp:=zp+yp; x:=x+dx;
    yq := 10 * (cos(x)-1)*(cos(x)-1);  zq:=zp+yp;
    x1 := x0 + Round(x * mx);         y1 := y0 - Round(yp * my);
    z1 := y0 - Round(zp * mz);        SetPenWidth(3);
    if k > 1 then
      begin
        SetPenColor(clBlack);
        Line(x1, y1, x2, y2); Line(x1, z1, x2, z2);
      end;
      x2 := x1; y2 := y1; z2:=z1;
  end;
  SetPenWidth(1);
  Line(100, 56, 445, 56);  yd:=zq/60;
  ye:=10*(1.5*pd+sin(2*pd)/4-sin(pd));
  ed:=100*(ye-yd)/ye;
  var t1:=new TextABC(650,130,11,'Error');
  var t2:=new TextABC(650,150,11, 'of approximation');
  Str(ed:4:3,s);
  var t3:=new TextABC(650,170,11,s+' %');
end;
//===============Main===============
begin
  CoordSys;  Energy;
  var t1:=new TextABC(400,30,12,'Pulse');
  var t2:=new TextABC(400,190,12,'In tegral');
 end.
```

Listing 3.1

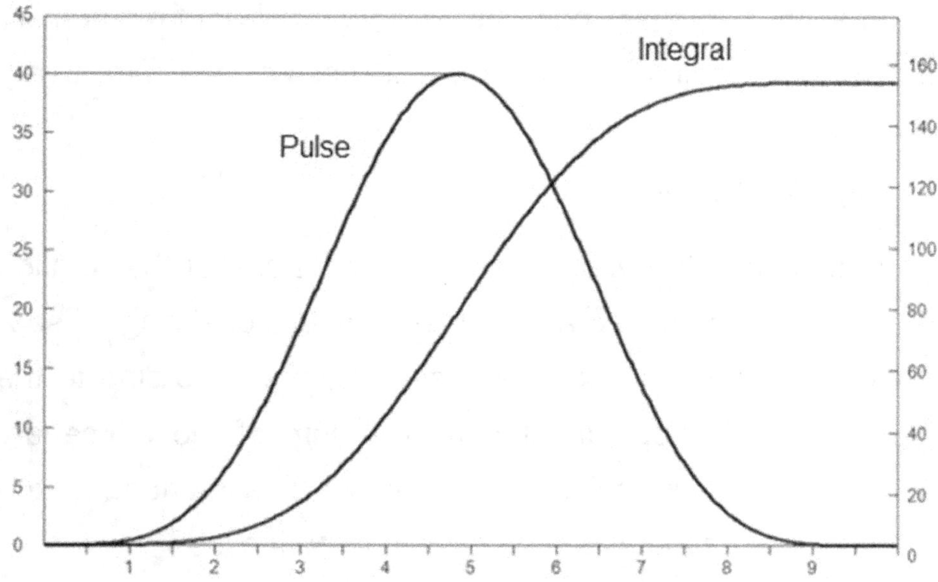

Figure 24. The integration of the pulse
Error of approximation is 0.407%

The numerical integration can be applied to calculate the square of an irregular figure or to calculate the volume of a complex form body filled with liquid. A reciprocal task is being solved as well for instance calculating liquid's level using its volume in the vessel.

Example 3.2.

Calculate the volume of the vessel which has surface of revolution with radius R(x) dependent on the current hight x which is the distance between the vessel bottom and the level of liquid:

$$R(x) = \frac{25\sin\left(\frac{x}{8} + 0.3\right)}{1 + 0.006x^2} + \sqrt{1 + 0.04x^2}$$

where $0 \le x \le 34.4$ (the vessel hight is 34.4 cm)

Solution.

Imagine the vessel consists of grate number of elementary discs with the radius R(x) and the thickness dx = 1 mm. Every disc volume is

$$dV = \pi R^2(x)*dx$$

In this case the vessel outline in the plane of the rotation axis is approximated by a step-like contour consisting of segments with the length dx. The vessel volume according to the rectangle rule is approximated by the sum of the elemenary disc volumes. Program Prog_3_2 (listing 3.2) is intended for the numerical integration. It contains two subroutines:

- "CoordSys" that draws the Cartesian axes;

- "Integral" that calculates the volume, draws the vessel contour and the integral graphic (fig.25) and estimates the approximation error.

The exact integration is possible using the formula R(x) but at our opinion it is a waste of time. It is better to investigate how the volume is approximated depending on the step value dx. Five variants are shown in the table 1. It seems the vessel volume is near V ≈ 12.75 litre with the error 0.7% if dx = 1 mm.

Table 1

Volume approximations

dx, mm	Volume, litre	dx, mm	Volume, litre
0.2	12.7444	1.5	12.7570
0.5	12.7460	2.0	12.7591
1.0	12.7495		

Program Prog_3_2;
Uses GraphABC, ABCobjects;

Procedure CoordSys;
var j, x: integer; **var** d: string;

```pascal
begin
  SetWindowSize(850, 500); SetPenStyle(psClear);
  SetPenColor(clBlack);        SetPenWidth(1);
  Line(100, 40, 790, 40);      Line(790, 40, 790, 440);
  Line(100, 240, 790, 240);    Line(100, 440, 790, 440);
  Line(100, 40, 100, 440);
  for j := 1 to 16 do
     begin
        Line(100, 440 - 25 * j, 108, 440 - 25 * j);
     end;
  for j := 1 to 17 do
     begin
        Line(100 + 40 * j, 430, 100 + 40 * j, 440);
     end;
  for j := 0 to 16 do
   begin
     Line(784, 440-25*j, 792, 440-25 * j);
   end;
  for j := 0 to 8 do
     begin
        x := 4 * j; d := IntToStr(x);
        var t := new TextABC(96 + 80 * j, 445, 12, d);
     end;
  for j := 0 to 8 do
     begin
        x := 4 * j-16; d:= IntToStr
        var t := new TextABC(72, 430 - 50 * j, 12, d);
     end;
  for j := 0 to 8 do
     begin
        x := 2 * j; d:= IntToStr(x);
        var t := new TextABC(800, 430 - 50 * j, 12, d);
     end;
end;

Procedure Integral;
  const x0=100; y0=240; w0=440; pi=3.14158;
        mx=20; my=10; mw=25;
  var k, x1, x2, y1, y2, z1, z2, w1, w2, m1, m2, m3, m4: integer;
  var x, r, dx, v, dv, r1, r2, r3, r4: real;
  var vv: string;
  begin
```

```pascal
  x:=0; dx:=0.1; v:=0; k:=0;
  While x<34.5 do
   begin
    r:=25*sin(0.125*x+0.3)/(1+0.006*x*x) + sqrt(1+0.04*x*x)-1;
    x:=x+dx;
    v:=v+pi*r*r*dx/1000;
    x1:=x0+Round(x*mx);     y1:=y0-Round(r*my);
    z1:=y0+Round(r*my);     w1:=w0-Round(v*mw);
    SetPenWidth(3);         SetPenColor(clBlue);
    k:=k+1;
    if k>1 then Line(x1, y1, x2, y2);
    if k>1 then Line(x1, z1, x2, z2);
    SetPenColor(clRed);
    if k>1 then Line(x1, w1, x2, w2);
    x2:=x1;  y2:=y1;   z2:=z1; w2:=w1;
   end;
   Str(v:5:4,vv);
   var t:=new TextABC(380,10,12,
          'Vessel volume V = '+vv+'  litre');
  SetPenColor(clBlue);    Line(788, 210, 788, 270);
  SetPenWidth(5);         Line(x0+2, 165, x0+2, 315);
  SetPenColor(clBlack);   SetPenWidth(2);
  SetPenStyle(psDot);     Line(100, 202, 260, 202);
  Line(260, 60, 260, 240);
 end;
//================ Main ================
begin
 CoordSys;  Integral;
 var t1:=new TextABC(600,95,12,'Integral V(x)');
 var t2:=new TextABC(150,180,12,'Level x');
 var t3:=new TextABC(270,130,12,'R(x)');
 var t4:=new TextABC(100,10,12,'cm');
 var t5:=new TextABC(740,10,12,'cub. dm');
end.
```

Listing 3.2

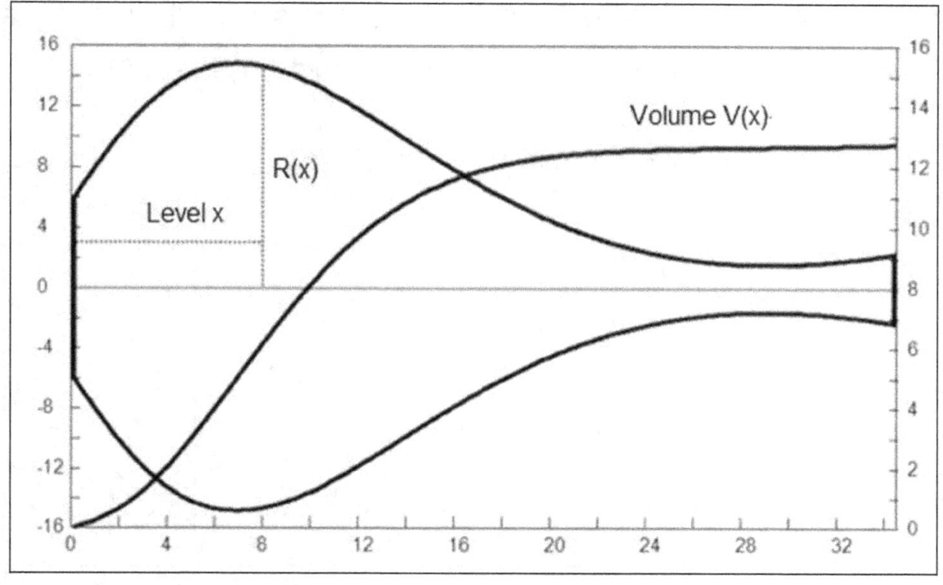

Figure 25. The vessel volume

Example 3.3.

The radius R of the inner surface of the tall wineglass is dependent on the distance h between the bottom and the liquid level (fig 26):

$$R(h) = 2.4h/(1+0.7h); \quad 0 \le h \le 9 \text{ cm}$$

Write the program for the next operations:
- to receive the liquid volume from the user;
- to check the permissable volume (not more then 160 cm^3);
- to show the liquid level growth in real time;
- to display the given value and the final level.

Solution.

The program Prog_3_3 (listing 3.3) contains two subroutines:

- "Coppa" that draws the empty wine glass;

- "Level" that receives the given liquid volume, computes the level growth and shows how the glass is filled in real time due to slowing down the process by the function Sleep (10).

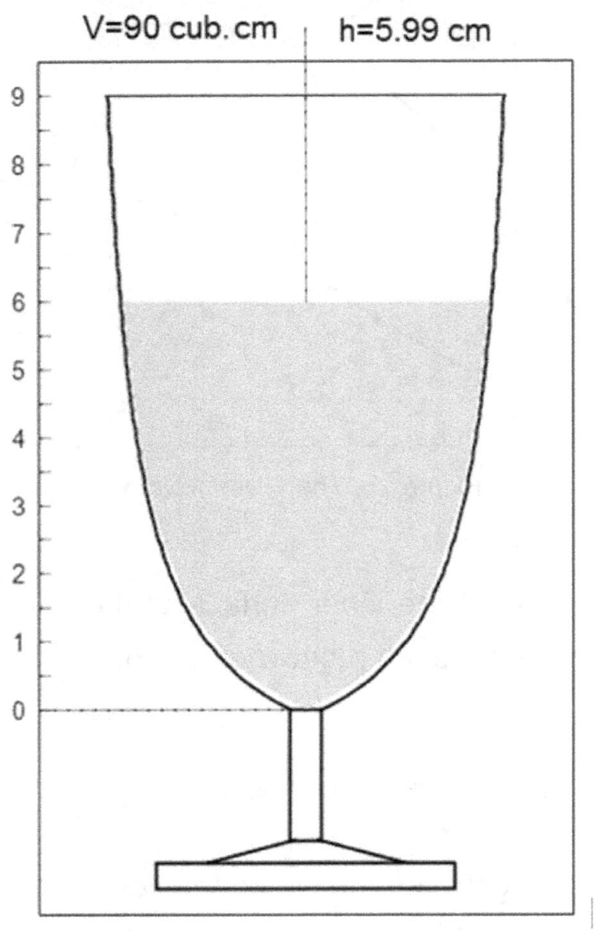

Figure 26. The tall wineglass

Program Prog_3_3;
Uses GraphABC, ABCobjects;

Procedure Coppa;
  const x0=300; y0=400; mx=40; my=40
  var k, x1, x2, y1, y2, z1, z2, d: integer;
  var h, dh, r: real;
  var s: string;
  begin
    SetWindowSize(600,580);   SetPenStyle(psClear);
    SetPenWidth(1);   SetPenColor(clBlack);

```pascal
    Line(140,20,460,20);        Line(140,520,460,520);
    Line(140,20,140,520);       Line(460,20,460,520);
    SetPenStyle(psDashDotDot);   Line(300,0,300,420);
    SetPenStyle(psSolid);        SetPenWidth(2);
    Rectangle(290,y0,310,y0+78); Line(240,490,294,476);
    Line(306,476,360,490);       SetPenWidth(2);
    Rectangle(210,490,390,506);  SetPenWidth(1);
    SetPenStyle(psDashDotDot);   Line(140,y0,300,y0);
    for k:=0 to 19 do
      begin
        Line(140,y0-20*k,148,y0-20*k);
      end;
    for k:=1 to 10 do
      begin
        d:=k-1; s:=IntToStr(d);
        var t:=new TextABC(124,y0-40*k+32,11,s);
      end;
//----------------------- form -----------------------
    SetPenWidth(3);  SetPenStyle(psSolid);
    h:=0;   dh:=0.1;
    for k:=1 to 91 do
      begin
        h:=h+dh;   r:=2.4*h/(1+0.7*h);
        x1:=x0+Round(r*mx);   z1:=x0-Round(r*mx);
        y1:=y0-Round(h*my)+4;
        if k>1 then
          begin
            Line(x1,y1,x2,y2);   Line(z1,y1,z2,y2);
          end;
        x2:=x1; y2:=y1; z2:=z1;
      end;
    SetPenWidth(2);   Line(180,40,420,40);
  end;

Procedure Level(vm: real);
  const x0=300; y0=400; pi=3.14158; mx=40; my=40;
  var k, x1, x2, y1, y2, z1, z2: integer;
  var dr, r, v, dv, h, dh: real;
  var s1, s2 : string;
  begin
    SetPenColor(clWhite);   Rectangle(10,540,580,590);
    dh:=0.01;  v:=0; k:=0;     Sleep(10);
```

```
    while v<vm-0.1 do
      begin
       h:=h+dh;         r:=2.4*h/(1+0.7*h);
       dv:=pi*r*r*dh;   v:=v+dv;
       x1:=x0+Round(r*mx)-2;  z1:=x0-Round(r*mx)+2;
       y1:=y0-Round(h*my);      SetPenColor(clRed);
      if k>1 then
          begin
            Line(x1,y1,x0,y2); Line(z1,y1,x0,y2);
          end;
          x2:=x1; y2:=y1; z2:=z1; k:=k+1;  Sleep(10);
      end;
    Str(vm:3:0,s1);
    var t1:=new TextABC(170,0,12, 'V = '+s1+' cub. cm');
    Str(h:3:2,s2);
    var t2:=new TextABC(320,0,12, 'h = '+s2+' cm');
   end;
//=====================Main=====================
begin
  var vm: real;
  Coppa;
  var t1:=new TextABC(15,550,12,
          'Input here liquid level from 30 to 160 cub. cm');
  readln(vm);
  if vm<161 then Level(vm);
  if vm>160 then
    begin
     SetPenColor(clWhite);
     Rectangle(10,530,590,595);
     var t2:=new TextABC(10,550,12,'Inadmissible
            number. Close the picture and repeat start');
    end;
end.
```

Listing 3.3

Chapter 4. FUNCTION INTERPOLATION

4.1. Interpolation methods

There are fields of investigation where experimental data equisition entailes great waste of time, materials or finances. In such cases the investigator is to be content with very small quantity of the goal function values and is in need to produce the missing data by means of some mathematical methods.

The problem is formulated by the next way. The experiment gives some values of the investigated function $y = f(x)$ at the argument boundaries x_k ($k = 0, 1, 2, \ldots$). The missing information inside the interval is recovered analytically by the interpolation of the known values y_k. The recover process as well as its result are called the approximations to the function $y = f(x)$. The points x_k are called the interpolation knots. The function given by the knot array is called the grid one. The grid may be either uniform (if $x_k - x_{k-1} = $ const) or non-uniform.

Any quantity of lines can be drawn through two neighbour points. It is difficult to select a function these points belong to. It is known that some optimum functions exist only. During the last 250 years several methods based on the polynomial approximation were suggested for solution of this problem. Three methods are demanded more frequently. These are the Newton difference polynomial, the Lagrange polynomial and the cubic spline.

The interpolation methods are estimated at view-point of the accuracy and the convergence. The knot points in the educational examples usually are taken from a known function $f(x)$ and are used for the polynom coefficients calculation. Hence, it is easy to compute the approximation error $R(x) = f(x) - P_a(x)$. In practice function $f(x)$ is always unknown. In such case the interpolation error is defined on the analogy of the numerical integration error.

4.2. Lagrange polynomial

The Lagrange approximation is the sum of the products:

$$P_n(x) = \sum_{k=0}^{n} f(x_k) L_{n,k}(x)$$

where

$$L_{n,k}(x) = \prod_{\substack{j=0 \\ j \neq k}}^{n} \frac{(x - x_j)}{(x_k - x_j)}$$

We'll find the polynomial as the sum of five addends:

$$P(x) = y_0 {*} L_0(x) + y_1 {*} L_1(x) + y_2 {*} L_2(x) + y_3 {*} L_3(x) + y_4 {*} L_4(x)$$

The Lagrange coefficients are computed by the formulae:

$$L_0(x) = \frac{(x - x_1)(x - x_2)(x - x_3)(x - x_4)}{(x_0 - x_1)(x_0 - x_2)(x_0 - x_3)(x_0 - x_4)};$$

$$L_1(x) = \frac{(x - x_0)(x - x_2)(x - x_3)(x - x_4)}{(x_1 - x_0)(x_1 - x_2)(x_1 - x_3)(x_1 - x_4)};$$

$$L_2(x) = \frac{(x - x_0)(x - x_1)(x - x_3)(x - x_4)}{(x_2 - x_0)(x_2 - x_1)(x_2 - x_3)(x_2 - x_4)};$$

$$L_3(x) = \frac{(x - x_0)(x - x_1)(x - x_2)(x - x_3)}{(x_3 - x_0)(x_3 - x_1)(x_3 - x_2)(x_3 - x_4)};$$

$$L_4(x) = \frac{(x - x_0)(x - x_1)(x - x_2)(x - x_3)}{(x_4 - x_0)(x_4 - x_1)(x_4 - x_2)(x_4 - x_3)}.$$

It is easy to notice the recurrence in the formulae. It allows to introduce the intermediate variables:

1) denom(n) – denominator of the polynomial $L_n(x)$; n = 0, 1, 2, 3, 4;

2) a(n) = y(n)/denom(n).

The Lagrange algorithm gives an elegant mathematical solution of the interpolation problem. The regularity of expressions is comfortable for the code developing at the programming language.

Remark.

The limited volume of the monograph does not allow to illustrate here another interpolation methods. It is possible to see the Newton polynomial and cubical spline in the book [2].

Example 4.1.

Recover unknown function using its knot points from the table 2. These points belong to the check function:

$$y(x) = 0.3 + 5exp(-0.2x)*sin(x/16)$$

Table 2

The grid function

k	x(k)	y(k)	k	x(k)	y(k)
0	0.30	1.453	3	1.40	2.162
1	0.55	2.365	4	1.75	1.806
2	0.9	2.698			

Solution.

The program Prog_4_1 (listing 4.1) has subroutines "CoordSys" for drawing the Cartesian axes and "Interpol" that is intended for the next actions:

- to calculate the Lagrange coefficients, the approximate polynomial values and the intermediate values of the check function;

- to estimate the interpolation error;

- to draw graphic (fig.27).

```
Program Prog_4_1;
Uses GraphABC, ABCobjects;

Procedure CoordSys;
  var j, v: integer;  var x: real;  var d: string;
begin
  SetWindowSize(840, 500); SetPenStyle(psClear);
  SetPenColor(clBlack);       SetPenWidth(1);
  Line(140, 20, 780, 20);       Line(780, 20, 780, 440);
  Line(140, 440, 780, 440);   Line(140, 20, 140, 440);
  Line(140, 300, 780, 300);   SetPenWidth(1);
  SetPenColor(clBlack);
  for j := 1 to 14 do
    begin
        Line(140, 440 - 28 * j, 148, 440 - 28 * j);
    end;
  for j := 1 to 17 do
    begin
      Line(140 + 36 * j, 430, 140 + 36 * j, 440);
    end;
  for j := 0 to 9 do
    begin
        x := 0.2 * j; Str(x:2:1,d);
        var t := new TextABC(134 + 71 * j, 445, 12, d);
    end;
  for j := 0 to 7 do
    begin
        x := 0.4 * j; Str(x:2:1,d);
        if x>0 then
          begin
            var t := new TextABC(112, 430 - 55 * j, 12, d);
          end;
    end;
  for j := 1 to 17 do
    begin
        Line(140 + 36 * j, 300, 140 + 36 * j, 310);
    end;
  for j := 2 to 10 do
    begin
```

```pascal
      Line(770, 450 - 25 * j, 780, 450 - 25 * j);
    end;
  for j := 0 to 7 do
    begin
      x := 0.5 * j-2.5; Str(x:2:1,d);
      if x>-1.5 then
      begin
        var t := new TextABC(790, 540 - 50 * j, 12, d);
      end;
    end;
end;

Procedure Interpol;
const x0=140; y0=440; e0=300;
mx=358; my=140; me=100;
var j, k, x1, x2, y1, y2, z1, z2, e1, e2: integer;
var v, w, dn, a: array[1..5] of real;
var p, p1, p2, p3, p4, p5: real;
var x, dx, y, z, er: real;
begin
  v[1]:=0.30; w[1]:=1.453; v[2]:=0.55; w[2]:=2.365;
  v[3]:=0.90; w[3]:=2.698; v[4]:=1.40; w[4]:=2.162;
  v[5]:=1.75;  w[5]:=1.806;
  dn[1]:=(v[1]-v[2])*(v[1]-v[3])*(v[1]-v[4])*(v[1]-v[5]);
  dn[2]:=(v[2]-v[1])*(v[2]-v[3])*(v[2]-v[4])*(v[2]-v[5]);
  dn[3]:=(v[3]-v[1])*(v[3]-v[2])*(v[3]-v[4])*(v[3]-v[5]);
  dn[4]:=(v[4]-v[1])*(v[4]-v[2])*(v[4]-v[3])*(v[4]-v[5]);
  dn[5]:=(v[5]-v[1])*(v[5]-v[2])*(v[5]-v[3])*(v[5]-v[4]);
  for k:=1 to 5 do
    begin
      a[k]:=w[k]/dn[k]; x1:=x0+Round((v[k])*mx);
      y1:=y0-Round(w[k]*my);
      SetPenStyle(psSolid); SetPenWidth(2);
      if k<16 then
        begin
          SetPenColor(clBlack); Circle(x1,y1, 7);
        end;
    end;
  x:=v[1]; dx:=0.005;
  for k:=1 to 294 do
    begin
      p1:=a[1]*(x-v[2])*(x-v[3])*(x-v[4])*(x-v[5]);
```

```
    p2:=a[2]*(x-v[1])*(x-v[3])*(x-v[4])*(x-v[5]);
    p3:=a[3]*(x-v[1])*(x-v[2])*(x-v[4])*(x-v[5]);
    p4:=a[4]*(x-v[1])*(x-v[2])*(x-v[3])*(x-v[5]);
    p5:=a[5]*(x-v[1])*(x-v[2])*(x-v[3])*(x-v[4]);
    p:=p1+p2+p3+p4+p5;
    z:=70*sin(x/15)/(1+x*x*x)+0.3*x;
    er:=((p-z)/2.75)*100;   x:=x+dx;
    x1:=x0+Round(x*mx);   y1:=y0-Round(p*my);
    z1:=y0-Round(z*my);    e1:=e0-Round(er*me);
    if k>1 then
      begin
        SetPenStyle(psSolid); SetPenColor(clBlack);
        SetPenWidth(4);          Line(x1,y1,x2,y2);
        SetPenWidth(2);          SetPenStyle(psDot);
        Line(x1,z1,x2,z2);       SetPenWidth(3);
        SetPenStyle(psSolid);  Line(x1,e1,x2,e2);
      end;
    x2:=x1; y2:=y1;  z2:=z1; e2:=e1;
  end;
end;
//=============== Main ================
begin
  CoordSys; Interpol;        SetPenWidth(2);
  SetPenColor(clBlack);      SetPenStyle(psDot);
  Line(248,242,248,300);    Line(338,112,338,300);
  Line(463,66,463,300);    Line(642,140,642,300);
  Line(768,180,768,300);
  var t1:=new TextABC(790,160,12,'%');
  var t2:=new TextABC(110,0,12,'Polynomial');
  var t3:=new TextABC(535,160,12,'Error');
  var t4:=new TextABC(550,65,12,'P(x)');
  var t5:=new TextABC(540,100,12,'f(x)');
  var t6:=new TextABC(710,145,12,'f(x)');
  var t7:=new TextABC(700,180,12,'P(x)');
end.
```

Listing 4.1

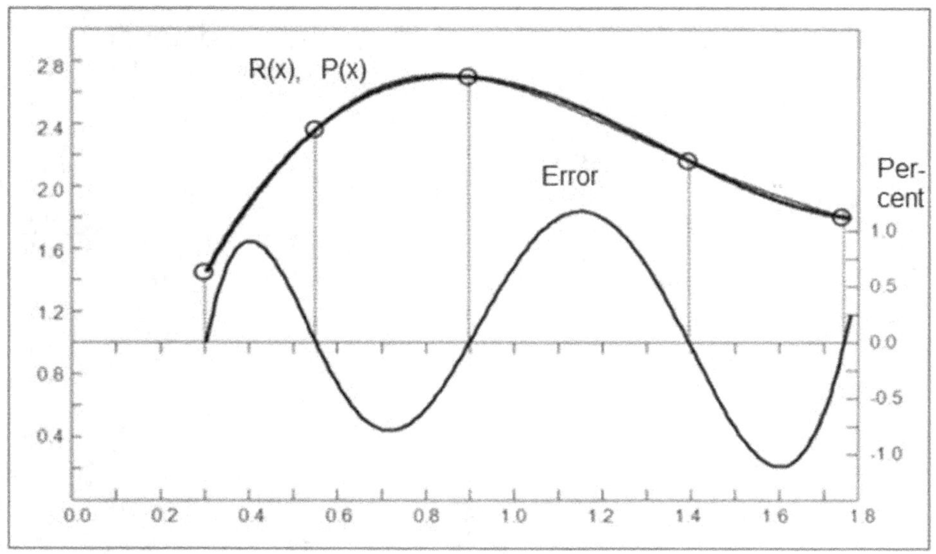

Figure 27. Interpolation with Lagrange polynomial

Example 4.2.

Recover function using its knot points from the table 3. These points belong to the check function:

$$y(x) = -0.12x^4 + 1.1x^3 - 5x + 2.5$$

Table 3

The grid function

k	x(k)	y(k)
0	-2.4	-4.688771
1	-1.2	6.350368
2	0.0	2.500000
3	1.2	1.848032
4	2.4	1.725088

Solution plan is built on analogy of the previous example. The program Prog_4_2 (listing 4.2) contains two subroutines 'CoordSys' and 'Interpol'. The results are shown at the fig.28.

```
Program Prog_4_2;
Uses GraphABC, ABCobjects;
Procedure CoordSys;
  var   j, v, yf: integer;
  var   x: real;
  var   d: string;
```

```pascal
begin
  SetWindowSize(850, 500);          SetPenStyle(psClear);
  SetPenColor(RGB(1, 1, 1));        SetPenWidth(1);
  Line(140, 20, 790, 20);           Line(140, 20, 140, 412);
  Line(140, 244, 790, 244);         Line(140, 412, 790, 412);
  Line(465, 20, 465, 412);          Line(790, 20, 790, 412);
  SetPenWidth(1);                   SetPenColor(RGB(1, 1, 1));
  for j := 1 to 14 do
     begin
       Line(458, 440 - 28 * j, 465, 440 - 28 * j);
     end;
  for j := 1 to 26 do
     begin
       Line(140 + 25 * j, 245, 140 + 25 * j, 252);
     end;
  for j := 1 to 13 do
     begin
       x := 0.4 * j- 2.8;     Str(x:2:1,d);
       var t := new TextABC(106 + 50 * j, 256, 11, d);
     end;
   for j := 1 to 3 do
     begin
       yf := 2* j;      d:=IntToStr(yf);
       var t := new TextABC(444, 236 - 56 * j, 11, d);
     end;
   for j := 1 to 2 do
     begin
       yf := -2* j;      d:=IntToStr(yf);
       var t := new TextABC(440, 236 + 56 * j, 11, d);
     end;
   for j := 2 to 16 do
     begin
       Line(782, 388 - 18 * j, 790, 388 - 18 * j);
     end;
   for j := 0 to 7 do
     begin
       yf :=2 * j-6;       d:=IntToStr(yf);
       var t := new TextABC(800, 340 - 36 * j, 11, d);
     end;
end;
```

```pascal
Procedure Interpol;
const x0=465; y0=244; e0=100;  mx=125; my=28; me=18;
var j, k, x1, x2, y1, y2, z1, z2, e1, e2: integer;
var v, w, dn, a: array[1..5] of real;
var p, p1, p2, p3, p4, p5, x, dx, y, z, er: real;
begin
  v[1]:=-2.4;  w[1]:=-4.6887712;
  v[2]:=-1.2;  w[2]:= 6.350368;
  v[3]:= 0.0;  w[3]:= 2.500;
  v[4]:= 1.2;  w[4]:=-1.848032;
  v[5]:= 2.4;  w[5]:= 1.725088;
  dn[1]:=(v[1]-v[2])*(v[1]-v[3])*(v[1]-v[4])*(v[1]-v[5]);
  dn[2]:=(v[2]-v[1])*(v[2]-v[3])*(v[2]-v[4])*(v[2]-v[5]);
  dn[3]:=(v[3]-v[1])*(v[3]-v[2])*(v[3]-v[4])*(v[3]-v[5]);
  dn[4]:=(v[4]-v[1])*(v[4]-v[2])*(v[4]-v[3])*(v[4]-v[5]);
  dn[5]:=(v[5]-v[1])*(v[5]-v[2])*(v[5]-v[3])*(v[5]-v[4]);
  for k:=1 to 5 do
     begin
       a[k]:=w[k]/dn[k];
       x1:=x0+Round((v[k])*mx);
       y1:=y0-Round(w[k]*my);
       SetPenStyle(psSolid);
       SetPenWidth(2);
       if k<16 then
         begin
           SetPenColor(RGB(1,1,1));
           Circle(x1,y1, 6);
         end;
     end;
  x:=v[1];   dx:=0.01;
  SetPenStyle(psSolid);
  for k:=1 to 490 do
    begin
      p1:=a[1]*(x-v[2])*(x-v[3])*(x-v[4])*(x-v[5]);
      p2:=a[2]*(x-v[1])*(x-v[3])*(x-v[4])*(x-v[5]);
      p3:=a[3]*(x-v[1])*(x-v[2])*(x-v[4])*(x-v[5]);
      p4:=a[4]*(x-v[1])*(x-v[2])*(x-v[3])*(x-v[5]);
      p5:=a[5]*(x-v[1])*(x-v[2])*(x-v[3])*(x-v[4]);
      p:=p1+p2+p3+p4+p5;
      z:=2.5-5*x+1.1*x*x*x-0.12*x*x*x*x;
      er:=((p-z)/6.1)*1000000;
```

```pascal
    x:=x+dx;
    x1:=x0+Round(x*mx);    y1:=y0-Round(p*my);
    z1:=y0-Round(z*my);    e1:=y0-Round(er*me);
    if k>1 then
      begin
        SetPenWidth(4);        SetPenColor(clBlue);
        Line(x1,z1,x2,z2);     SetPenColor(clBlack);
        SetPenWidth(2);
        if x>-1.3 then  Line(x1,e1,x2,e2);
      end;
    x2:=x1; z2:=z1; e2:=e1;
  end;
end;
//================= Main =================
begin
  CoordSys;  Interpol;        SetPenWidth(3);
  SetPenColor(clBlack);       SetPenStyle(psDot);
  Line(315,65,315,246);       Line(615,246,615,290);
  Line(765,200,765,246);
  var t0:=new TextABC(386,30,12,'Polynom');
  var t1:=new TextABC(420,50,12,'axis');
  var t2:=new TextABC(667,40,12,'The value is ');
  var t3:=new TextABC(670,60,12,'to be multiplied');
  var t4:=new TextABC(670,80,12,'by 0.0001%');
  var t5:=new TextABC(240,70,12,'P(x)');
  var t6:=new TextABC(610,160,12,'Error');
end.
```

Listing 4.2

The task peculiarity is in the source of the knot points. The points belong to the congener polynomial function y(x). That is why the maximum interpolation error is less then 0.001% and the curves P(x) and y(x) are indistinguishable at the diagram.

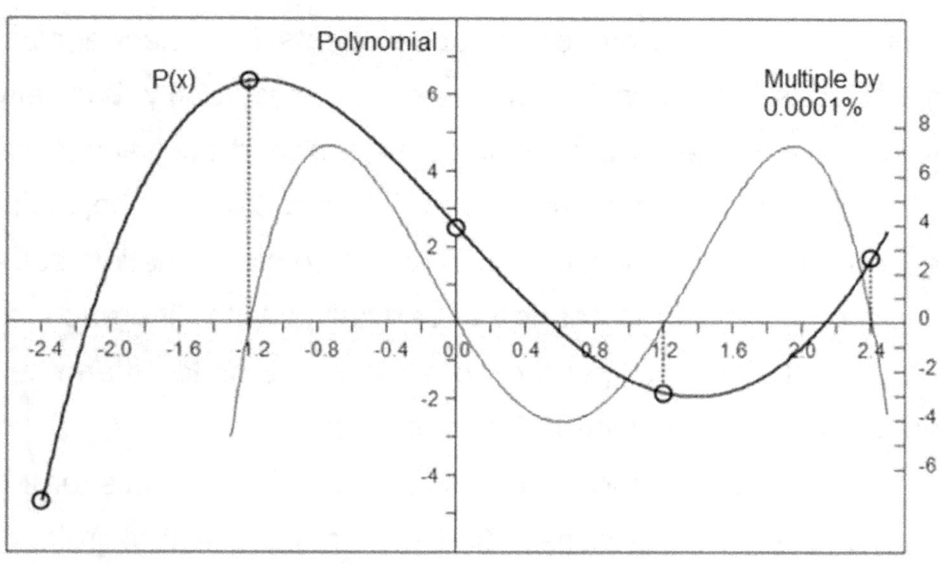

Figure 28. Recovering of the polynomial function

Chapter 5. DIFFERENTIAL EQUATIONS

5.1. Numerical solution of differential equations

A differential equation describes very well the control system behaviour in real time. It is necessary to find a function x(t) that denotes the output coordinate of the controlled object state at any moment. The equation sets the dependance of function x(t) and its derivatives on the input drive-signal u(t) applied to the object. The differential equation is called ordinary if its elements are multiplied by constant coefficients. The exact analytical solutions are accessible in the simplest cases only, basically for the homogeneous equations of the first and the second order. In practice the numerical methods are used in the main. The numerical methods of ordinary differential equation solutions are based on the change of derivatives with difference relations which allow to get the equation in the finite differences and to solve it by the algebraical methods.

The solution methods are discussed in the manuals for the Caushy task. It is necessary find the solution of the ordinary first-order equation:

$$\frac{dy(x)}{dx} = f(x, y)$$

The solution is found in the interval [x0, x0 + z] divided at n segments with the legth h = z/n. In so manner the knot points of the grid function are got. The difference task is solved at every subinterval. The Euler algorithm gives the recurrent equation:

$$y_{k+1} = y_k + hf(x_k, y_k), \quad 0 \le k \le n-1$$

It is considered that the Euler method has a theoretical interest only because of low accuracy. There are two ways to increase the solution accuracy – either to diminish the value h or to use the modified method:

$$y_{k+1} = y_k + 0.5h[f(x_k, y_k) + f(x_{k+1}, y_{k+1})]$$

Among methods of the Cauchy task solution the variants of the Runge-Kutta method are mentioned most often. One of its is called "the predictor-corrector" method. Essence of the method is in two stages. First, the rough prediction is made at every subinterval:

$$\tilde{y}_{k+1} = y_k + f(x_k, y_k)h.$$

Second, the predicted value is made more precise by the formula:

$$y_{k+1} = y_k + \frac{h}{2}[f(x_k, y_k) + f(x_{k+1}, \tilde{y}_{k+1})].$$

The method provides the second-order accuracy. The forth-order accuracy may be achieved due to the doubling of the operation quantity:

$$y_{k+1} = y_k + \frac{h}{6}(m_1 + 2m_2 + 2m_3 + m_4),$$

where

$$m_1 = f(x_k, y_k); \qquad m_2 = f\left(x_k + \frac{h}{2}, y_k + \frac{h}{2}m_1\right);$$

$$m_3 = f\left(x_k + \frac{h}{2}, y_k + \frac{h}{2}m_2\right); \qquad m_4 = f(x_k + h, y_k + hm_3)$$

The exact solution of the differential equation is not known beforehand as a rule. Hence the accuracy of the approximate solution is estimated by means of the double calculation method as it is met in the practice of the numerical integration.

The non-homogeneous differential equations with the accidental drive-signal at the right part are the most frequently in the control systems. In such case the problem can not be formulated in terms of Cauchy. One is to approximate the derivative with the first-order finite difference and to achieve the demanded accuracy by diminishing the step size of time increment (time quantum step).

5.2. The first-order equation

The canonical expression is used in the authomatic control theory for the non-homogeneous ordinary first-order differential equation:

$$T\frac{dx(t)}{dt} + x(t) = u(t), \qquad (5.1)$$

where T – the time-constant of the controlled object.

The parameter T in time units reflects proportonally inertial properties of the object. It may be a body mass, heat capacity, production volume, number of workers, an investment and so on.

Let the drive-signal is a semi-infinite leap:

$$u(t) = \begin{cases} 0, & t < 0 \\ 1, & t \geq 0 \end{cases}$$

The object reaction under zero start conditions is the known expression:

$$x(t) = 1 - \exp\left(-\frac{t}{T}\right), \qquad t > 0.$$

Output variable x(t) aspires to 1 in infinity but one may consider the process is over with x(t) ≈ 0.95.

The solution becomes complicated significally when the function u(t) is not so simple as the leap or it is impossible in general with analytical methods when u(t) involves the random elements. In such case one ought to use the numerical method. For the purpose it is necessary to change the continious functions in expression (5.1) with the grid functions at discrete time t = Δt*k where k is natural numbers.

The digital model of the control system deals with the past and current values of variable. Hence, the left approximation of derivative is possible only:

$$\frac{dx(t)}{dt} \approx \frac{x(k) - x(k-1)}{\Delta t}, \qquad (5.2)$$

We'll substitute the derivative (5.2) into the equation (5.1) and will have the algebraical equation:

$$(T/\Delta t)*[x(k) - x(k-1)] + x(k) = u(k) \qquad (5.3)$$

Using denote b = Δt/T one gets the solution of the equation (5.3) relatively the output coordinate

$$x(k) = b*u(k) + (1 - b)*x(k-1), \quad k = 0, 1, 2, \qquad (5.4)$$

In so manner the differential equation is transformed into an adequate algebraical recurrent equation in finite differences. The recurrence is explained by the fact that the current value of the output coordinate is dependent on the its previous value.

Objects described with the equation (5.1) are called inertial and may be met in many spheres of acivity. The inertialness may be mechanical, thermal, electrical, magnetical, financial, industrial and so on. The smoothing-iron is a typical example. It has an artifically enlarged iron mass to keep given temperature. The electrical heating is switched on instantly, but the demanded temperature is achieved gradually in some minutes. After the heating is switched out the iron temperature aspires to the air temperature in the room also gradually.

Any aircraft and any ship are inertial objects. The instant changing of the rudder position forces slow turning of the velocity vector and slow trajectory changing. The battle plane can make the turn in some seconds, for instance, but the sea tanker takes new direction of the movement in several minutes.

It was seen the same rule in all above-mentioned examples: the exponential changing of the output coordinate $x(t)$ takes place. The next example confirms the rule.

Example 5.1.

Find reaction of an abstract inertial object forced by the pulse $u(t)$ that has finite break points:

$$u(t) = \begin{cases} 0, & 0 \leq t \leq 1; \\ 23 - 7.1*(t - 1), & 1 \leq t \leq 8; \\ 0, & t > 8 \end{cases}$$

Find two solutions with step $dt = 0.1$ using the parameter values

$T_1 = 0.3$ and $T_2 = 1.5$.

Solution.

1. Calculate values of the coefficients in the equation (5.4):

$$b_1 = dt / T_1; \quad b_2 = dt / T_2$$

2. Develop the program Prog_5_1 (listing 5.1) containing subroutines "CoordSys" and "Equat". The first draws the Cartesian coordinate system. The second subroutine is intended for the next actions:

- to calculate values of the drive-signal $u(t)$;

- to find values of the output function $x(t)$;

- to draw graphics (fig.29).

```pascal
Program Prog_5_1;
Uses GraphABC, ABCobjects;

Procedure CoordSys;
  const x0=100;
  var j, k, xc, yc: integer;
  var x: real;
  var s: string;
  begin
    SetWindowSize(840,500);      SetPenStyle(psClear);
    SetPenColor(clBlack);        SetPenWidth(3);
    Line(140, 20, 140, 440);     Line(140, 226, 784, 226);
    SetPenWidth(1);              Line(140, 20, 784, 20);
    Line(140, 440, 784, 440);    Line(784, 20, 784, 440);
    for k:=1 to 30 do
       begin
          Line(140+20*k, 224, 140+20*k, 234);
       end;
    for k:=1 to 13 do
       begin
          Line(132, 450-32*k, 140, 450-32*k);
       end;
    for k:=2 to 16 do
       begin
          x:=1.0*(k-1);    Str(x:1:0,s);
```

```pascal
          if x<5 then
          begin
             var t:= new TextABC(96+40*k,238,12,s);
          end;
          if x>4 then
          begin
             var t:= new TextABC(94+40*k,200,12,s);
          end;
      end;
    for k:=1 to 13 do
      begin
         yc:=4*(k-1)-24;    s:=IntToStr(yc);
         var t:= new TextABC(102,440-32*k,12,s);
      end;
end;


Procedure  Equat;
const x0=140; y0=226; mx=45; my=8;
var k, x1, x2, y1, y2: integer;
var b, dt, t, t1, t2: real;
var u, y: array[1..200] of real;
begin
   t1:=0.8;  t2:=5.0;
   dt:=0.2;  t:=0;     u[1]:=0;   u[2]:=0;   u[3]:=0;
   For k:=1 to 65 do
     begin
        t:=t+dt;
        if t<0.6 then u[k]:=0;
        if t>0.5 then u[k]:=23-5.8*(t-0.8);
        if t>8 then u[k]:=0;
        x1:=x0+Round(t*mx);  y1:=y0-Round(u[k]*my);
        if k>2 then
          begin
             SetPenWidth(2);  Line(x1,y1,x2,y2);
          end;
        x2:=x1; y2:=y1;
     end;
   b:=0.26;  y[1]:=0;  t:=0;  x2:=0;  y2:=0;
   for k:=2 to 68 do
     begin
      t:=t+dt;
```

```pascal
      y[k]:=b*u[k]+(1-b)*y[k-1];
      x1:=x0+10+Round(t*mx);  y1:=y0-Round(y[k]*my);
      SetPenWidth(3);          SetPenColor(clBlack);
      if k>2 then
        begin
          Line(x1,y1,x2,y2);
        end;
      x2:=x1; y2:=y1;
    end;
  b:=0.14;  y[1]:=0;  t:=0;  x2:=0;  y2:=0;
  for k:=2 to 68 do
    begin
     t:=t+dt;
     y[k]:=b*u[k]+(1-b)*y[k-1];    x1:=x0+10+Round(t*mx);
     y1:=y0-Round(y[k]*my);      SetPenWidth(3);
     SetPenStyle(psDot);         SetPenColor(clBlack);
     if k>2 then
        begin
          Line(x1,y1,x2,y2);
        end;
      x2:=x1; y2:=y1;
    end;
end;
//========= Main ==================
begin
  CoordSys;
  Equat;
  var t1:=new TextABC(450,80,12,
         'Equation in finite differences:');
  var t2:=new TextABC(450,105,12,
        ' y(k) =  b*u(k) + (b - 1)*y(k-1)');
  var t3:=new TextABC(200,40,12,'u(k)');
  var t4:=new TextABC(170,310,12,
         'Reaction of the controlled object:');
  var t5:=new TextABC(170,335,12,'with T = 0.8 -  solid line');
  var t6:=new TextABC(170,355,12,'with T = 5.0 -  dotted line');
end.
```

Listing 5.1

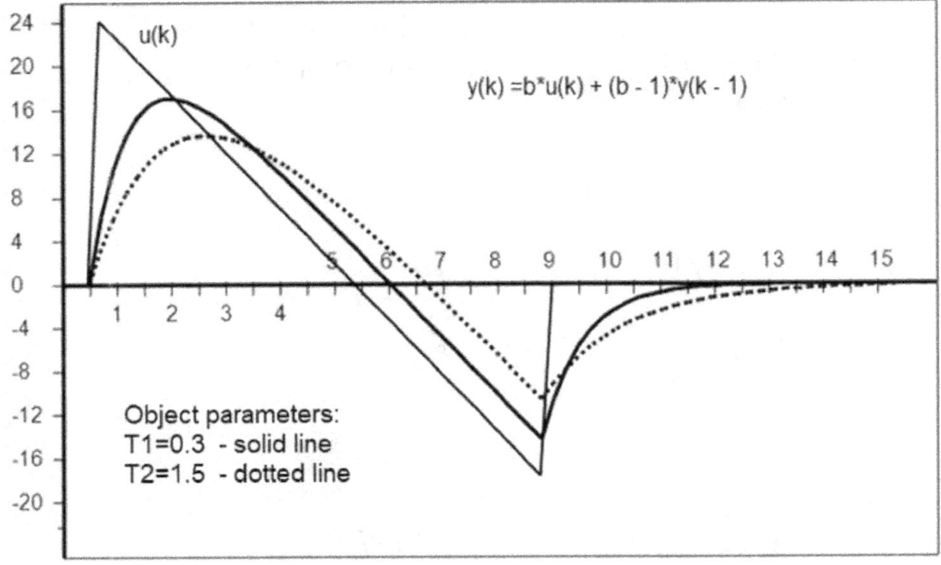

Figure 29. Solution of the differential
first-order equation

Conclusion.

The object with the parameter T = 0.3 sec. begins repeating input signal u(t) exactly in 2.5 seconds with the constant delay 0.2 sec.

If T = 1.5 sec., then the transition process is over in 3.5 seconds and its delay becomes 1.0 sec.

5.3. The second-order equation

The ordinary non-homogeneous second-order differential equation in canonical form is:

$$T_1^2 \frac{d^2 x(t)}{dt^2} + T_2 \frac{dx(t)}{dt} + x(t) = u(t), \qquad (5.5)$$

where T_1 and T_2 – time-constants of the investigated object.

The exact solution is possible if the drive-signal u(t) has unit leap form. A lot of solutions exist depending on values of the pair T_1 and T_2. Every solution belongs either to aperiodic or oscillation mode.

Exact solution is need as an etalon for the accuracy estimation of an approximated one. Let us use the Laplace method. The Laplace transform of the differential equation (5.5) gives the characteristic equation:

$$p^2 + \frac{T_2}{T_1^2} p + \frac{1}{T_1^2} = 0 \qquad (5.6)$$

where p – Laplace operator.

The roots of equation (5.6):

$$p_{1,2} = -\frac{T_2}{2T_1^2} \pm \sqrt{D},$$

$$D = \left(\frac{T_2}{2T_1^2}\right)^2 - \frac{1}{T_1^2}.$$

The roots are real if determinant $D > 0$. In such case the unit leap function u(t) gives the aperiodic solution:

$$x(t) = 1 - A_1 \exp(p_1 t) + A_2 \exp(p_2 t) \qquad (5.7)$$

where A_1 and A_2 are the constants of integration:

$$A_1 = \frac{T_2}{T_2 - T_1}; \qquad A_2 = \frac{T_1}{T_1 - T_2}.$$

The roots become complex if $D < 0$. It is a true sign that oscillations occur in the solution:

$$x(t) = 1 - \frac{\exp(-\gamma\omega_0 t)}{\sqrt{1-\gamma^2}} \sin\left(\omega_0 t\sqrt{1-\gamma^2} + arctg\frac{\sqrt{1-\gamma^2}}{\gamma}\right), \quad (5.8)$$

Frequency of oscillation: $\omega_0 = \dfrac{1}{T_1}$ Fading coefficient: $\gamma = \dfrac{T_2}{2T_1}$

The exponent argument in the expression (5.8) may occure both negative and positive. In the first case the oscillaton is fading. Positive argument gives oscillation with arising amplitude. It is a reason to investigate the stability of the object behaviour.

Now it is time to study approximated numerical solution of the equation (5.5). We'll transform it into the algebraical equation in finite differences by means of the left approximations of both derivatives:

$$\frac{T_1^2}{T^2}[x(k) - 2x(k-1) + x(k-2)] + \frac{T_2}{T}[x(k) - x(*k-1)] + x(k) = u(k) \quad (5.9)$$

where T – time quantum step.

We'll execute grouping of variables in equation (5.10):

$$a_0 x(k) - a_1 x(k-1) + a_2 x(k-2) = u(k), \quad (5.10)$$

where:

$$a_0 = \frac{T_1^2}{T^2} + \frac{T_2}{T} + 1; \qquad a_1 = 2\frac{T_1^2}{T^2} + \frac{T_2}{T}; \qquad a_2 = \frac{T_1^2}{T^2}.$$

Solution of the equation (5.10) relatively the output coordinate is:

$$x(k) = Au(k) + Bx(k-1) - Cx(k-2) \qquad (5.11)$$

where

$$A = \frac{1}{a_0}; \qquad B = \frac{a_1}{a_0}; \qquad C = \frac{a_2}{a_0}.$$

Let us remember that exact solution is expressed (depending on the the values of T_1 and T_2) in two modes – aperiodic or oscillation. The mode is predicted by the determinant D. The numerical finite difference method suggests the single solution (5.11). It is true for any pair of T_1 and T_2. Hence, there is falling away a necessity to calculate any indirect sign of the solution stability.

Example 5.2.

Study behaviour of the controlled object using the second-order differential equation with time-constants T_1 = 0.11 and T_2 = 0.04. Find its reaction on the drive-signal having the rectangle form with amplitude A = 10 and duration 5 sec. Repeat the experiment with another values T_1 = 0.12 and T_2 = 0.12. In both cases take time step dt = 0.02.

Solution.

Develop the program Prog_5_2 (listing 5.2) which contains three subroutines:

- "CoordSys" for drawing the coordinate axes;
- "Signal" for drawing the input function;
- "Equat" for approximating of solution.

The subroutines "Equat" is intended for the next actions:

- to receive values T1 and T2;

- to calculate values of the equation coefficients;

- to solve the differential equation;

- to draw two output functions (fig.30).

```
Program Prog_5_2;
Uses GraphABC, ABCobjects;
Procedure CoordSys;
 const x0=100;
 var j, k, xc, yc: integer;
 var x: real;
 var s: string;
 begin
   SetWindowSize(850,500);   SetPenStyle(psClear);
   SetPenColor(clBlack);     SetPenWidth(1);
   Line(140, 0, 140, 434);   Line(140, 324, 800, 324);
   Line(140, 0, 800, 0);     Line(140, 434, 800, 434);
   Line(800, 0, 800, 434);
   for k:=1 to 20 do
     begin
       Line(140+32*k, 324, 140+32*k, 332);
     end;
   for k:=0 to 19 do
     begin
       Line(132, 434-22*k, 140, 434-22*k);
     end;
   for k:=2 to 11 do
     begin
       x:=1*(k-1);   Str(x:2:0,s);
       var t:= new TextABC(68+64*k,334,11,s);
     end;
   for k:=1 to 10 do
     begin
       yc:=2*(k-1)-4;   s:=IntToStr(yc);
       var t:= new TextABC(112,448-44*k,12,s);
     end;
 end;

Procedure Signal;
const x0=140; y0=324; mx=62.5; my=22;
var k, x1, x2, y1, y2: integer;
var amp, a, b, c, dt, t, u: real;
```

```pascal
begin
  amp:=10;            dt:=0.01; t:=0;
  SetPenWidth(2);   SetPenColor(clBlack);
  For k:=1 to 800 do
    begin
      t:=t+dt;
      if (t<0.2) or (t>5) then u:=0;
      if (t>0.2) and (t<5.1) then u:=amp;
      x1:=x0+Round(t*mx);     y1:=y0-Round(u*my);
      if k>1 then
        begin
          Line(x1,y1,x2,y2);
        end;
      x2:=x1; y2:=y1;
    end;
end;

Procedure  Equat(t1,t2:real);
const x0=140; y0=324; mx=160; my=22;
var k, x1, x2, y1, y2: integer;
var amp, a, b, c, dt, t, a0, a1, a2: real;
var u, y: array[1..900] of real;
begin
  amp:=10;  dt:=0.02;  t:=0;
  u[1]:=0;     u[2]:=0;     u[3]:=0;
    for k:=1 to 250 do
        begin
            t:=t+dt;
            if k<2 then u[k]:=0;
            if k>2 then u[k]:=amp;
            if k>99 then u[k]:=0;
            x1:=x0+Round(t*mx);   y1:=y0-Round(u[k]*my);
      end;
  a0:=(t1*t1)/(dt*dt)+ t2/dt+1;   a1:=(2*t1*t1)/(dt*dt)+t2/dt;
  a2:=(t1*t1)/(dt*dt);                a:=1/a0; b:=a1/a0; c:=a2/a0;
  y[1]:=0;  t:=0;  x2:=0;  y2:=0;
  SetPenWidth(3);                SetPenColor(clBlue);
  for k:=3 to 200 do
    begin
      t:=t+dt;                   y[k]:=a*u[k]+b*y[k-1]-c*y[k-2];
      x1:=x0+8+Round(t*mx);  y1:=y0-Round(y[k]*my);
```

```pascal
  if k>3 then
    begin
      Line(x1,y1,x2,y2);
    end;
    x2:=x1; y2:=y1;
  end;
end;
//===================== Main =======================
var t1,t2: real;
begin
  CoordSys;  Signal;
  t1:=0.11;  t2:=0.04;   Equat(t1,t2);
  t1:=0.12;  t2:=0.12;   Equat(t1,t2);
  var v1:=new TextABC(540,15,12,
          'Equation in finite differences:');
  var v2:=new TextABC(530,40,12,
          ' y(k) =  A*u(k) + B*y(k-1) - C*y(k-2)');
  var v3:=new TextABC(150,80,12,'u(k)');
  var v4:=new TextABC(220,10,11,'T1 = 0.11;   T2 = 0.04');
  var v5:=new TextABC(280,50,11,'T1 = 0.12;   T2 = 0.12');
  SetPenWidth(1);   SetPenColor(clBlack);
  Line(256,92,280,64);
end.
```

Listing 5.2

Conclusion.

1. Both solutions are of the oscillation mode. Oscillations are fading during three periods if parameters have values T1 = 0.11 and T2 = 0.04. Function y(k) intersects the levels 1 and 0 six times. Such reaction of the object is called "over-regulation". The control system is to be damped.

2. Regulation process becomes optimum if parameters have values T1 = T2. The single swing of the output function is less then 30%. Such an error is permissible.

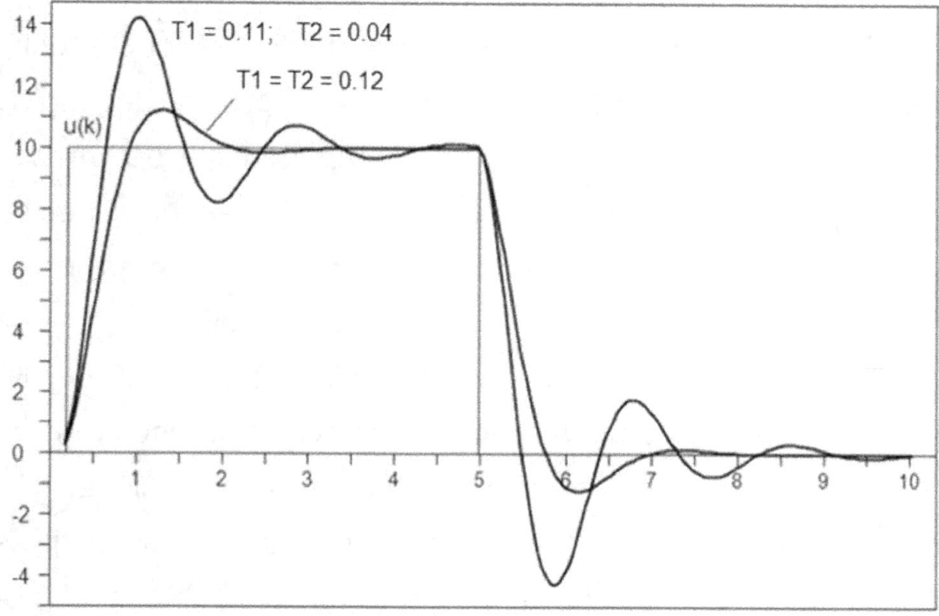

Figure 30. Solution of the second-order differential equation
y(k)=A*u(k)+B*y(k-1)-C*y(k-2)

Differential second-order equations are often met in physics and engineering where mathematical models use velocity and acceleration of movement. Such equations describe behaviour of many objects.

These are for instance:

- mechanism with a spring,

- resonant electric circuit;

- high TV tower swinged by the wind,

- automatic control system of guided missile,

- sea tanker swimming on waves and so further.

Chapter 6. DIGITAL FILTERS

6.1. Statement of the problem

Results of experimental measurements are often registered as numerical rows or a number flow. Both are depending on time. The flow usually has two components – determined (regular) and random one. The first part describes the expected process. The second component is considered as an error or a hindrance (noise) that disturbes observation results.

There are many sources of the noise. It depends on observed process nature. Natural and artifical hindrances in the radio communication channels are the reasons of signal random fluctuations. Heat noise in semiconductor transistor gives birth to data disturbance in the radio-receiver and in photo-receiver with strong signal amplification. Departure from technological norms is a reason of size deviations in a serial production of machine detailes. Random processes are inevitable in macro- and micro-economic owing to great number of unforeseen events in a country.

Observed data are to be previously processed before coming to some important decision. The preliminary processing is called data filtering. Data row is a mixture of useful signal and noise. The mixture can be analysed either in time or in frequence space. In the second case one investigates the mixture spectrum and developes an algorithm of signal selection and noise suppression. The aim of preliminary processing is to encrease the relation of the signal to noise by amplitude or by power. In economic models the data preliminary treatment is used to detect a trend of an economic process and to prepare observed data to the next stage of the mathematical prognosis.

Signal and noise separation is made in the frequence space by means of filter that is transparent for the harmonics in the given interval. Noise usually consists of more high-frequence harmonics than signal does. In such case the filter is needy to pass the slow oscillations and to impede quick ones. In time space the mentioned filter acts as an inertial procedure.

An economical index can not have quick changes like a random noise has. It means that an inertial procedure is effective only to select slow signal of business activity and to suppress the quick noise oscillations. Such procedure is called smoothing.

There are several smoothing algorithms:
- arithmetic average in sliding window;
- exponential;
- by means of the second-order low-pass filter;
- median alignment.

One must know main properties and peculiarities of the investigated process also statistical characteristics of its random component. It is necessary for a scientifically grounded smoothing algorithm. In addition, one ought to know methods of random process modelling.

The termin "noise" is used already here to denote a random component of the investigated input mixture. The mixture is called additive if it is the algebraical sum of a "pure" signal and a noise. Sometimes the noise influences upon one of the "pure" signal parameters – an amplitude, a frequence or a phase. In this case an influence is described as a mathematical multiplication. Such mixture is called multiplicative.

The important characteristic of a random variable is a density function. There are several modes of density – uniform probable, normal, binomial, Poisson's and so further. The Pascal ABC has the library function Random which returns equally probable value "s" from the interval [0, 1]. The function is often used in the examples of this monograph. Numbers of real type produced by function Random are transforming further into natural numbers.

A number array resulted from an experiment is called a random sample. Coincidence sample number in an interval of a number axes is an event frequence in this interval. Frequence distribution is drawn graphically as a histogram.

If an array size is arising infinitely and an interval length aspires to 0 then the histogram transforms into a probability density function.

Sum of all frequency values along all number intervals gives an increasing function named a probability integral.

6.2. Extract from the digital filters theory

Let us consider that a digital filter is an algorithm of transformation of a time number sequence $x(kT)$, where $k = 1, 2, 3,$ ... and $T = const$, into the sequence $y(kT)$, associated with $x(kT)$ by means of the definite differences equation. The sequence $x(kT)$ is called input, the $y(kT)$ is called output. Now we have two grid functions of time with the constant quantum step of current time at discrete moments $t = kT$. Further we'll use brief denotes

$$x(kT) = x(k); \quad y(kT) = y(k). \qquad (6.1)$$

A filter with parameters not depending on time is described by the N-order equation [6]:

$$a_0 y(k) + a_1 y(k-1) + \ldots + a_{N-1} y(k-N+1) + a_N y(k-N) =$$
$$= b_0 x(k) + b_1 x(k-1) + \ldots + b_{n-1} x(k-N+1) + b_N x(k-N),$$

where a_k, b_k – filter constant coefficients.

The goal of filter design is selection of such coefficient values which provide the demanded quality of the input flow transformation.

Solution of the equation (6.1) relatively the current value of the output variable is

$$y(k) = \frac{1}{a_0}\left[\sum_{i=0}^{N} b_i x(k-i) - \sum_{i=1}^{N} a_i y(k-i) \right]. \qquad (6.2)$$

So far as y(k) depends not only on the input flow elements but also on the previous values y(k - i) of the output flow, the equation (6.2) is called "recurrent" and the filter is "recursive". If $a_0 = 1$ and $a_i = 0$ for i > 1 then the filter becomes "nonrecursive".

Equation in finite differences is reduced to an algebraical one by means of the mathematical apparatus of z-transform for the grid functions.

Z-transform formulae are got from a discrete Laplace transformation by means of new complex variable:

$$z = \exp(s) \qquad (6.3)$$

where

$$s = \sigma + j\omega; \quad \sigma - \text{real part}; \quad \omega - \text{circular frequency}.$$

Let us consider that the input discrete numerical flow is got from the continious function x(t) by superposition of the pulse infinite small duration function δ(t) repeating with quantum step T:

$$X*(s) = \sum_{k=-\infty}^{\infty} x(kT)z^{-k} = X(z) \qquad (6.4)$$

We'll transform function (6.4) by Laplace:

$$X*(s) = L\big[x*(t)\big] = \sum_{k=-\infty}^{w} x(kT)\exp(-ksT)$$

We'll substitute variable (6.3) in expression (6.4):

$$X*(s) = \sum_{k=-\infty}^{\infty} x(kT)z^{-k} = X(z) \qquad (6.5)$$

The operation (6.5) is called a two-sided z-transform of function x(k). The sequence of values with index k > 0 is interesting in real process only. Hence, one consideres all x(k) = 0 with k < 0 and in so manner one-sided z-transform is received.

There is a list of the very important z-transform properties:

1. Linearity:

$$a_1x_1(k) + a_2x_2(k) \longleftrightarrow a_1X_1(z) + a_2X_2(z)$$

2. Delay (time shift):

$$x(k-m) \longleftrightarrow z^{-m}X(z)$$

3. Time-reversal:

$$x(-k) \longleftrightarrow X(z^{-1})$$

The convergence of two grid function is the expression:

$$g(k) = \sum_{k=0}^{N} x(N-k)y(k) = \sum_{k=0}^{N} x(k)y(N-k).$$

The convergence transform of the pointed functions is found as the product of its transforms:

G(z) = X(z)Y(z)

The limit of the discrete function x(k) is defined by the formula:

$$\lim_{k \to \infty} x(k) = \lim_{z \to 1}(z-1)X(z)$$

The digital filter is considered usually as "a black box" with the input number flow x(k) and the output one y(k). The properties of the box are described completely by its transfer function. It is easy to get the output function transform as the product of the input function transfom multiplied by the transfer function of the filter. The models of all automatic control systems are based on this fundamental principle.

Let use the z-transform properties to reduce the formula of the digital filter (DF) transfer function. First, we denote the DF reaction on the sample single test pulse as the number sequence w(k). It is called weight function of the filter. On the analogy of the expression (6.4) we'll write:

$$y(k) = \sum_{k=-\infty}^{\infty} x(i)w(k-i)$$

Z-transform of the DF reaction on an arbitrary input flow x(k) is:

$$Y(z) = Z[y(k)] = \sum_{k=-\infty}^{\infty}\left[\sum_{i=-\infty}^{\infty} x(i)w(k-i)\right]z^{-k} \qquad (6.6)$$

We'll do the substitution m = k – i at the formula (6.6). The result is the expression (6.7).

$$Y(z) = \sum_{m=-\infty}^{\infty}\sum_{i=-\infty}^{\infty} x(i)w(m)z^{-(m+i)} = \sum_{m=-\infty}^{\infty}\left[\sum_{i=-\infty}^{\infty} x(i)z^{-i}\right]w(m)z^{-m} =$$

$$= \sum_{m=-\infty}^{\infty} X(z)w(m)z^{-m} = X(z)H(z), \qquad (6.7)$$

where H(z) – transfer function of the digital filter:

$$H(z) = \sum_{m=-\infty}^{\infty} w(m)z^{-m}$$

It follows that:

$$H(z) = \frac{Y(z)}{X(z)}$$

It means that the DF transfer function is computed as a relation of the output function z-transform to the z-transform of the input function. Now we'll set the relationship of the function H(z) with the coefficients of the difference equation. For the purpose we'll do z-transform of the expression (6.1) with zero start conditions:

$$a_0 Y(z) + a_1 z^{-1}Y(z) + \ldots + a_{N-1}z^{-(N-1)}Y(z) + a_N z^{-N}Y(z) =$$
$$= b_0 X(z) + b_1 z^{-1}X(z) + \ldots + b_{N-1}z^{-(N-1)}X(z) + b_N z^{-N}X(z) \qquad (6.8)$$

We'll solve the equation (6.8), using the transfer definition:

$$Y(z) = \frac{1}{a_0} * \frac{b_0 + b_1 z^{-1} + ... + b_{N-1} z^{-(N-1)} + b_N z^{-N}}{1 + a_1 z^{-1} + ... + a_{N-1} z^{-(N-1)} + a_N z^{-N}} ;$$

$$Y(z) = \frac{1}{a_0} * \frac{P(z)}{Q(z)} * X(z) = H(z)X(z)$$

It means that the DF transfer function can be presented as the relation of the polynomials $P(z)$ and $Q(z)$ formed with the coefficients of the filter equations. The roots of the equation $P(z)$ are called filter zero's ($z1$, $z2$, ...); the roots equation $Q(z)$ are called filter pole's ($p1$, $p2$, ...).

We'll pick up the normalized multiplier $M = b_0/a_0$ in the transfer function and will have the second variant of its record:

$$H(z) = M \frac{(z - z_1)(z - z_2)...(x - z_N)}{(z - p_1)(z - p_2)...(z - p_N)} \qquad (6.9)$$

The expression (6.9) may be transformed also into the sum of rational fractions. In this case the complex filter algorithm becomes as parallel or parallel-sequential junction of the simple links called basic. The important conclusion flows out of (6.9): the transfer function of the high-order filter may be obtained as the product of several first-order and second order transfer functions. Hence, the high-order filter may be sinthesized as the successive junction of first-order and second-order filters.

The transition function of the digital filter is its reaction on the test semi-infinite leap:

$$x(k) = \begin{cases} 0, & k < 0 \\ 1, & k \geq 0 \end{cases} \qquad (6.10)$$

Its z-transform is:

$$X(z) = \frac{z}{z-1}$$

It is easy to reduce the filter transition function using the transfer function H(z):

$$Y(z) = \frac{z}{z-1} H(z)$$

The weight function of the filter coincides with its reaction on the single drive-signal:

$$x(k) = \begin{cases} 0, & k < 0; \\ 1, & k = 0; \\ 0, & k \geq 1. \end{cases} \qquad (6.11)$$

The formulae for analysis of the digital filter in the frequence space are reduced from its transfer function by means of substitution:

$$z = \exp(j\omega T).$$

The digital filter is described in the frequent space with the magnitude-frequency $A(\omega)$ and phase-frequency $\varphi(\omega)$ responses. Its are a module and an argument of the transfere function:

$$H(j\omega T) = A(\omega)\exp[\varphi(j\omega T)]$$

As far as digital filters are investigated usually in the time space we omit here a discussion about the frequence responces. Notion about DF can not be complete without knowledge of the frequence properties however. There is an opportunity to see the examples of the frequence analysis in the chapter 8 of the monograph.

The constant-coefficient filters are called linear. Such filters are used effectively for smoothing of random fluctuations with a normal distribution. The smoothing quality is estimated by the statistical parameter which is computed as a relation of the noise quadratic mean of the input and output numerical sequences:

$$R = \frac{\sigma(x)}{\sigma(y)}.$$

The integral parameter R is a quality measure in static when a mathematical expectation of the input signal $M[x(k)]$ = const. In practice such sequences are met rarely however. It is necessary to pay attention to a dynamic accuracy of the filter as well. It is measured by the lag (delay, time-shift) of the output sequence.

Two accuracy estimations are discrepant. Rise of the static quality is accompanied by the lag increase. True, it is not so important in the economic investigations where data are processed aposteriory and the delay may be compensated. But it is impossible in the automatic control system working in the real time.

Strong smoothing means a loss of the higher harmonics that becomes apparent by the amplitude decrease. One may to multiply the output signal by the scale coefficient nearly 1.05 ... 1.15, when it is necessary.

6.3. Ariphmetical average smoothing

Let the coefficient values in the equation (6.1) are:

$a_0 = 1$, $a_k = 0$ for $k >= 1$;

$b_k = 1/N$ for $k >= 0$.

The input sequence x(k) is smoothing with the filter that is described by the equation:

$$y(k) = \frac{1}{N+1} \sum_{i=0}^{N} x(k-i) \qquad (6.12)$$

In accordance with the expression (6.12) the digital filter sums up the current and the N previous values x(k) with the equal weight values

$w = 1/(N +1)$

Value A = N + 1 is called as filter aperture. It is the width of the window for observation of the input sequence (fig.31).

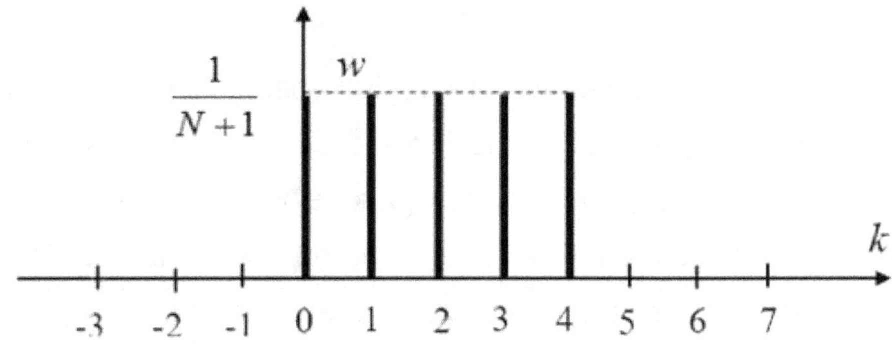

Figure 31. Weight function of the filter

The indexes $k \geq 0$ denote 5 weight coefficients of the current and 4 past numbers x(k). The future elements x(k) with k < 0 and old elements with k > 4 are not seen by the filter. The sequence x(k) is being observed in the window that is sliding along the time axis. The aperture size is constant.

The filter structure (fig.32) contains three type operations:

T — one-step delay at the discrete time T;

"+" — algebraical addition;

"x" — multipication by the weight coefficient.

Let the z-transform of the equation (6.12) is:

$$Y(z) = \frac{1}{N+1}\left[X(z) + z^{-1}X(z) + \ldots + z^{-N}X(z)\right]$$

The filter transit function is resulted when both equation parts are divided by X(z):

$$H(z) = \frac{1}{N+1}\sum_{k=0}^{N} z^{-k}$$

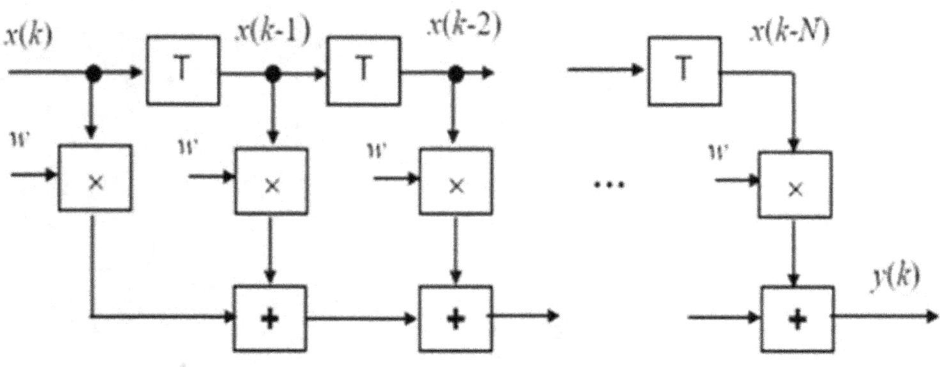

Figure 32. Structure of the equal-weight filter

As far as the output values y(k) are not depended on its previous values y(k-1), the equal-weight filter is non-recoursive. It belongs to the class of filters with a finite pulse diagram. Sometimes the filter is mentioned in the literature with other names reflecting some properties, for instance, mean arithmetical smoothing filter, sliding average filter.

The equal-weight filter is easy recognized by its reaction on the standard test drive-signal:

$$x(k) = \begin{cases} 0, & k < 0 \\ 1, & k \geq 0 \end{cases}$$

The amplitude of output sequence y(k) rises by steps w = 1/A up to level 1 (fig.33). The aperture is more, the step is less, and the step quantity is more (the filter inertialness is more).

The filter passes from the state 0 to the state 1 with delay A steps. It is considered in pactice that the transitional process is over when the level 2/3 is reached. Hence the dynamic error (i.e. the delay) is estimated as d = 2*A/3.

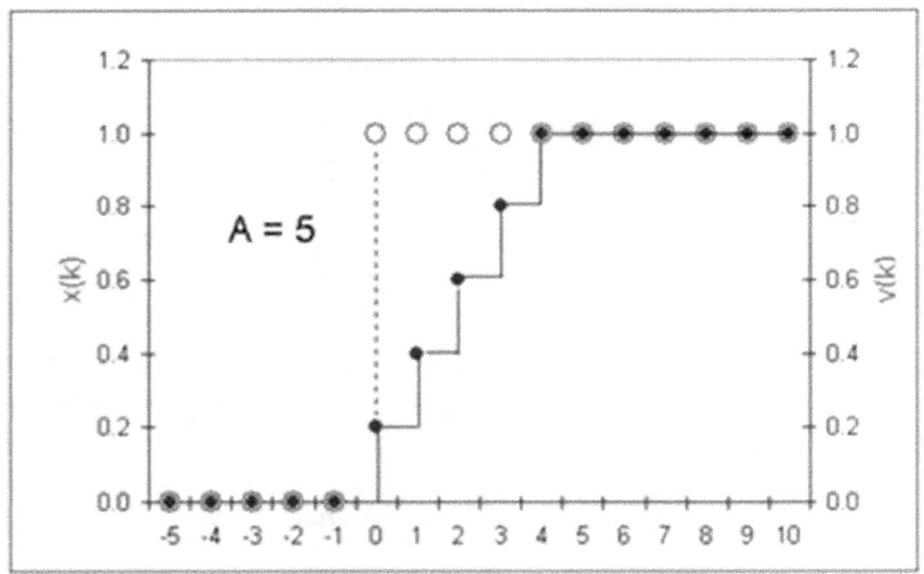

Figure 33. The transition function
of the equal-weight filter

Now we'll see examples of the equal-weight smoothing.

Example 6.1.

The input numerical flow has 30 random elements with parameters:

- the mathematical expectation me = 20;

- the random deviation -6 < rd < 6.

Develop the program of smoothig by the equal-weight filter with the aperture A = 5.

Solution.

The program Prog_6_1 (listing 6.1) consists three subroutines:

- "CoordSys" draws the rectangle coordinates;

- "Noise" is modelling the input flow;

- "Mid_Arithm" calculates the output number sequence, estimates quality and draws the graphic (fig.34).

Remark.

The noise decrease coefficient takes random value 1.4 < R < 2.9 when the program is run repeatedly.

```
Program Prog_6_1;
Uses GraphABC, ABCobjects;
  var u, z: array[1..30] of real;
  var sig1, sig2, R: real;
Procedure CoordSys;
  var  j, x: integer;
  var  d: string;
begin
  SetWindowSize(850, 450);  SetPenStyle(psClear);
  SetPenColor(clBlack);     SetPenWidth(1);
  Line(80, 30, 800, 30);    Line(800, 30, 800, 370);
  Line(80, 370, 800, 370);  Line(80, 30, 80, 370);
  SetPenWidth(2);           SetPenColor(clBlack);
  for j := 1 to 9 do
    begin
      Line(80, 380 - 50 * j, 88, 380 - 50 * j);
    end;
```

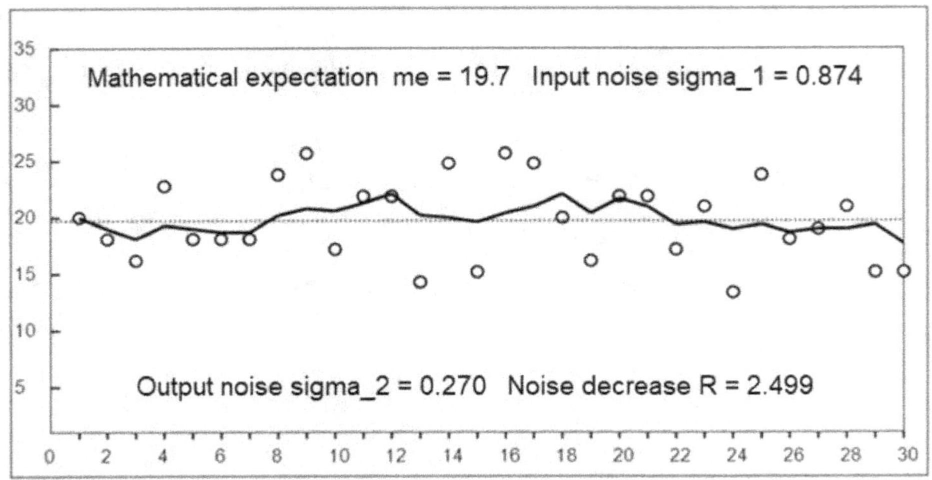

Figure 34. Equal-weight smoothing with the aperture A = 5

```pascal
for j := 1 to 30 do
   begin
        Line(80 + 24 * j, 370, 80 + 24 * j, 376);
   end;
 for j := 0 to 15 do
  begin
    x := 2 * j; d := IntToStr(x);
    var t := new TextABC(75 + 48 * j, 382, 12, d);
  end;
 for j := 1 to 10 do
  begin
    x := 5 * j; d := IntToStr(x);
    var t := new TextABC(50, 370 - 50 * j, 12, d);
  end;
end;

Procedure Noise;
 const M=30; x0=80; y0=370; mx=10; my=9.5;
 var  k, x1, x2, y1, y2: integer;
 var v: array[1..30] of real;
 var  x, y, z, noise, su, ss, nm, ds, sds: real;
 var s1, s2 :string;
 begin
  SetPenColor(clBlack);  SetPenWidth(1);
  ss:=0;
  for k := 1 to 30 do
   begin
```

```pascal
      v[k]:=Round(Random(14)-7);
      u[k] := 20+v[k]; ss:=ss+u[k];
      x1:=x0+24*k;
      y1:=y0-Round(u[k]*my);
      SetPenWidth(2);          SetPenColor(clBlack);
      SetBrushColor(ClWhite); SetPenStyle(psSolid);
      Circle(x1,y1,5);
    end;
  nm:=ss/30;              y1:=y0-Round(nm*my);
  SetPenWidth(2);         SetPenColor(clBlue);
  SetPenStyle(psDot);   Line(x0,y1,800,y1);
  Str(nm:3:1,s1);
  for k:=1 to 30 do
    begin
      ds:=u[k]-nm; sds:=sds+ds*ds;
    end;
  sig1:=sqrt(sds)/30; Str(sig1:4:3,s2);
  var t1:=new TextABC(200,40,12,
        'Mathematical expectation = '+s1);
  var t2:=new TextABC(480,40,12,
        'Input noise: sigma_1 = '+ '+s2);
  end;

Procedure Mid_Arithm(A: integer);
 const x0=80; y0=370; mx=10; my=9.5;
 var  k, x1, x2, y1, y2, z1, z2: integer;
 var ss, ds, zm, sds: real;
 var s1, s2, s3: string;
 begin
  SetPenColor(clBlack); SetPenWidth(2);
  z[1]:=u[1];
  z[2]:=(u[1]+u[2])/2;
  z[3]:=(u[1]+u[2]+u[3])/3;
  z[4]:=(u[1]+u[2]+u[3]+u[4])/4;
  ss:=0;
  for k:=5 to 30 do
   begin
    z[k]:=(u[k]+u[k-1]+u[k-2]+u[k-3]+u[k-4])/5;
    ss:=ss+z[k];
   end;
  for k:=1 to 30 do
   begin
```

```pascal
    x1:=x0+24*k;
    y1:=y0-Round(u[k]*my);
    z1:=y0-Round(z[k]*my);
    SetPenWidth(3);   SetPenColor(clRed);
    SetPenStyle(psSolid);
    if k>1 then Line(x1,z1,x2,z2);
    x2 := x1; z2:=z1;
  end;
 zm:=ss/25;
 for k:=1 to 30 do
   begin
      ds:=z[k]-zm; sds:=sds+ds*ds;
   end;
 sig2:=sqrt(sds)/30;   R:=sig1/sig2;    s1:=IntToStr(a);
 var t1:=new TextABC(300,0,12,
        'Equal-weight smoothing   A = '+s1);
 Str(sig2:4:3,s2);
 var t2:=new TextABC(200,340,12,
        'Output noise   sigma_2 =  '+s2);
 Str(R:4:3, s3);
 var t3:=new TextABC(480,340,12,
        'Noise decrease R = '+s3);
 end;
//====== Main ========
var A: integer;
begin
 CoordSys; Noise;
 A:=5;  Mid_Arithm(A);
end.
```

Listing 6.1

The static accuracy of the smoothing is rising with the aperture A > 5 but the lag is rising too in this case. On the contrary the lag may be decreased with the aperture A < 5 but at the cost of the static error increases. It can be seen at fig.35.

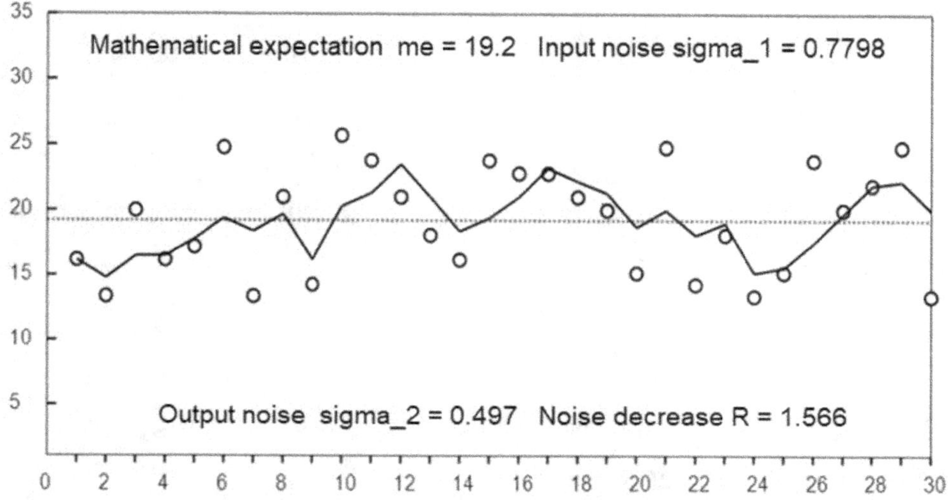

Figure 35. Equal-weight smoothing with the aperture A = 3

6.4. Exponential smoothing

The algorithm of the exponential smoothing has been already described above in the expression (5.4). Let us remember the inertial object equation:

$$y(k) = b*x(k) + (1 - b)*y(k - 1)$$

The current output value of the exponential filter accumulates theoretically the information about all previous values of the input numerical sequence but with weight coefficients that are decreasing in accordance with the exponential rule. One may be convinced of the fact using drive-signal in the form of a single pulse with the amplitude 1. The infinite number sequence of the weight coefficients $w(k)$ is resulted in this case:

$$w(0) = bx(0) + (1-b)y(-1) = b;$$
$$w(1) = (1-b)y(0) = b(1-b);$$
$$w(2) = (1-b)y(1) = b(1-b)^2;$$
$$...$$
$$w(k) = b(1-b)^k \qquad\qquad (6.13)$$

In practice the values w(k) coincide with the points of the exponent curve (fig.36)

$$s(k) = 0.3\exp(-k/m),$$

where m = 2.8076 – scale coefficient. The smoothing algorithm name reflects the curve type.

The exponent algorithm belongs to the filter class with the infinite pulse function. Its pulse function is finite indeed due to the limited length of computer code. On other hand the previous input values with indexes k < 12 have very small weight coefficients. It is the reason to be ignored completely. For instance, w(k) < 0.01 if k > 10. It means that out of date information exerts no influence upon the current output value y(k) and can be ignored.

The transition function (fig.37) is the filter reaction on the standard drive-signal

$$u(k) = \begin{cases} 0, & k < 0 \\ 1, & k \geq 0 \end{cases}$$

The first step size y(0) is equal to the parameter value b = 0.3. Difference x(k) – y(k) → 0 when k → ∞ (in practice k > 15 if b = 0.3). The b is less, the digital filter is more inertial.

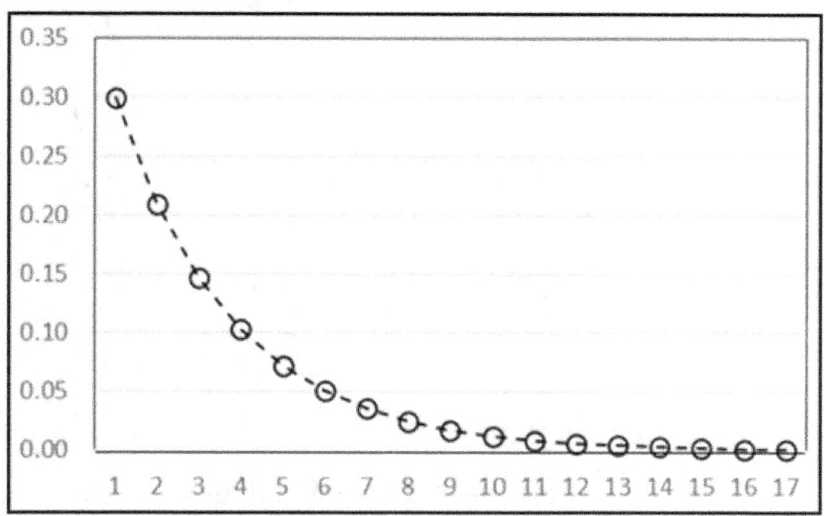

Figure 36. Weight function

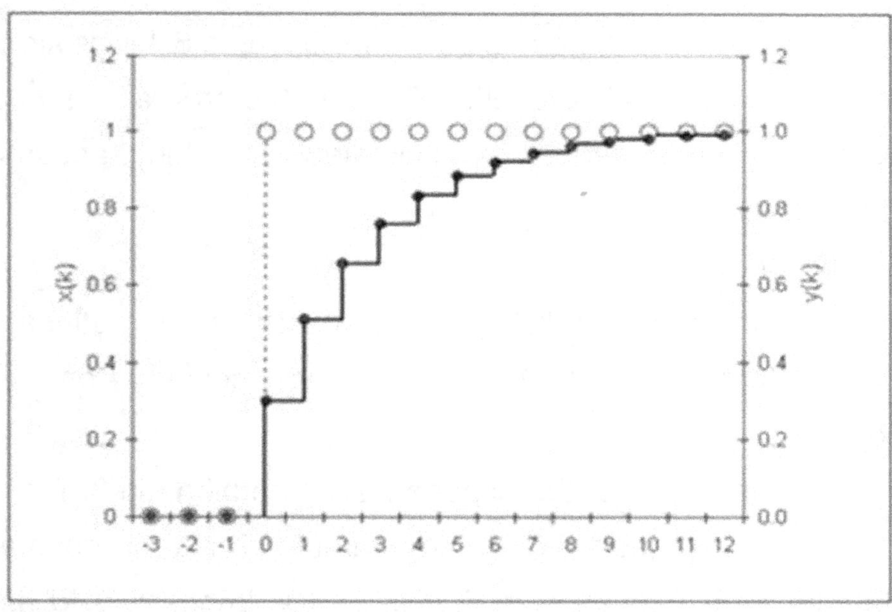

Figure 37. The transition response of exponential filter

The structure of the exponential filter reflects its recourciveness, i.e. using of the previous smoothed value for calculating of the current output value. It is shown by the feedback relation (fig.38).

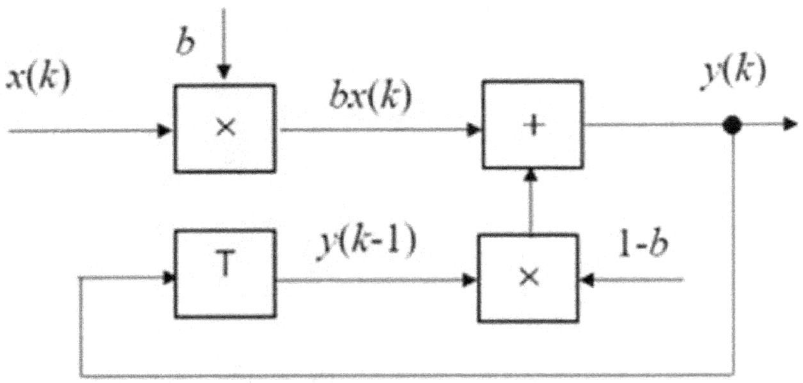

Fig.38. The exponential filter

Example 6.2.

The input flow of the exponential filter with the smoothing parameter b = 0.2 is the same as in the Example 6.1 (it is significant for the comparable results). Develop the program that calculates the output sequence, estimates smoothing accuracy and draws diagram.

Solution.

Program Prog_6_2 (listing 6.2) contains the subroutines:

- "CoordSys" and Noise that are copied from the Prog_6_1;

- "ExpSmooth": for the exponential smoothing (fig.39).

The input sequences in both examples have the constant mathematical expectation. In that case it is impossible to estimate the most significant property of the exponential filter which preferes the newest input values and depreciates the previous values that becomes obsolete. It is the main reason for wide using the exponential filters in the technical automatic control systems.

Results are shown on fig.39: mathematical signal expectation me = 19.5. Input noise sigma1 = 0.742; output noise sigma2 = 0.24. Nose decreasing R = 3.098.

```pascal
Program Prog_6_2;
Uses GraphABC, ABCobjects;

var u, z: array[1..30] of real;   var sig1,sig2, R: real;
Procedure CoordSys;
  var j, x: integer;    var d: string;
begin
  SetWindowSize(850, 450);  SetPenStyle(psClear);
  SetPenColor(clBlack);        SetPenWidth(1);
  Line(80, 30, 800, 30);       Line(800, 30, 800, 370);
  Line(80, 370, 800, 370);     Line(80, 30, 80, 370);
  SetPenWidth(2);              SetPenColor(clBlack);
  for j := 1 to 9 do
    begin
      Line(80, 380 - 50 * j, 88, 380 - 50 * j);
    end;
  for j := 1 to 30 do
    begin
      Line(80 + 24 * j, 370, 80 + 24 * j, 376);
    end;
  for j := 0 to 15 do
    begin
      x := 2 * j; d := IntToStr(x);
      var t := new TextABC(75 + 48 * j, 382, 12, d);
    end;
    for j := 1 to 10 do
    begin
      x := 5 * j; d := IntToStr(x);
      var t := new TextABC(50, 370 - 50 * j, 12, d);
  end;
end;

Procedure Noise;
  const M=30; x0=80; y0=370;
  mx=10; my=9.5;
  var k, x1, x2, y1, y2: integer;
  var v: array[1..30] of real;
```

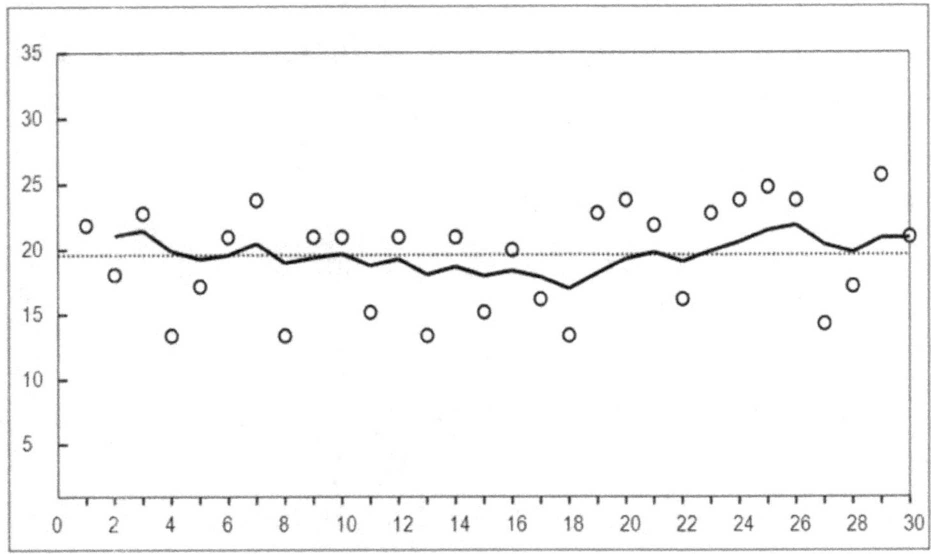

Figure 39. The exponential smoothing with the parameter b = 0.2.

```
var x, y, z, noise, su, ss1, nm, ds, sds: real;
var s1, s2 :string;
begin
  SetPenColor(clBlack);  SetPenWidth(1);
  ss1:=0;
  for k := 1 to 30 do
     begin
       v[k]:=Round(Random(14)-7);
       u[k] := 20+v[k];  ss1:=ss1+u[k];  x1:=x0+24*k;
       y1:=y0-Round(u[k]*my);  SetPenWidth(2);
       SetPenColor(clBlack);    SetBrushColor(ClWhite);
       SetPenStyle(psSolid);    Circle(x1,y1,5);
     end;
    nm:=ss1/30;              y1:=y0-Round(nm*my);
    SetPenWidth(2);          SetPenColor(clBlue);
    SetPenStyle(psDot);  Line(x0,y1,800,y1);
    Str(nm:3:1,s1);
    for k:=1 to 30 do
      begin
        ds:=u[k]-nm; sds:=sds+ds*ds;
      end;
    sig1:=sqrt(sds)/30; Str(sig1:4:3,s2);
    var t1:=new TextABC(200,40,12,
         'Mathematical expectation = '+s1);
```

```pascal
    var t2:=new TextABC(480,40,12,
          'Input noise: sigma_1 = '+s2);
  end;

 Procedure ExpSmooth(b: real);
  const x0=80; y0=370; mx=10; my=9.5;
  var  k, x1, x2, y1, y2, z1, z2: integer;
  var ss, ds, zm, sds, ss2: real;
  var s1, s2, s3: string;
  begin
   SetPenColor(clBlack);   SetPenWidth(2);
   z[1]:=u[1]; ss:=0;
   for k:=2 to 30 do
     begin
       x1:=x0+24*k;
       z[k]:=u[k]*b+z[k-1]*(1-b);  ss:=ss+z[k];
       y1:=y0-Round(u[k]*my);   z1:=y0-Round(z[k]*my);
       SetPenWidth(3);           SetPenColor(clBlack);
       SetPenStyle(psSolid);
       if k>2 then Line(x1,z1,x2,z2);
       x2 := x1; z2:=z1;
     end;
    zm:=ss/29;
   for k:=1 to 30 do
     begin
       ds:=z[k]-zm;  sds:=sds+ds*ds;
     end;
   sig2:=sqrt(sds)/30;  R:=sig1/sig2;
   Str(b:3:2,s1);
   var t1:=new TextABC(280,0,12,
          'Exponential smoothing b =  '+s1);
   Str(sig2:4:3,s2);
   var t2:=new TextABC(200,340,12,
          'Output noise sigma_2 =  '+s2);
   Str(R:4:3,s3);
    var t3:=new TextABC(480,340,12,
          'Noise decrease  R =  '+s3);
  end;
//========= Main ======
var b: real;
begin
 CoordSys; Noise;
```

  b:=0.2;   ExpSmooth(b);
end.

Listing 6.2

6.5. Median alignment

The initial numerical sequence firstly ought to be aligned and only then smoothed during the prevent processing. The matter is fact that the homogeneity of the observation data is demanded in many control systems but it is desturbed often by the abnormal deflections of some input values. In result an anomalous splash or downfall can arise. The reason of the random numerical deviation may be a single pulse hindrance in the radio channel or the ordinary misprint in the digital text. In so manner the value appears that is many times more then the mathematical expectation or otherwise it is many times less.

The equal-weight or exponential filters cannot suppress the anomalous deflections owing their inertialness. The effective method to exclude the extraordinary interference has been developed many years ago in the technology of the digit picture processing [9], but is being unknown in the economic modelling. This aligning method is not included even into the standard library of Microsoft Excel up to date (October 2019).

The problem solution, as it often happens, was evident long ago for specialists who ever used the function "Median". The indication of an anomaly is very simple: in the input sequence of several adjacent values the non-standard amplitude can never be median. Hence, it is necessary to compare the values $x(k - j)$ that are observed in the sliding window, and to award median value to the current output element $y(k)$:

$$y(k) = \operatorname*{med}_{j}\big[x(k - j)\big], \qquad j = 0,1,..., N$$

The comparison of the adjacent values may be correct in the case only when the filter aperture is odd, for instance A = 3, 5, 7. Of course, the dynamic delay arises with big values A. Hence, values 3 or 5 are recommended. The median filter is capable also to suppress at the same time the train of the sequence with the normal distribution, i.e. to suppress elements which have the module more then 3σ.

Example 6.3.

Develop the alignment model for the numeric sequence with length 30 integer elements u(k), three being abnormal. The sequence mathematical expectation M = 10, the random deviations are in the interval (-2; 2). Set the model abnormal values u[7] = 19, u[17] = 2, u[27] = 18. The program is to detect logically the anomalies in the sliding window with the apperture A = 3 and to carry out the substitution u[k-1] or u[k-2] depending on which of them becomes the median.

Solution.

The program Prog_6_3 (listing 6.3) contains three subroutines:

- "CoordSys" that draws the rectangle axes;

- "Noise" that models the sequence with three abnormal values;

- "Levelling" that alignes the input flow (fig.40).

```
Program Prog_6_3;
Uses GraphABC, ABCobjects;

var u: array[1..30] of real;
var v: array[1..30] of real;
Procedure CoordSys;
```

```pascal
  var j, x: integer;   var d: string;
begin
  SetWindowSize(850, 500);   SetPenStyle(psClear);
  SetPenColor(cLBlack);        SetPenWidth(1);
  Line(100, 40, 820, 40);     Line(820, 40, 820, 440);
  Line(100, 440, 820, 440); Line(100, 40, 100, 440);

  for j := 1 to 10 do
    begin
     Line(100, 440 - 40 * j, 108, 440 - 40 * j);
    end;
  for j := 0 to 15 do
    begin
     x := 2 * j;  d := IntToStr(x);
     var t := new TextABC(96 + 48 * j, 445, 12, d);
    end;
  for j := 1 to 10 do
    begin
     x := 2 * j; d := IntToStr(x);
     var t := new TextABC(75, 430 - 40 * j, 12, d);
  end;
end;

Procedure Noise;
  const x0=100; y0=440; mx=10; my=20;
  var k, x1, x7, x17, x27, y1, y2, y7, y17, y27: integer;
  begin
   for k := 1 to 30 do
     begin
        u[k] := 9+Random(30)/9;
        x1:=x0+24*k;   y1:=y0-Round(u[k]*my);
        if (k=7) or (k=17) or (k=27) then y1:=y0;
        SetPenColor(clBlack);     SetPenWidth(2);
        SetBrushColor(ClWhite);  SetPenStyle(psSolid);
        Circle(x1,y1,5);            SetPenWidth(1);
        Line(x1,y0,x1,y1);
     end;
     u[7]:=19; u[17]:=2; u[27]:=18;
     x7:=x0+24*7;        y7:=y0-Round(u[7]*my);
     x17:=x0+24*17;     y17:=y0-Round(u[17]*my);
     x27:=x0+24*27;
     y27:=y0-Round(u[27]*my);
```

```pascal
      SetPenWidth(4);        SetPenColor(clBlue);
      Line(x7,y0,x7,y7);     Line(x17,y0,x17,y17);
      Line(x27,y0,x27,y27);  SetPenWidth(2);
      Circle(x7,y7,6);       Circle(x17,y17,6);
      Circle(x27,y27,6);
   end;
 Procedure Levelling;
   const x0=100; y0=440; mx=10; my=20;
   var  k, x1, x2, y1, y2, z1, z2: integer;
   var  a, b, c, med, maxi, x, y,z: real;
   begin
    SetPenColor(clBlack);   SetPenWidth(3);
    v[1]:=u[1];  v[2]:=(u[1]+u[2])/2;
    for k:=3 to 30 do
      begin
        a:=u[k-2];   b:=u[k-1];   c:=u[k];
        if (a>b) and (a<c) then med:=a;
        if (a>c) and (a<b) then med:=a;
        if (b<a) and (b>c) then med:=b;
        if (b>a) and (b<c) then med:=b;
        if (c<a) and (c>b) then med:=c;
        if (c>a) and (c<b) then med:=c;
        v[k]:=med;
      end;
      for k:=1 to 30 do
      begin
        x1:=x0+24*k;
        z1:=y0-Round(v[k]*my);
        SetPenWidth(4);    SetPenColor(clRed);
        If k>1 then Line(x1,z1,x2,z2);
        x2:=x1;  z2:=z1;
      end;
 end;
//====== Main ====
begin
  CoordSys;  Noise;
  Levelling;
end;
```

Listing 6.3

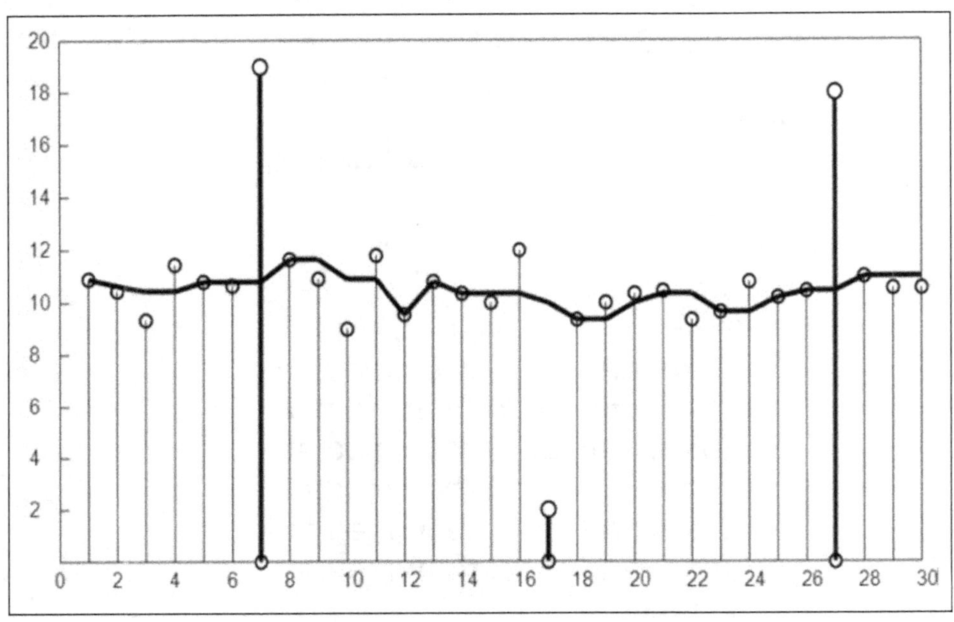

Figure 40. The median alignment of the numerical flow

Conclusion.

The median filter has deleted all anomalies in the input sequence and made it homogeneous. The anomalies are excluded ideally. The "suspicious" elements u[7], u[17], u[27] are detected and substituted by the elements u[5], u[16] and u[25] accordingly.

Chapter 7. EXTRAPOLATION

7.1. Methods of numerical prediction

At all times the economic success of mankind was depending on ability to look in the future. History, religion and literary monuments testify to this fact. The agriculture and the sea transportation were needy to know the weather forecast in some days and weeks ahead. The producers of material values and the merchants were keenly dependent on the expected customer demand. It is not mere chance to appear the words "predict", "prophecy", "foretell", "foresee". The word "prognosis" is produced from Grecian "gnosis" (knowledge).

The centuries-old experience accumulated in written and oral form is the basis for prediction. Every nation has the real omen system used to form hypotheses relatively the visible future events. In all times the prediction was considered as an art but in the last 20-30 years the so-called soothsayers attending to the "high educated clients" are to use the new quasi-scientific termins such as "positive and negative energy", "bio-fields", "information flows" and so further. The russian scientist D. Mendeleev, known as the author of the periodical element system and the metrology originator in Russia, denoted a frontier between the art and the science 130 years ago. According to him, a science begins when measurements start.

The descrete mathematic gives the quantitative predictions based on the measurement results that engenders the empiric numerical rows registered in the finite time interval. One may predict with some probability the future value of this row using the retrospective value sequence. Such prediction is called single-point prognosis. If the value boundaries are prognosed, then prediction is called the interval prognosis. The prediction truth can be proved by the fact only i.e. a posteriory. That is why any prediction is true as a stochastic parameter. The empiric values are disturbed by the random deviations.

The numeric row is considered as a reliable source of data for the prediction if the next conditions are satisfied:

- representativeness (statistical sufficiency);
- stability (conforming to natural laws);
- homogeneity (all elements belong to the same row).

Not every numeric row can be used for a prediction at once. It is necessary to continue the observations if the row is too short or unstable. The row is to be aligned previously if it contains some abnormal elements. Further the row is to be smoothed by any method.

The extrapolation is one of the forecast methods used for solution of many problems in practice. There are some examples:

1. The forestalling of the input signal deviation is calculated in the automatic control system, for instance, using the velocity and acceleration values.

2. The aim trajectory is extrapolated to compute the optimum angle values of the gun tube for the shot moment.

3. The extrapolation of the aim trajectory is calculated duiring the guiding of the missile.

4. The purchase of raw materials is planning in the microeconomic using extrapolation methods.

5. The demographic prognosis is needy in macroeconomic to form the state budget.

7.2. Methods of continuous function restoration

The restoration of a continious function using its discrete values is the task reverse the analog-to-digit transforming of the function. It is necessary to join the discrete points of the sequence in so manner that a broken line would be got with small deflections from the origin smooth function. The problem is how to predict the value of the function one step ahead.

Let us define the task more exact. The input numerical row contains values y(k) that follow with constant time intervals T. Every interval T is to be divided at m parts with the step dt = T/m and filled with calculated values z(j), j = 1, 2, ..., m. There are three methods used most often to compute z(j).

Zero-order extrapolation

The algorithm is based on the hypothesis that the unknown function preserves the current value y(k) in the interval T. It is supposed the first derivative of the function (i.e. velocity) is equal to 0. The extrapolation error arises if the hypothesis is not true and the error is to be estimated.

In accordance with the hypothesis the value y(k) is to be remem-bered up to the moment of getting new value y(k+1). The method is attractive due to the technological simplicity. The device called "pick-detector" makes sampling and keeping of the instant function value.

Remark.

Here and further the control function is used to produce the input numerical row y(k):

$$y(t) = 20*\exp(-0.8t)*[1 - \cos(0.6t)]^2 + 0.3t,$$

where t – continuous time in the interval (0; 10).

The function y(t) is computed with the step $\Delta t = T/m = 0.02$.

Example 7.1.

The input sequence has 25 numbers that are formed with the control function in the discrete time intervals T. Develop the program to restore the function by the method of the zero-order extrapolation.

Solution.

Program Prog_7_1 (listing 7.1) contains two subroutines "CoordSys" and "Extra_0". The first one draws the coordinate axes. The subroutine "Extra_0" solves the next tasks:

- to compute the control function values;

- to generate the input number sequence with time step 0.4;

- to draw the control function and the step graph (fig.41);

- to estimate the extrapolation error.

```
Program Prog_7_1;
Uses GraphABC, ABCobjects;

var z: array[1..600] of real;
Procedure CoordSys;
  const x0=100;   var k: integer;
  var x, y: real;
  var s: string;
  begin
    SetWindowSize(800,550);  SetPenStyle(psClear);
    SetPenColor(clBlack);       SetPenWidth(1);
    Line(140, 0, 140, 486);      Line(140, 300, 770, 300);
    Line(140, 0, 770, 0);         Line(140, 430, 770, 430);
    Line(770, 0, 770, 486);      Line(140, 486, 770, 486);
    for k:=1 to 21 do
      begin
        Line(140+30*k, 300, 140+30*k, 290);
```

```pascal
      end;
    for k:=6 to 20 do
      begin
        Line(140, 400-20*k, 146, 400-20*k);
      end;
    for k:=2 to 11 do
     begin
       x:=1*(k-1); Str(x:2:0,s);
       var t:= new TextABC(74+60*k,304,11,s);
     end;
    for k:=4 to 10 do
     begin
       y:=0.4*(k-1)-0.8; Str(y:3:1,s);
       var t:= new TextABC(110,414-40*k,11,s);
     end;
    for k:=6 to 13 do
      begin
        Line(134, 598-21*k, 140, 598-21*k);
      end;
    for k:=3 to 10 do
     begin
       y:=1*(k-1)-4; Str(y:2:0,s);
       var t:= new TextABC(116,526-21*k,10,s);
     end;
end;

Procedure Extra_0;
  const x0=140; y0=300; z0=430;  mx=60; my=70; me=4.5;
  var j, k, x1, x2, y1, y2, z1, z2, u1, u2: integer;
  var x, dx, y, ee: real;
  begin
   x:=0; dx:=0.02; j:=1;
   z[1]:=20*(1-cos(0.6*x))*(1-cos(0.6*x))*
   exp(-0.8*x)+0.5*x;
   z1:=y0-Round(z[1]*my);   SetPenWidth(3);
   SetPenColor(clBlue);        Line(x0,z1,x0+30,z1);
   for k:=1 to 500 do
   begin
    x:=x+dx;     y:=20*(1-cos(0.6*x))*(1-cos(0.6*x))*
    exp(-0.8*x)+0.3*x;
    if (k mod 25 = 0) then
      begin
```

```pascal
    z[k]:=y;                    z1:=y0-Round(z[k]*my);
    SetPenWidth(3);             SetPenColor(clBlue);
    Line(x1,z1,x2,z2);          Line(x1,z1,x1+30,z1);
  end;
  x1:=x0+Round(x*mx);         y1:=y0-Round(y*my);
  if (k mod 25)=0 then
    begin
     j:=j+1; z[j]:=y;
    end;
  ee:= 100*(y-z[j])/3.8;      u1:=z0-Round(ee*me);
  if k>1 then
    begin
      SetPenWidth(3);           SetPenColor(clBlack);
      SetPenStyle(psSolid);  SetPenColor(clBlack);
      Line(x1,y1,x2,y2);         Line(x1,u1,x2,u2);
    end;
    x2:=x1; y2:=y1;            z2:=z1; u2:=u1;
    ee:=y1-z1;
  end;
end;
//==================== Main ================
var s: string;
begin
  CoordSys; Extra_0;
  var t1:=new TextABC(350,200,12, 'Extrapolation');
  var t2:=new TextABC(370,370,12, 'Error, %');
end.
```

Listing 7.1

Conclusions.

1. The extrapolation error is less than 5 percent with the given interval value T.

2. The most error arises where the first derivative of control function has maximun.

3. In accordance with the hypothesis the extrapolation accuracy increases when the change velocity of the control function decreases.

4. The error sign change is in the extremum points of the control function.

5. The evident conditions of the extrapolation error decrease:

- to diminish the time quantizing step T;

- to limit the change velocity of the control function.

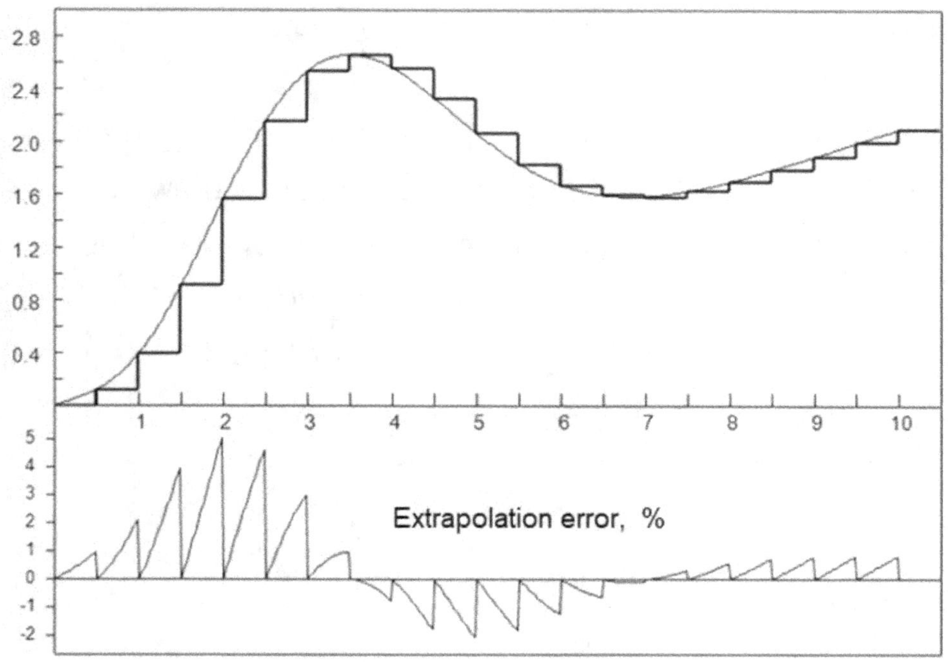

Figure 41. The zero-order extrapolation

The first-order extrapolation

The hypothesis is choosed: the change velocity of the function is constant in every interval of the extrapolation:

$$v(k) = [y(k) - y(k-1)] / T = const.$$

The approximating function has linear dependence on the time:

$$z(j) = y(k-1) + v(k)*T/m, \quad j = 1, 2, \ldots, m$$

Example 7.2.

Develop the program that approximates the control function with the broken line formed of the linear segments. The incline grade of the segment is equal to the velocity v(k).

Remark.

As far as the velocity is not known in the first interval T, it is considered to be equal 0 and z[2] = y[1] = 0. The linear extrapolation is beginning at the second interval.

Solution.

Program 7_2 (listing 7.2) contains three subroutines "CoordSys", "Func" and "Extrapol". The first one draws the rectangle axes, the second generates the input row y(k). The subroutine "Extrapol" is intended for the next actions:

- to measure the change velocity of the input flow;

- to realize the extrapolation;

- to estimate the extrapolation error;

- to draw graphic (fig.42).

```
Program Prog_7_2;
Uses GraphABC, ABCobjects;

var  z,v: array[0..600] of real;

procedure CoordSys;
const  x0 = 100;
var  k: integer;
var  x, y: real;
var  s: string;
begin
  SetWindowSize(800, 550);  SetPenStyle(psClear);
  SetPenColor(clBlack);     SetPenWidth(1);
  Line(140, 0, 140, 506);   Line(140, 300, 770, 300);
  Line(140, 0, 770, 0);     Line(140, 420, 770, 420);
  Line(770, 0, 770, 506);   Line(140, 506, 770, 506);
  for k := 1 to 21 do
```

```pascal
    begin
      Line(140 + 30 * k, 300, 140 + 30 * k, 290);
    end;
  for k := 6 to 20 do
    begin
      Line(134, 470 - 34 * k, 140, 470 - 34 * k);
    end;
  for k := 2 to 11 do
    begin
      x := 100 * (k - 1);    Str(x:2:0, s);
      var t := new TextABC(70 + 60 * k, 304, 11, s);
    end;
  for k := 4 to 8 do
    begin
      y := 1 * (k - 1) - 3;   Str(y:1:0, s);
      var t := new TextABC(120, 564 - 68 * k, 11, s);
    end;
  for k := 6 to 14 do
    begin
      Line(134, 630 - 21 * k, 140, 630 - 21 * k);
    end;
  for k := 3 to 11 do
    begin
      y := 2 * (k - 1) - 12;
      Str(y:2:0, s);
      var t := new TextABC(116, 556 - 21 * k, 10, s);
    end;
end;

Procedure Func;
const  x0 = 140;  y0 = 300;  mx = 60;  my = 70;
var  j, k, n, x1, x2, y1, y2, z1, z2: integer;
var  x, dx, y, w: real;
begin
  x := 0; dx := 0.01; j:=0;
  for k := 1 to 1000 do
    begin
      x := x + dx;
      y := 20 * (1 - cos(0.6 * x)) * (1 - cos(0.6 * x)) *
           exp(-0.8 * x) + 0.3 * x;
      x1 := x0 + Round(x * mx);
      y1 := y0 - Round(y * my);
```

```pascal
      if k > 1 then
        begin
          SetPenWidth(2);         SetPenColor(clBlack);
          SetPenStyle(psSolid); SetPenColor(clBlack);
          Line(x1, y1, x2, y2);
        end;
      x2 := x1;y2 := y1;
  end;
end;

Procedure Extrapol;
const  x0 = 140;  y0 = 300;  z0 = 420;
mx = 60;  my = 70;  mv=1; me = 10.5;
var j, k, x1, x2, y1, y2, w1, w2, z1, z2: integer;
var x, y, dx, w, vv, xp, xq, yp, yq, yc, ee : real;
  begin
  dx:=0.02;  SetPenWidth(2);  SetPenColor(clBlue);
  x:=0;         dx:=0.5;              z[0]:=0;
  for k:=1 to 20 do
    begin
    x:=x+dx; x1:=x0+Round(x*mx);
    y := 20 * (1 - cos(0.6 * x)) * (1 - cos(0.6 * x)) *
                            exp(-0.8 * x) + 0.3 * x;
    z[k]:=y;                    y1:=y0-Round(y*my);
    SetPenStyle(psSolid);    SetPenColor(clBlack);
    SetBrushColor(clWhite); SetPenWidth(2);
    Circle(x1,y1,4);
  end;
  for k:=1 to 20 do
    begin
    v[k]:=(z[k]-z[k-1])/0.50;
    end;
        SetPenColor(clBlue);  SetPenWidth(3);
        dx:=0.05;
  for k:=1 to 10 do
    begin
    x:=dx*k;
    yc := 20 * (1 - cos(0.6 * x)) * (1 - cos(0.6 * x)) *
                            exp(-0.8 * x) + 0.3 * x;
    y:=0;
    ee:=100*(y-yc)/4.4;     x1:=x0+Round(x*mx);
    y1:=y0-Round(y*my);   z1:=z0-Round(ee*me);
```

```pascal
    if k>1 then
      begin
        Line(x1,y1,x2,y2);
        Line(x1,z1,x2,z0);
      end;
   x2:=x1; y2:=y1; z2:=z1;
  end;
  z1:=y0-Round(z[1]*my);   Line(x1,y1,x2,z1);
 for k:=1 to 10 do
   begin
    x:=0.5+dx*k;
    yc := 20 * (1 - cos(0.6 * x)) * (1 - cos(0.6 * x)) *
                      exp(-0.8 * x) + 0.3 * x;
    y:=v[1]*x;                 ee:=100*(y-yc)/4.4;
    x1:=x0+Round(x*mx);  y1:=y0-Round(y*my);
    z1:=z0-Round(ee*me);
    if k>1 then
      begin
        Line(x1,y1,x2,y2);    Line(x1,z1,x2,z2);
        Line(x1,z1,x2,z0);
      end;
  for j:=2 to 19 do
    begin
     for k:=1 to 10 do
      begin
       x:=0.5*j+dx*k;
       yc := 20 * (1 - cos(0.6 * x)) * (1 - cos(0.6 * x)) *
                      exp(-0.8 * x) + 0.3 * x;
       y:=z[j]+v[j]*dx*k;
       ee:=100*(y-yc)/4.4;      x1:=x0+Round(x*mx);
       y1:=y0-Round(y*my);   z1:=z0-Round(ee*me);
       if k>1 then
       begin
         Line(x1, y1, x2, y2);  Line(x1, z1, x2 z2);
         Line(x1, z1, x2, z0);
       end;
        x2:=x1; y2:=y1; z2:=z1;
     end;
    z1:=y0-Round(z[j+1]*my); Line(x1,y1,x1,z1);
  end;
end;
```

```
//===========================Main=================
var  s: string;
begin
  CoordSys;
  Func;    Extrapol;
  var t1 := new TextABC(320, 210, 12,  'Extrapolation');
  var t2 := new TextABC(310, 470, 12,  'Extrapolation
         error,  %');
end.
```

Listing 7.2

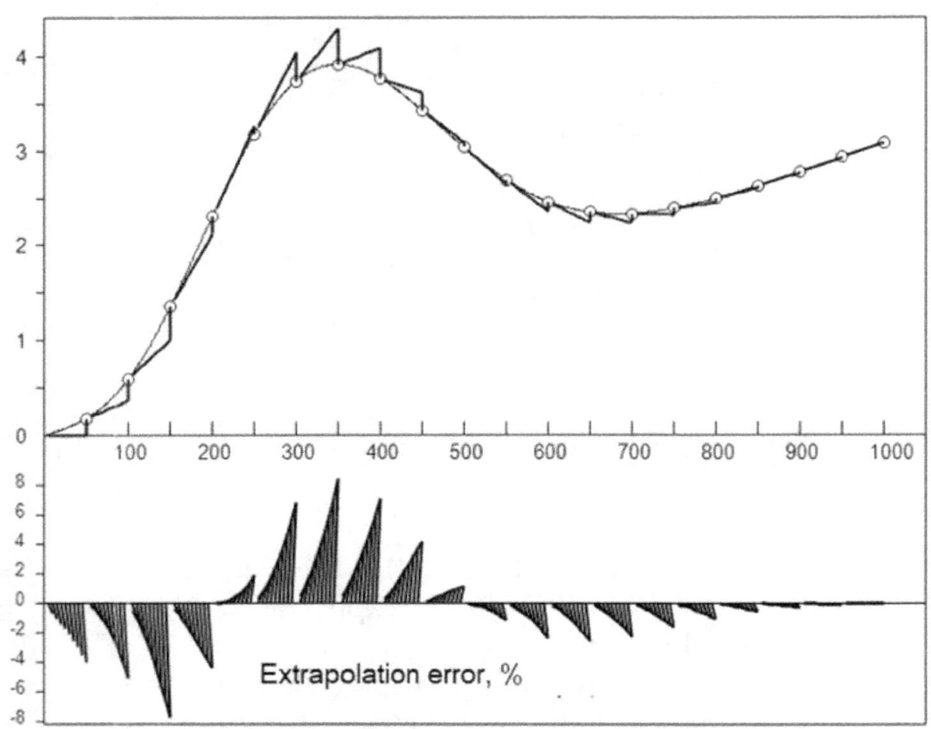

Figure 42. The linear extrapolation

Conclusions.

1. The extrapolation error is near 8% in the intervals where the acceleration is maximum.

2. The linear extrapolation works off the settled velocity. The error aspires to 0 if the velocity becomes constant. It is the

important advantage of the first-order extrapolation over above-mentioned prediction method.

3. For increasing the accuracy of the linear extrapolation it is necessary to decrease the time quantum step and to limit the acceleration of the input numerical flow.

The second-order extrapolation

The hypothesis is taken: the numerical flow acceleration is constant in the interval T. The mathematical formulae of the assumption:

$$v(k) = [y(k) - y(k - 1)] / T; \qquad \text{- velocity;}$$

$$a(k) = [v(k) - v(k - 1)] / T = \text{const.} \quad \text{- acceleration}$$

The future value of the approximation is calculated:

$$y(k + 1) = y(k) + v(k)^*(T\ /m) + a(k)^*(T / m)^2.$$

Hence, the prediction becomes non-linear. The second degree polynom is planned, namely the parabola. On this reason the extrapolation is called parabolic as well.

Example 7.3.

Develop the program that approximates the control function with parabolic segments in the intervals of the grid function used in the previous example.

Remarks.

1. The flow velocity is not defined in the first interval. One ought to consider:

$$v(k=1) = 0; y(k) = y(t = 0).$$

2. The flow acceleration is not known in the second interval. Hence, the linear extrapolation hypothesis is to be used.

3. The parabolic hypothesis is being realized in the next intervals.

Solution.

Program Prog_7_3 (listing 7.3) contains the subroutines "CoordSys" for drawing the rectangle axes and "Func" for computing the control function. The third subroutine "Predict" is to do the next actions:

- to generate the input numerical sequence;

- to calculate the flow parameters - velocity and acceleration;

- to approximate the function;

- to estimate the extrapolation error;

- to draw the graphic (fig.43).

```pascal
Program Prog_7_3;
Uses GraphABC, ABCobjects;

var  a,v,z: array[0..600] of real;
procedure CoordSys;
const x0 = 100;
var  k: integer;
var  x, y: real;
var  s: string;
begin
  SetWindowSize(800, 550);  SetPenStyle(psClear);
  SetPenColor(clBlack);     SetPenWidth(1);
  Line(140, 0, 140, 520);   Line(140, 300, 770, 300);
  Line(140, 0, 770, 0);     Line(140, 400, 770, 400);
  Line(770, 0, 770, 520);   Line(140, 520, 770, 520);
for k := 1 to 21 do
   begin
     Line(140 + 30 * k, 300, 140 + 30 * k, 290);
   end;
 for k := 6 to 20 do
   begin
     Line(134, 470 - 34 * k, 140, 470 - 34 * k);
   end;
```

```pascal
    for k := 2 to 11 do
      begin
        x:= 100 * (k - 1); Str(x:2:0, s);
        var t := new TextABC(70 + 60 * k, 304, 11, s);
      end;
  for k := 4 to 8 do
   begin
     y := 1 * (k - 1) - 3;    Str(y:1:0, s);
     var t := new TextABC(120, 564 - 68 * k, 11, s);
   end;
   for k := 6 to 14 do
     begin
        Line(134, 631 - 21 * k, 140, 631 - 21 * k);
     end;
   for k := 2 to 10 do
     begin
       y := 1 * (k - 1) - 6;    Str(y:2:0, s);
       var t := new TextABC(116, 540 - 21 * k, 10, s);
     end;
 end;

Procedure Func;
const x0 = 140; y0 = 300; mx = 60; my = 70;
var j, k, n, x1, x2, y1, y2, z1, z2: integer;
var x, dx, y, w: real;
begin
 x := 0; dx := 0.01; j:=0;
 for k := 1 to 1000 do
   begin
     x := x + dx;
     y := 20 * (1 - cos(0.6 * x)) * (1 - cos(0.6 * x))
                          * exp(-0.8 * x) + 0.3 * x;
     x1 := x0 + Round(x * mx);   y1 := y0 - Round(y * my);
     if k > 1 then
       begin
         SetPenWidth(2);         SetPenColor(clBlack);
         SetPenStyle(psSolid);   SetPenColor(clBlack);
         Line(x1, y1, x2, y2);
       end;
     x2 := x1;y2 := y1;
 end;
end;
```

```pascal
Procedure Predict;
const x0 = 140; y0 = 300; z0 = 400;
mx = 60; my = 70; mv=1; me = 20;
var j, k, x1, x2, y1, y2, w1, w2, z1, z2: integer;
var x, y, dx, w, vv, xp, xq, yp, yq, yc, ee : real;
  begin
  dx:=0.02;    SetPenWidth(2);  SetPenColor(clBlue);
  x:=0; dx:=0.5; z[0]:=0;
for k:=1 to 20 do
    begin
    x:=x+dx;    x1:=x0+Round(x*mx);
    y := 20 * (1 - cos(0.6 * x)) * (1 - cos(0.6 * x))
                        * exp(-0.8 * x) + 0.3 * x;
    z[k]:=y;                   y1:=y0-Round(y*my);
    SetPenStyle(psSolid);    SetPenColor(clBlack);
    SetBrushColor(clWhite); SetPenWidth(2);
    SetPenColor(clRed);      Circle(x1,y1,4);
  end;
  for k:=1 to 20 do
  begin
    v[k]:=(z[k]-z[k-1])/0.50;
  end;
  for k:=2 to 20 do
  begin
    a[k]:=(v[k]-v[k-1])/0.50;
  end;
  SetPenStyle(psSolid);  SetBrushColor(clWhite);
  SetPenColor(clBlue);   dx:=0.05;
  for k:=1 to 10 do
    begin
    x:=dx*k;
    yc := 20 * (1 - cos(0.6 * x)) * (1 - cos(0.6 * x)) *
                        exp(-0.8 * x) + 0.3 * x;
    y:=0; ee:=100*(y-yc)/4.4;
    x1:=x0+Round(x*mx); y1:=y0-Round(y*my);
    Circle(x1,y1,1);        z1:=z0-Round(ee*me);
    if k>1 then              Line(x1,z1,x2,z0);
    x2:=x1; y2:=y1; z2:=z1;
    end;
    z1:=y0-Round(z[1]*my);  Line(x1,y1,x2,z1);
    for k:=1 to 10 do
```

```pascal
    begin
      x:=0.5+dx*k;
      yc := 20 * (1 - cos(0.6 * x)) * (1 - cos(0.6 * x)) *
                              exp(-0.8 * x) + 0.3 * x;
      y:=v[1]*x;                ee:=100*(y-yc)/4.4;
      x1:=x0+Round(x*mx);   y1:=y0-Round(y*my);
      Circle(x1,y1,1);         z1:=z0-Round(ee*me);
      if k>1 then  Line(x1,z1,x2,z0);
      x2:=x1; y2:=y1; z2:=z1;
    end;
  z1:=y0-Round(z[2]*my);      Line(x1,y1,x1,z1);
  j:=2;
  for j:=3 to 20 do
    begin
      for k:=1 to 10 do
        begin
          x:=0.5*j+dx*k;
          yc := 20 * (1 - cos(0.6 * x)) *(1 - cos(0.6 * x)) *
                              exp(-0.8 * x) + 0.3 * x;
          ee:=100*(y-yc)/4.4;  y:=z[j]+v[j]*dx*k+a[j]*dx*dx*k*k;
          x1:=x0+Round(x*mx);    y1:=y0-Round(y*my);
          Circle(x1,y1,1);          z1:=z0-Round(ee*me);
          if k>1 then
            begin
              Line(x1,z1,x2,z0);
            end;
          x2:=x1; y2:=y1; z2:=z1;
        end;
      z1:=y0-Round(z[j+1]*my);
      Line(x1,y1,x1,z1);
    end;
end;
//=======================Main=======================
var  s: string;
begin
  CoordSys;  Func;  Predict;
  var t1 := new TextABC(320, 210, 12, 'Extrapolation');
  var t2 := new TextABC(310, 470, 12, 'Extrapolation error, %');
end.
```

Listing 7.3

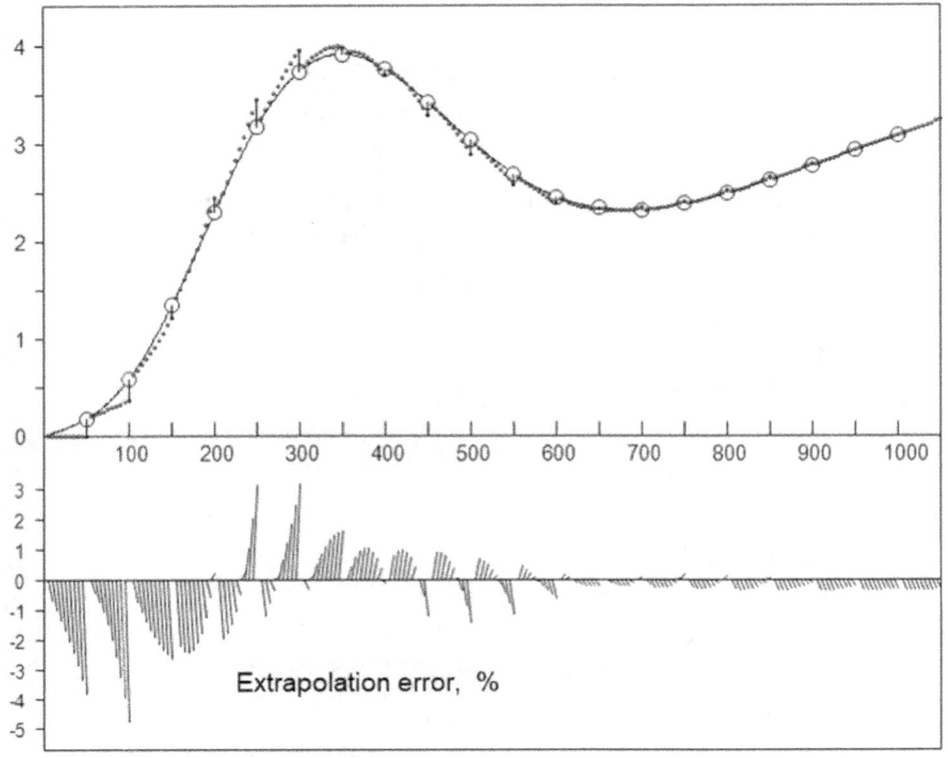

Figure 43 The parabolic extrapolation

Conclusions.

1. The function restoration error is less than 2% with the rational limits of the numerical flow acceleration.

2. The big errors are possible in the initial intervals only owing to the start vagueness.

3. The parabolic extrapolation provides the restoration accuracy two times more than the linear method and four times more than the zero-order extrapolation (other things being equal).

7.3. Models of numerical rows prediction

The predictions of events at several steps ahead are the very valuable. There are three types of the forecasts depending on the time remoteness in the future – short-term, average-term and long-term. These termines are not defined in the mathematics and are to be explained as far as it is important for studying the behaviour of the real control systems.

The fact is that every object has its own "inertial mass" measured in tons, dollars, quantity of workers and in other units. For instance, the short-term prognosis of the sea tanker movement is the prediction of the trajectory in the nearest 10 minutes, but for the quick patrol boat the same 10 minutes mean the long-term prognosis. The grocery owner can forecast his income a year ahead with a small probability but one year is a too short time for plans of the state tax service. That is why one ought at first to define the type of the investigated object (or the process) and to indicate its inertialness. Only if the said above has been done, one gets the real basis for the prediction terms classification (short, average or long).

An ambiguity of the verbal definitions is removed when the prognosed future time interval is measured with quantity of time quantum steps and the time measurement unit is indicated. Let us take values $x(k)$, $x(k-1)$, $x(k-2)$, … of the experimental data that are necessary and adequate to forecast the future values $x(k+1)$, $x(k+2)$, …, $x(k+n)$.

The input values have in general two components – determined and random ones. No future value can be known exactly and it cannot be predicted with an absolute accuracy. By the way, Winston Churchill is considered as an author of the statement about political foresight: "The politician skill is an ability to predict event and to explain the reason if this prediction has not been true". Something like is met in the mathematical prognosis. One selects at first the forecast model and then estimates the prediction quality. As distinct from politicians and other foretellers, however, the mathematician is obliged to produce evidence of his prognosis before the prognosed time interval will begin. It is not difficult to suspect that the above-mentioned condition entails a definite risk.

There are two variants of the prognosis model.

1. Approximation of the input numerical sequence with a trend function.

2. Extrapolation of the input numerical sequence with a polynom.

The hypothesis is taken in both variants: the conformity to law detected in the input sequence will occur as well in the array of the future elements, in other words, the predicted values x(k+1), x(k+2), ..., x(k+n) will belong to the trend or the polynom.

The optimum trend function y(k) is to be found in so manner to provide the minimum deflection x(k) from y(k) at an average. The optimum criterion S is the average square:

$$S = \sum_{j=0}^{m} [y(k - j) - x(k - j)]^2 \rightarrow \min$$

The analytical synthesis worked up for linear trend only. The selection non-linear trend by method of the least squares is in the theory up till now. In such case the trend line is choosed from the known functions – hyperbolic, exponential, logarithmic. There are the wave processes in the economics, for instance, seasonal price fluctuatioins. The trigonometric functions are used in such cases.

7.4. The linear trend

K. Gauss and A. Legendre suggested the analytical method for calculating the linear trend parameters. Straight line takes optimum place between the points of the compact agregate $y(x_i)$ given in a table. The optimum criterion is the minimum value of the sum R:

$$R = \sum_{i=0}^{n} (y_i - z_i)^2, \qquad j = 0, 1, 2, \dots$$

The optimum approximation function $z(x_i)$ is described by the polynom:

$$z_i = \sum_{j=0}^{k} a_j x^j$$

Coefficients a_j are selected to provide the minimum value R. It means that the first derivative of the function $R(a_j)$ is to be zero. In this case the algebraic equation system is formed to calculate a_j. In order to minimize the quantity of calculations one reduces the approximate function to expressions:

$z(x) = a_0 + a_1 x$

$a_0[x] + a_1[xx] = [xy]$

Coefficients a_0 and a_1 are solution of the system for "n" points from the table function [8]:

$$a_0 n + a_1 \sum_{i=1}^{n} x_i = \sum_{i=1}^{n} y_i;$$

$$a_0 \sum_{i=1}^{n} x_i + a_1 \sum_{i=1}^{n} x_i^2 = \sum_{i=1}^{n} x_i y_i$$

More compact expressions are possible using Gauss denotes:

$$\begin{cases} a_0 n + a_1[x] = [y] \\ a_0[x] + a_1[xx] = [xy] \end{cases}$$

This method was popular long ago before the computer invention, but it may be used now to calculate coefficients with library functions in work list of the Microsoft Excel.

Still more effective technology of the problem solution provides the language Pascal ABC.

Example 7.4.

Develop the program to automatize two processes:

- finding optimum straight line for the array x[21] of random numbers;

- calculation the mathematical expectation at four steps ahead.

Remark.

It is necessary to produce at first the arising numerical sequence with the additive noise. The Microsoft Excel is recommended for number modelling.

Solution.

The program Prog_7_4 (listing 7.4) contains three subroutines:

1. "CoordSys" that draws rectangle axes;

2. "Array_21" that displays the array of random points;

3. "Find_AB" intended to the next actions:

- to select parameters of the equation x(k) = A*k + B;

- to calculate 4 values of the mathematical expectation;

- to draw graphic (fig.44).

```
Program Prog_7_4;
Uses GraphABC, ABCobjects;

var x: array[1..30] of real;
var A, B, sq : real;
var sqm: array[1..30] of real;

Procedure CoordSys;
  const x0=100;
  var j, k, xc, yc: integer;
  var s: string;
  begin
    SetWindowSize(800,500);   SetPenStyle(psClear);
    SetPenColor(clBlack);     SetPenWidth(1);
    Line(100, 24, 100, 440);  Line(100, 440, 724, 440);
    SetPenWidth(1);           Line(100, 24, 724, 24);
    Line(724, 24, 724, 440);
    for k:=1 to 26 do
      begin
        Line(100+24*k, 440, 100+24*k, 448);
      end;
    for k:=1 to 13 do
      begin
        Line(100, 440-30*k, 108, 440-30*k);
end;
    for k:=1 to 13 do
      begin
        xc:=2*k;  s:=IntToStr(xc);
        var t:=new TextABC(96+48*k, 452, 10, s);
      end;
    for k:=1 to 7 do
      begin
        yc:=2*(k-1);  s:=IntToStr(yc);
```

```pascal
      var t:=new TextABC(80, 490-60*k, 10, s);
    end;
end;
//---------------------------- Array ----------------------------
Procedure Array_21;
const x0=100; y0=440; mx=24; my=30;
var k, x1, x2, y1: integer;
begin
 x[1]:= 3.06;  x[6]:= 3.42;   x[11]:=8.60;   x[16]:= 6.27;
 x[2]:= 5.25;  x[7]:= 6.18;   x[12]:= 5.83;  x[17]:= 10.38;
 x[3]:= 2.79;  x[8]:= 5.07;   x[13]:= 7.85;  x[18]:= 9.08;
 x[5]:= 4.96;  x[9]:= 8.15;   x[14]:= 7.03;  x[19]:= 6.76;
 x[4]:= 2.40;  x[10]:= 6.87;  x[15]:= 5.76;  x[20]:= 6.97;
 x[21]:= 8.10;
 for k:=1 to 21 do
   begin
     x1:=x0+Round(k*mx);   y1:=y0-Round(x[k]*my);
     SetPenStyle(psSolid);  SetPenWidth(2);
     SetPenColor(clBlack);  SetBrushColor(clWhite);
     Circle(x1,y1, 4);
   end;
end;
Procedure Find_AB(a0,b0: real);
const x0=100; y0=440; mx=24; my=30;
var j, k: integer;
var y, d, sd: real;
var sqm: array[1..21] of real;
var x1, x2, y1, y2: integer;
B:=b0;
begin
 j:=1; A:=a0+0.01*j; sd:=0;
 for k:=1 to 21 do
   begin
     y:=A*k+B; d:=y-x[k];    sd:=sd+d*d; sq:=sqrt(sd);
     x1:=x0+Round(k*mx);  y1:=y0-Round(y*my);
     if k>1 then
       begin
         SetPenWidth(3);    SetPenColor(clBlack);
         Line(x1,y1,x2,y2);  SetPenWidth(1);
         SetPenColor(clBlack);
       end;
```

```pascal
      x2:=x1; y2:=y1;
    end;
  end;
//=========== Main =============
const x0=100; y0=440; mx=24; my=30;
var j, k, x1, y1: integer;   var a0,b0: real;
var sa, sb, ss, s1, s2, s3, s4: string;
begin
  CoordSys;  Array_21; b0:=3.035;
    for j:=1 to 20 do
      begin
        Rectangle(100,24,724,440);
        CoordSys;          Array_21; a0:=0.2+0.01*j;
        Find_AB(a0,b0);   sqm[j]:=sq;
        if (j>1) and (sqm[j] > sqm[j-1]) then break;
        Sleep(100);
      end;
  Str(A:3:2,sa);  Str(b0:3:2,sb);   Str(sqm[j]:4:3,ss);
  var t1:=new TextABC(210,50,12, 'Optimum line');
  var t2:=new TextABC(210,75,12,  'x(k) = A*k + B');
  var t3:=new TextABC(230,110,12,'A = '+sa);
  var t4:=new TextABC(230,135,12,'B = '+sb);
for k:=22 to 25 do
  begin
   x[k]:= A*k+b0;
   x1:=x0+Round(k*mx); y1:=y0-Round(x[k]*my);
   SetPenWidth(3);        Circle(x1,y1,5);
  end;
  var t5:=new TextABC(500,300,12,
          'Mathematical expectation');
  Str(x[22]:4:3,s1);   Str(x[23]:4:3,s2);
  Str(x[24]:4:3,s3);   Str(x[25]:4:3,s4);
  var t6:=new TextABC(540,330,12,'x[22] = '+s1);
  var t7:=new TextABC(540,350,12,'x[23] = '+s2);
  var t8:=new TextABC(540,370,12,'x[24] = '+s3);
  var t9:=new TextABC(540,390,12,'x[25] = '+s4);
end.
```

Listing 7.4

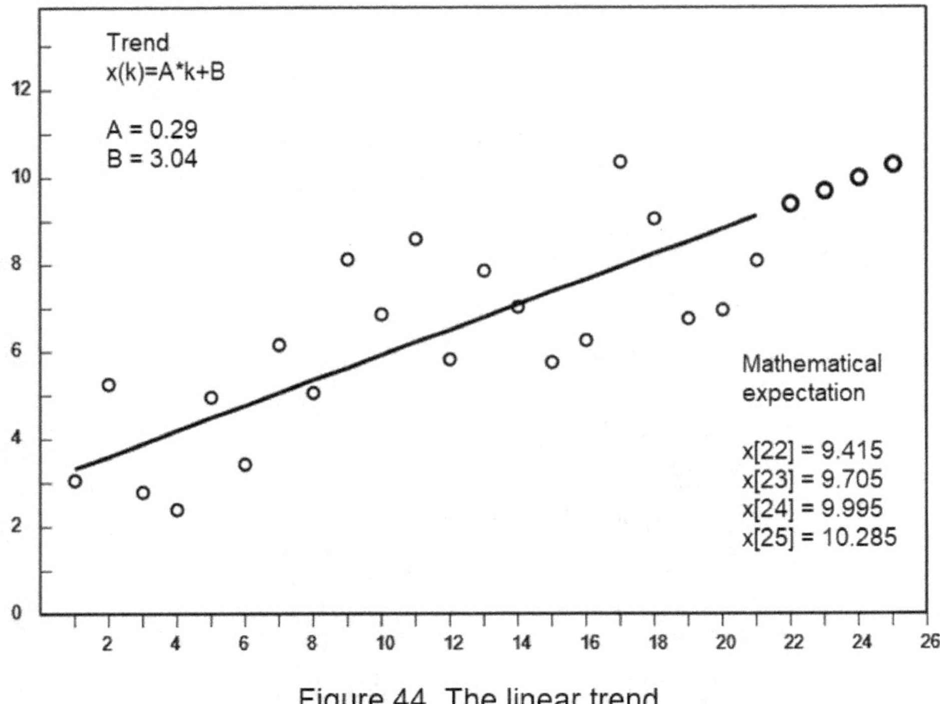

Figure 44. The linear trend

There is a possibility to automatize the selection of the parameter B in the linear trend equation. It is necessary to add the adequate external cycle into the program Find_AB. Here you are if you want.

7.5. Nonlinear trends

Let us consider two examples of nonlinear trends – exponential and parabolic.

Example 7.5. There is a rising numeric sequence disturbed by the noise deviations. Develop the program that selects the coefficients of the exponential trend:

$$y(k) = A*exp(-B*k)$$

and uses the trend for the short-term prognosis at 4 steps ahead.

Remark.

At first one ought to model the arising exponential sequence of integer numbers x(k) for k = 1, 2, …, 18. Every value x(k) is produced as a sum of regular and random parts.

Solution.

Program Prog_7_5 contains three subroutines:

1. "CoordSys" that draws the rectangle axes;

2. "Array_18" that displays the array x[18];

3. "Find_ABC" (listing 7.5) that is needed to fulfil the following actions:

    - to find the optimun coefficients for the trend y(k);

    - to calculate the expected values y(k);

    - to draw graphic (fig.45).

```pascal
Program Pro_7_5;
Uses GraphABC, ABCobjects;

var w: array[1..30] of real;
var A, B, sq : real;
var sqm: array[1..300] of real;

Procedure CoordSys;
  const x0=100;
  var j,k: integer;
  var xc, yc: integer;
  var s: string;
  begin
    SetWindowSize(800,500);
    SetPenStyle(psClear);    SetPenColor(clBlack);
    SetPenWidth(1);             Line(100, 0, 100, 440);
    Line(100, 440, 772, 440); SetPenWidth(1);
    Line(100, 0, 772, 0);       Line(772, 0, 772, 440);
    for k:=1 to 28 do
      begin
        Line(100+24*k, 440, 100+24*k, 448);
      end;
```

```pascal
    for k:=1 to 17 do
      begin
        Line(100, 440-25*k, 108, 440-25*k);
      end;
    for k:=1 to 15 do
      begin
        xc:=2*(k-1); s:=IntToStr(xc);
        var t:= new TextABC(48+48*k,452,12,s);
      end;
  for k:=1 to 8 do t
    begin
      yc:=10*k; s:=IntToStr(yc);
      var t:= new TextABC(70,430-50*k,12,s);
    end;
end;
//------------------------------ Array ------------------------------------
Procedure Array_18;
const x0=100; y0=440; mx=24; my=5;
var k,x1,x2,y1: integer;
begin
 w[1]:= 9.89;    w[6]:= 21.46;   w[11]:= 23.20;  w[16]:= 45.41;
 w[2]:= 7.86;    w[7]:= 29.13;   w[12]:= 30.28;  w[17]:= 55.40;
 w[3]:= 7.68;    w[8]:= 25.75;   w[13]:= 39.11;  w[18]:= 60.12;
 w[4]:= 17.87;  w[9]:= 24.19;   w[14]:= 33.07;
 w[5]:= 26.27;  w[10]:=29.22;   w[15]:= 34.78;

  for k:=1 to 18 do
    begin
     x1:=x0+Round(k*mx);   y1:=y0-Round(w[k]*my);
     SetPenStyle(psSolid);   SetPenWidth(2);
     SetPenColor(clBlack);   SetBrushColor(clWhite);
     Circle(x1,y1, 5);
    end;
end;
//----------------- Trend -----------------
Procedure Find_ABC(a0,b0: real);
const x0=100; y0=440; mx=24; my=5;
var j, k: integer;
var y, d, sd: real;
var sqm: array[1..21] of real;
var x1, x2, y1, y2: integer;
begin
```

```pascal
      sd:=0;
      for k:=1 to 18 do
        begin
          y:=a0*exp(b0*k);         d:=y-w[k];
          sd:=sd+d*d;              sq:=sqrt(sd);
          x1:=x0+Round(k*mx);   y1:=y0-Round(y*my);
          if k>1 then
            begin
              SetPenWidth(4);      SetPenColor(clBlack);
              Line(x1,y1,x2,y2);     SetPenWidth(1);
              SetPenColor(clBlack);
            end;
          x2:=x1; y2:=y1;
        end;
end;
//============= Main ============
const x0=100; y0=440; mx=24; my=5;
var j, k, x1, y1:integer;
var a0, b0, sq1, sq2, dsq: real;
var sa, sb, ss, s1, s2, s3, s4: string;
begin
  CoordSys;
  Array_18;
  a0:=12; b0:=0.06; dsq:=20; j:=1;
  while dsq>0 do
    begin
      j:=j+1; Rectangle(100,0,772,440);
      CoordSys;    Array_18; b0:=b0+0.00002*j;
      Find_ABC(a0,b0);  sq1:=sq;
      if j>5 then dsq:=sq2-sq1;
      if j>5000 then break;
      Sleep(30); sq2:=sq1;
    end;
  A:=a0; B:=b0;         Str(A:3:2,sa);
  Str(b0: 4: 3, sb);       Str(sqm[j]:4:3,ss);
  var t1:=new TextABC(240,50,12,'Trend:');
  var t2:=new TextABC(210,75,12, 'x(k) = A*exp(B*k)');
  var t3:=new TextABC(235,110,12,'A = '+sa);
  var t4:=new TextABC(235,135,12,'B = '+sb);
```

```pascal
for k:=19 to 22 do
  begin
    w[k]:= A*exp(B*k);
    x1:=x0+Round(k*mx);   y1:=y0-Round(w[k]*my);
    SetPenWidth(3);          SetPenStyle(psSolid);
    SetPenColor(clBlack);  Circle(x1,y1,7);
  end;
  var t5:=new TextABC(520,300,12,
        'Mathemaical expectation');
Str(w[19]:4:3,s1); Str(w[20]:4:3,s2);
Str(w[21]:4:3,s3); Str(w[22]:4:3,s4);
  var t6:=new TextABC(550,330,12,'x[19] = '+s1);
  var t7:=new TextABC(550,350,12,'x[20] = '+s2);
  var t8:=new TextABC(550,370,12,'x[21] = '+s3);
  var t9:=new TextABC(550,390,12,'x[22] = '+s4);
end.
```

Listing 7.5

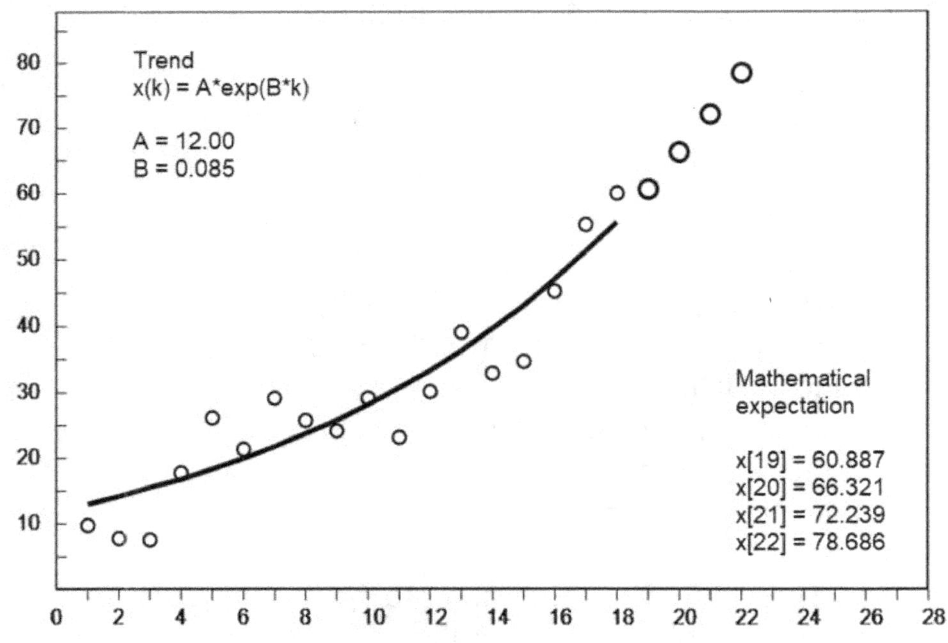

Figure 45. The exponential trend

Example 7.6.

There is a rising (with decreasing velocity) numeric sequence disturbed by the noise deviations. Develop the program that selects the coefficients of the trend:

$$y(k) = C + B*k - A*k^2$$

and uses the trend for the short-term prognosis at 4 steps ahead.

Remark.

At first one ought to model the parabolic sequence of integer number $x(k)$ for $k = 1, 2, \ldots, 18$. Every value $x(k)$ is produced as a sum of regular and random parts.

Solution.

Program Prog_7_6 (listing 7.6) contains subroutine "CoordSys" for drawing the rectangle axes and "Array_18" that fulfils the next actions:

- displays the input values $x(k)$;

- selects the optimun coefficient values for the trend $y(k)$;

- calculates the expected values $y(k)$;

- draws graphic (fig.46).

```
Program Pro_7_6;
Uses GraphABC, ABCobjects;

var w: array[1..22] of real;

Procedure CoordSys;
 const x0=100;
 var j, k, xc, yc: integer;
 var s: string;
 begin
   SetWindowSize(870,500);  SetPenStyle(psClear);
   SetPenColor(clBlack);       SetPenWidth(3);
   Line(140, 44, 140, 440);    Line(140, 440, 830, 440);
```

```pascal
    SetPenWidth(1);              Line(140, 44, 830, 44);
    Line(830, 44, 830, 440);
  for k:=1 to 22 do
    begin
        Line(140+30*k, 440, 140+30*k, 448);
    end;
   for k:=1 to 19 do
    begin
        Line(140, 440-20*k, 150, 440-20*k);
    end;
  for k:=1 to 12 do
  begin
      xc:=2*(k-1); s:=IntToStr(xc);
     var t:= new TextABC(76+60*k,452,12,s);
  end;
  for k:=1 to 10 do
    begin
        yc:=2*(k-1); s:=IntToStr(yc);
        var t:= new TextABC(116,472-40*k,12,s);
    end;
end;

Procedure Array_18;
const x0=140; y0=440; mx=30; my=20;
var j, k, x1, x2, y1, y2, z1, z2: integer;
var a, b, c, x, y, dy, z, ss, sq1, sq2, dsq: real;
begin
  w[1]:= 4.21;     w[10]:= 15.08;
  w[2]:= 5.22;     w[11]:= 13.97;
  w[3]:= 7.87;     w[12]:= 15.33;
  w[4]:= 10.97;    w[13]:= 17.26;
  w[5]:= 9.50;     w[14]:= 14.27;
  w[6]:= 11.77;    w[15]:= 16.10;
  w[7]:= 10.66;    w[16]:= 15.12;
  w[8]:= 12.18;    w[17]:= 15.27;
  w[9]:= 15.64;    w[18]:= 16.50;
  for k:=1 to 18 do
    begin
      x1:=x0+Round(k*mx);   y1:=y0-Round(w[k]*my);
      SetPenStyle(psSolid);  SetPenWidth(2);
      if k<19 then
       begin
```

```pascal
            SetPenColor(clBlack); Circle(x1,y1, 5);
        end;
      end;
a:=0.07; b:=1.75; c:=3.0; dsq:=10; sq2:=50;
while dsq>0 do
  begin
      a:=a-0.001; ss:=0; x1:=0;
      for k:=1 to 18 do
         begin
             x:=k; y:=c+b*k-a*k*k;
             dy:=abs(y-w[k]); ss:=ss+dy*dy;
         end;
         sq1:=sqrt(ss); dsq:=sq2-sq1;  sq2:=sq1;
end;
a:=a+0.0001;  b:=1.62; c:=3.0; dsq:=10;  sq2:=50;
while dsq>0 do
   begin
      b:=b+0.01; s:=0;  x1:=0;
      for k:=1 to 18 do
         begin
             x:=k;                y:=c+b*k-a*k*k;
             dy:=abs(y-w[k]);   ss:=ss+dy*dy;
         end;
         sq1:=sqrt(ss); dsq:=sq2-sq1; sq2:=sq1;
end;
  b:=b-0.001; c:=2.0;  dsq:=10;  sq2:=50;
  while dsq>0 do
     begin
        c:=c+0.1;  ss:=0; x1:=0;
        for k:=1 to 18 do
           begin
               x:=k;    y:=c+b*k-a*k*k;
               dy:=abs(y-w[k]); ss:=ss+dy*dy;
           end;
           sq1:=sqrt(ss); sq:=sq2-sq1; sq2:=sq1;
       end;
      c:=c+0.01;
      for k:=1 to 18 do
         begin
             x1:=x0+mx*k;
             z:=c+b*k-a*k*k;    z1:=y0-Round(z*my);
```

```pascal
        SetPenWidth(4);   SetPenColor(clBlack);
        if k>1 then Line(x1,z1,x2,z2);
        x2:=x1; z2:=z1;
      end;
    for k:=19 to 22 do
      begin
        x1:=x0+mx*k;  z:=c+b*k-a*k*k;  w[k]:=z;
        z1:=y0-Round(z*my);
        SetPenStyle(psSolid); SetPenWidth(4);
        SetPenColor(clBlack); Circle(x1,z1, 6);
      end;
  end;
  writeln('  A = ',a:3:2);  writeln('  B = ',b:3:2);
writeln('  C = ',c:3:2);
end;
//=============== Main ================
var s1, s2, s3, s4: string;
begin
 CoordSys;  Array_18;
 var t1:=new TextABC(140,10,12,
        'Trend:  y(x) = C + B*x - A*x^2');
 var t2:=new TextABC(200,60,12,'A = 0.06');
 var t3:=new TextABC(200,85,12,'B = 1.74');
 var t4:=new TextABC(200,110,12,'C = 3.01');
 var t5:=new TextABC(505,300,12,
        'Mathematical expectation');
 Str(w[19]:4:2,s1); Str(w[20]:4:2,s2);
 Str(w[21]:4:2,s3); Str(w[22]:4:2,s4);
 var t6:=new TextABC(550,330,12,'y[19] = '+s1);
 var t7:=new TextABC(550,350,12,'y[20] = '+s2);
 var t8:=new TextABC(550,370,12,'y[21] = '+s3);
 var t9:=new TextABC(550,390,12,'y[22] = '+s4);
end.
```

Listing 7.6

Conclusion.

The decrease of the row is predicted in despite of observed growth. It became possible due to detected slowing down of the growth. The parabolic trend was built truthful.

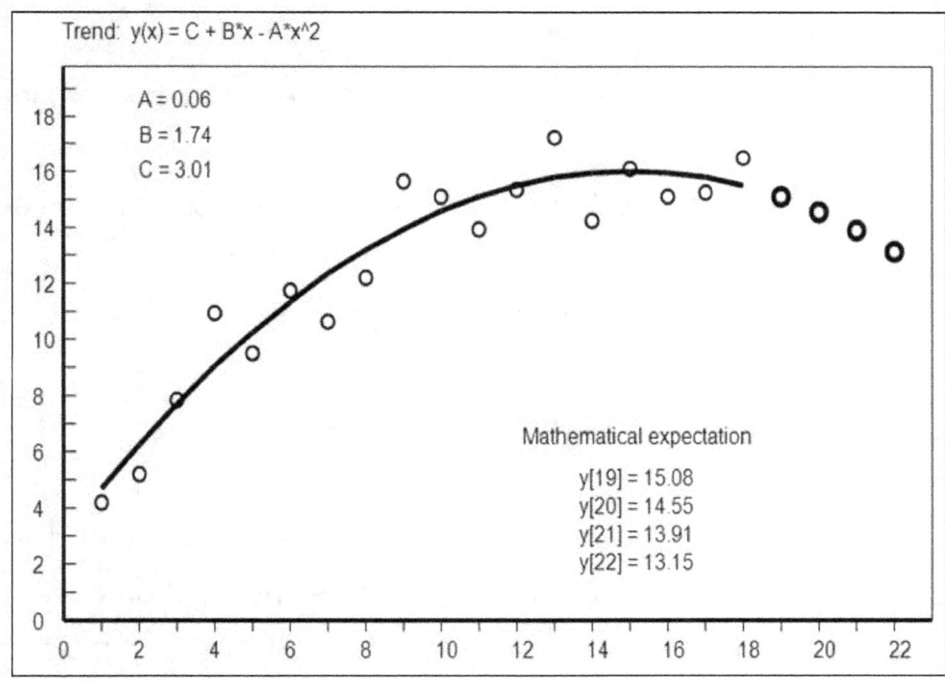

Figure 46. The parabolic trend

Chapter 8. DYNAMIC MODELS

8.1. Problems and methods of modelling

Any modern technical or economical complex project of the control system goes through a computer modelling. Computer simulation is intended:

- to shorten the projecting duration;

- to diminish the project cost;

- to exclude any possible project mistakes;

- to check the system behaviour under external random infuencies;

- to estimate how the system will work in extreme situatons.

Three model types are known:

- physical (or natural);

- semi-physical (semi-natural);

- mathematical.

The physical modelling has the centuries-old history. Designers of multistory buildings, bridges, ships, hydrotechnical constructions are using till now scale models of the objects in form of small size copies. There is a special theory of similarity that provides adequacy of the scale model. It is evident that a natural model is unique always. Any constructive change makes to create the new example of the model but it is too expensive and demands additional time.

The semi-physical model has two parts. The real system construction is reproduced by the first part. The second one is the computer program that simulates more complicated and changable system part. Such method is effective for pilot trainers and space simulators. For instance, the jet plane cabin is real but the flight program is modelled with the computer software. In so manner the pilot trainer is made also for the river navigation. The wheel house is natural but the ship movement is simulated. The picture of the navigating channel or the lock chamber is displayed on the computer monitor.

The mathematical model describes the control system with a limited completeness which is sufficient for the problem solution. Such method of approach demands the know-how to formulate the essential parts of the problem. Some insignificant details can be omited without any damage for the model. It has been already shown in the Chapter 1 (see Example 1.1).

The general property of many mathematical models is dependence of the observed parameters on the current time. That is why such models are called dynamic. A mathematic modelling has an important peculiarity: it does not demand the exact solution of the equations or the equation systems. As a rule, similar solutions occure impossible owing to a great dimensionality of variables and arrays.

The main idea of the dynamic modelling is to present the system as an agregate consisting of the links (type and non-standard) joined into the multi-contour scheme with positive and negative feedbacks. Designer is to describe the behaviour of every link and to work out the equation system in finite differences.

Now the dynamic models are being introduced step-by-step into the educational process of the economic high school of the Russian Federation. The computer trainers are used everywhere. Now it is possible to investigate the dymamic of many proceess, for instance:

- economical (production, finances, price formation, budget);

- social;

- demographical.

Some dynamic models are shown further.

8.2. A physical pendulum

The physical pendulum is a hard body that oscillates, drived by gravitation, around the horizontal axis which does not go through the body gravity center. The pendulum, drived out of the equilibrium state, performs fading oscillations. The reason of the attenuation is the energy dissipation owing to the air resistance and the friction in the hanging point. Let the pendulum parameters are:

$L = 20$ - length;

$s = 0.04$ - fading coefficient;

$V = 14$ - start lihear velocity.

We use well known formulae:

1. Angular frequency:

$$w = \sqrt{\frac{9.81}{L}}$$

2. The horizontal deflection from the vertical line:

$x(t) = A_{max}*exp(-s*t)*sin(w*t)$,   where $A_{max} = V / w$

3. The elevating over the rest point:

$$y(t) = L - \sqrt{L^2 - x(t)^2}$$

The pendulum model contains two programs.

1. Program Prog_8_1 (listing 8.1)

- calculates functions x(t) and y(t);

- draws graphic (fig.47).

```
Program Prog_8_1;
Uses ABCobjects, GraphABC;
Procedure CoordSys;
 const x0=100;
 var j, k, x: integer;  var xc, yc: integer;  var s: string;
 begin
   SetWindowSize(850,500);  SetPenStyle(psClear);
   SetPenColor(clBlackj);      SetPenWidth(2);
   Line(80, 16, 80, 456);        Line(80, 280, 800, 280);
   SetPenWidth(1);                Line(80, 16, 800, 16);
   Line(80, 456, 800, 456);    Line(800, 16, 800, 456);
```

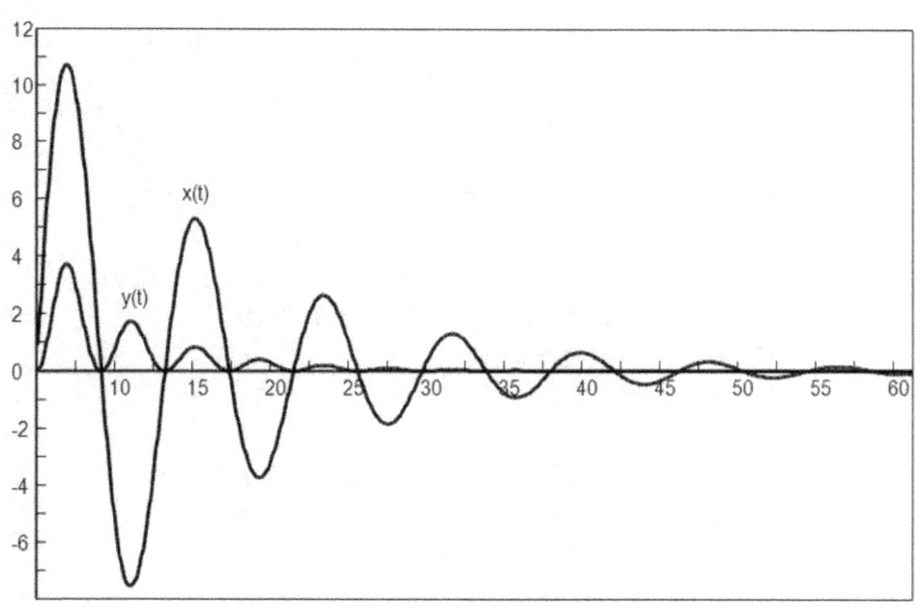

Figure 47. The pendulum oscillations

```
   for k:=1 to 22 do
     begin
         Line(80+32*k, 272, 80+32*k, 280);
     end;
    for k:=0 to 22 do
     begin
         Line(88, 456-22*k, 80, 456-22*k);
     end;
```

```pascal
  for k:=2 to 12 do
   begin
      x:=5*k; s:=IntToStr(x);
      var t:= new TextABC(12+64*k,284,12,s);
   end;
   for k:=1 to 10 do
   begin
      yc:=2*(k-1)-6; s:=IntToStr(yc);
      var t:= new TextABC(60,448-44*k,12,s);
   end;
end;
Procedure TXY;
const g=9.81; x0=80; y0=280; mt=12; my=20;
var k, x1, x2, y1, y2, z1, z2: integer;
var t1, t2, a0, a1, a2, A, s, R, L, V, w, w0, t, dt: real;
var b, x, dx, y, dy: real;
begin
   V:=10.0;  s:=0.08;     L:=19.0;   w0:=sqrt(g/L);
   w:=sqrt(w0*w0-s*s);  A:=V/w; dt:=0.1;
  for k:=0 to 600 do
   begin
     t:=dt*k;          x:=A*exp(-s*t);          b:=w*t;
     x:=x*sin(b);     dy:=L-sqrt(L*L-x*x);    y:=L-dy;
     x1:=x0+Round(t*mt);     y1:=y0-Round(x*my);
     z1:=y0-Round(dy*my);   SetPenWidth(3);
     SetPenColor(clBlack);
     if k>1 then
        begin
          Line(x1,y1,x2,y2);  Line(x1,z1,x2,z2);
        end;
     x2:=x1; y2:=y1; z2:=z1;
     Sleep(2);
   end;
end;
//============= Main ============
begin
 CoordSys;  TXY;
 var t1:=new TextABC(200,135,12,'x(t)');
 var t2:=new TextABC(150,215,12,'y(t)');
end.
```

Listing 8.1

2. Program Prog_8_2 (listing 8.2) emulates the pendulum oscilla-tions in real time and shows the fading swings (fig.48).

```pascal
Program Prog_8_2;
Uses ABCobjects, GraphABC;
const R=15;

Procedure Picture;
  const x0=100;  var k, x: integer;  var s: string;
  begin
    SetWindowSize(700,500);   SetBrushColor(ClWhite);
    SetPenColor(ClBlack);        Line(30,10,670,10);
    Line(30,480,670,480);        Line(30,10,30,480);
    Line(670,10,670,480);
//--------------------- Hanging point----------------------------
    SetPenWidth(6); Line(240,20,300,20);
    SetPenWidth(4); Line(254,20,270,40);
    Line(270,40,286,20); SetBrushColor(clBlack);
Circle(270, 42, 8);
//--------------------- Axis  X ---------------------
    SetPenWidth(2);   Line(50,440,510,440);
    for k:=1 to 24 do
      begin
        Line(50+20*(k-1),440,50+20*(k-1),432);
      end;
    for k:=1 to 13 do
      begin
        x:=2*k-14; s:=IntToStr(x);
        var t:=new TextABC(39*k-4,450,12,s);
      end;
//--------------------- Start position ---------------------
    SetPenColor(ClBlack);    SetPenWidth(3);
    Line(270,40,270,400);
    SetBrushColor(clBlack);  Circle(270, 410, R);
```

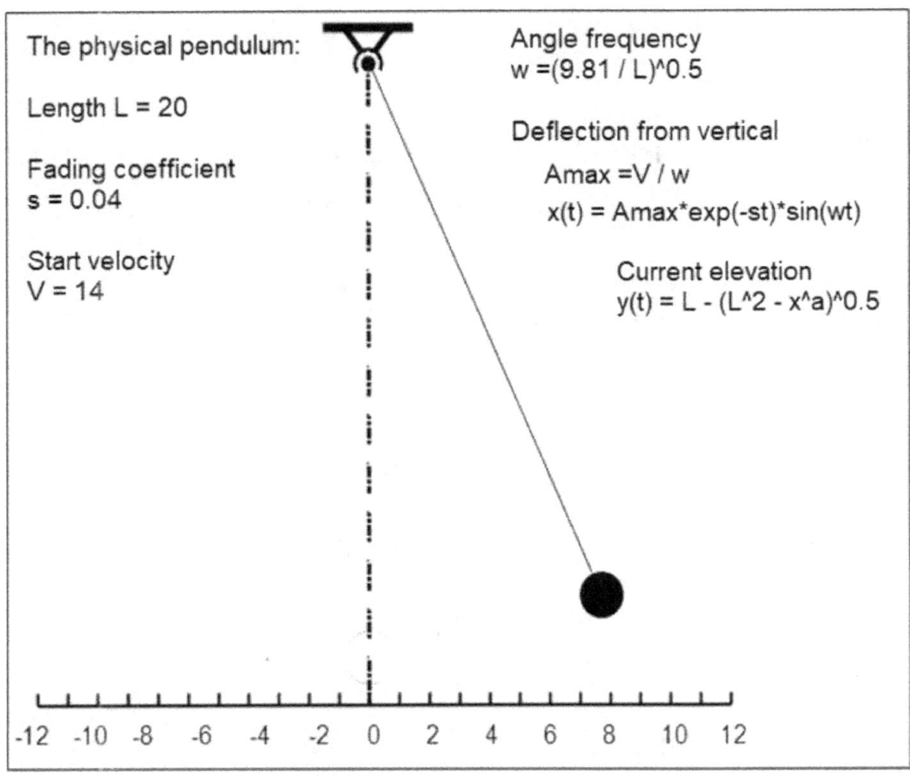

Figure 48. The pendulum swings in real time

```
//------------------- Analitical information----------------------------
  var t1:=new TextABC(40,15,12,'The physical pendulum');
  var t2:=new TextABC(40,40,11,'Length L = 20');
  var t3:=new TextABC(40,65,11,'Fading');
  var t4:=new TextABC(40,85,11,'coefficient');
  var t5:=new TextABC(40,105,11,'s = 0.04');
  var t6:=new TextABC(390,15,12,'Parameters:');
  var t7:=new TextABC(490,15,12, 'Angle frequency:');
  var t8:=new TextABC(490,40,11, 'w = (9.81 / L)^0.5');
  var t9:=new TextABC(390,60,11,
  'Deflection from vertical line:');
  var t10:=new TextABC(410,80,11,
  'Maximum:  Amax = V / w');
  var t11:=new TextABC(410,105,11,
       'Current: x(t) = Amax*exp(-st)*sin(wt)');
  var t12:=new TextABC(450,130,11, 'Current ekevation:');
  var t13:=new TextABC(450,150,11,
  ' y(t)= L - (L^2 - x^a)^0.5 ');
end;
```

```pascal
Procedure TXY(V: integer);
const g=9.81; x0=270; y0=410; mx=13; my=10;
var j, k, x1, y1, N, z: integer;
var A, s, L, w, w0, t, dt, b, x, dx, y, dy: real;
begin
  s:=0.04; L:=20.0;       w0:=sqrt(g/L);
  w:=sqrt(w0*w0-s*s);     A:=V/w; dt:=0.3;
  var t1:=new TextABC(40,135,11,'Start');
  var t2:=new TextABC(40,155,11,'velocity:');
  var t3:=new TextABC(40,180,11, 'V = '+IntToStr(V));
  N:=250+20*V;
  for k:=0 to N do
    begin
      t:=dt*k; x:=A*exp(-s*t);        b:=w*t; x:=x*sin(b);
      dy:=L-sqrt(L*L-x*x);            y:=L-dy;
      x1:=x0+Round(x*mx);             y1:=y0-Round(dy*my);
      SetPenStyle(psSolid);           SetPenWidth(1);
      SetPenColor(clBlack);           Line(270,40,x1,y1);
      SetPenWidth(1);                 SetPenColor(clWhite);
      SetBrushColor(clBlack);         Circle(x1, y1, R);
      Sleep(80);                      SetPenStyle(psSolid);
      SetPenWidth(3);                 SetPenColor(clWhite);
      Line(270,40,x1,y1);             SetPenColor(ClWhite);
      SetBrushColor(clWhite);         Circle(x1, y1, R);
      SetBrushColor(clBlack);         Circle(270, 42, 6);
      SetPenWidth(2);                 SetPenColor(clBlack);
      SetPenStyle(psDashDot);         Line(270,40,270,440);
      for j:=1 to 12 do
        begin
          z:=10*j; SetPenWidth(2);  SetPenColor(clWhite);
          SetBrushColor(clWhite);   Circle(270, 44+30*j, 5);
        end;
    end;
  SetPenColor(ClBlack);   SetPenWidth(1);
  Line(270,40,270,400);   SetBrushColor(clBlack);
  Circle(270, 410, R);
end;
'================-====== Main ==============
Var V: integer;
begin
  Picture; Sleep(1000); V:=14;
```

```
  if V<16 then TXY(V); if V>15 then
    begin
      var t1:=new TextABC(40,150,12,
      'Inputed V ='+IntToStr(V));
      var t2:=new TextABC(40,170,12, 'Change');
      var t3:=new TextABC(40,190,12, 'value !');
      var t4:=new TextABC(40,210,12, 'Admitted');
      var t5:=new TextABC(40,230,12, '5 < V < 15');
    end
end.
```

Listing 8.2

8.3. A steam-engine model

The steam-engine (fig.49) transforms a forward-recurrence motion of the piston into a rotatory movement of the fly-wheel. The piston is bound with the wheel by means of the rod and the crank. The piston is drived by steam.

The valve switches the direction of the steam movement in the work cylinder. The more a steam pressure, the more the fly-wheel rotation speed.

The rod-crank mechanisms were used in the reciprocating water-pumps long ago before the steam-engine invention. Nowadays this principle is being used in the internal-combustion engines.

The program Prog_8_3 (listing 8.3) draws steam-engine scheme and simulates its work in real time.

```
Program Prog_8_3;
Uses ABCobjects, GraphABC;
const x0=220; y0=180; mx=2.5; my=2.5;

Procedure StartPos (R,L,P: real);
var x1, y1, z1, u1, r1,r 2: integer;  var x, y, a, b, z, u:real;
begin
  SetWindowSize(800,400);    SetPenStyle(psClear);
  SetPenColor(clBlack);
```

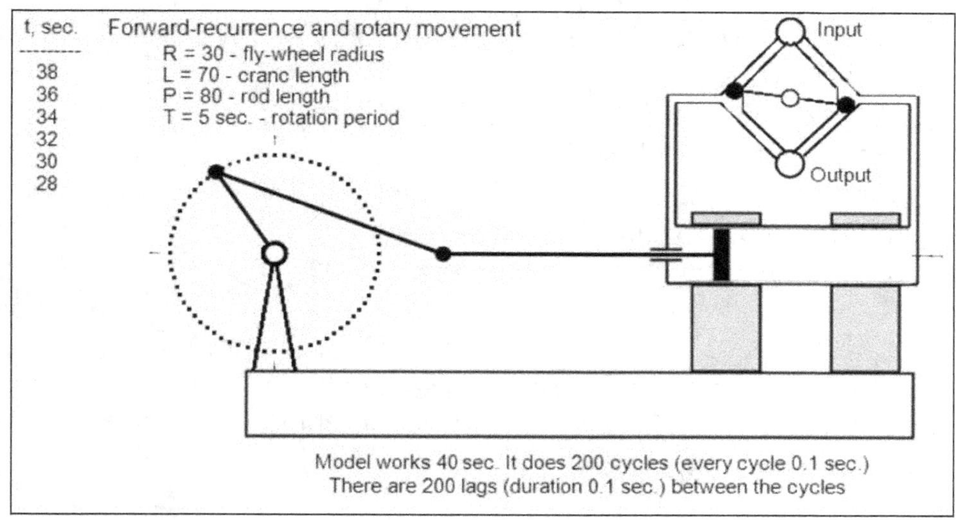

Figure 49. The steam-engine model

```
//------------------------fly-wheel------------------------
  SetPenStyle(psSolid);              r1:=Round(R*mx);
  Circle(x0,y0,r1);
//-------------------work cylinder-------------------------
  SetPenStyle(psSolid);              SetPenWidth(2);
  Line(508,y0-22,676,y0-22);   Line(502,y0+22,682,y0+22);
  SetPenWidth(2);                    Line(502,y0-122,502,y0-4);
  Line(508,y0-22,508,y0-116);  Line(676,y0-22,676,y0-116);
  Line(682,y0-122,682,y0+22);  SetPenWidth(2);
  Line(502,y0+4,502,y0+22);    Line(490,y0-4,514,y0-4);
  Line(490,y0+4,514,y0+4);
//------------------------Valve----------------------------
  SetPenWidth(2);              Circle(590,10,10);
  Circle(590,110,10);         SetPenWidth(1);
  Line(502,58,540,58);        Line(508,64,540,64);
  Line(540,58,582,14);        Line(554,54,588,18);
  Line(540,64,580,106);       Line(556,70,586,100);
  Line(556,54,556,70);        Line(640,58,680,58);
  Line(640,64,676,64);        Line(600,14,640,58);
  Line(592,18,623,52);        Line(592,102,624,70);
  Line(600,104,640,64);       Line(624,52,624,70);
  SetBrushColor(clBlack);     Circle(550,55,6);
  Circle(630,65,6);            SetPenWidth(2);
  Line(550,54,630,65);        SetBrushColor(clWhite);
  Circle(590,60,6);
```

```pascal
  r2:=6; x:=R*cos(pi/4);      y:=R*sin(pi/4);
  x1:=x0+Round(x*mx);      y1:=y0-Round(y*my);
  a:=sqrt(L*L-y*y);         z:=x+a;
  z1:=x0+Round((z*mx));     SetPenWidth(3);
  SetBrushColor(clBlack);   //--------rod
  Circle(x1,y1,r2);         Circle(z1,y0,r2);
  Line(x0,y0,x1,y1);        //--------crank
  SetPenWidth(3);           Line(x1,y1,z1,y0);
//----------------------Rod and piston ------------------
  b:=x+a+p;                 u1:=x0+Round(b*mx);
  SetBrushColor(clWhite);   SetPenWidth(4);
  Line(z1,y0,u1,y0);        SetPenWidth(12);
  Line(u1,y0-20,u1,y0+20);
//----------------------------Axes----------------------------
  SetPenStyle(psDashDotDot);   SetPenWidth(1);
  Line(130,y0,700,y0);         Line(x0,90,x0,320);
//------------------------------Bed--------------------------
  SetPenStyle(psSolid);         SetPenWidth(2);
  Rectangle(x0-20,270,680,320);
  SetBrushColor(clLightGray);   SetPenWidth(2);
  Rectangle(520,y0-32,570,y0-22);
  Rectangle(620,y0-32,670,y0-22);
  Rectangle(520,y0+23,570,y0+90);
  Rectangle(620,y0+23,670,y0+90);
//--------------------Wheel support--------------------
  Line(x0,y0,x0-15,270);    Line(x0,y0,x0+15,270);
  SetBrushColor(clWhite);   Circle(x0,y0,8);
end;

Procedure Movement(R,L,P: real);
  const x0=220; y0=180; mx=2.5; my=2.5;
  var j, k, m, x1, x2, y1, y2, z1, z2, u1, u2, r2, rad: integer;
  var x, y, z, u, a, b, c, f, g, t, ph, betg, betr, xm, ym: real;
  var xm1, xm2, ym1, ym2: integer;
  var d: string;
  begin
    writeln('       t, sec.');   writeln('       ---------');
    r2:=6;  x1:=x0;           y1:=y0-Round(R*my);
    a:=sqrt(L*L-R*R);         z1:=x1+Round(a*mx);
    f:=0.2;   ph:=pi/4; m:=0;
    for k:=1 to 200 do
      begin
```

```pascal
t:=0.1*k; g:=2*pi*f*t;        x:=R*cos(g+ph);
y:=R*sin(g+ph);               c:=sqrt(L*L-y*y);
a:=x+c; b:=a+P;               x1:=x0+Round(x*mx);
y1:=y0-Round(y*my);           z1:=x0+Round(a*mx);
u1:=x0+Round(b*mx);           SetPenStyle(psClear);
SetPenWidth(3);               SetPenColor(clWhite);
SetBrushColor(clBlack); Circle(x1,y1,6);
Circle(z1,y0,6);              SetPenColor(clBlack);
SetPenStyle(psSolid);         SetPenWidth(3);
Line(x0,y0,x1,y1);            //----wheel
SetPenWidth(3);
Line(x1,y1,z1,y0);            //----crank
SetPenWidth(3);
Line(z1,y0,u1,y0);            //----rod
SetPenWidth(12);
Line(u1,y0-20,u1,y0+20);   //----piston
x2:=x1; y2:=y1; z2:=z1; u2:=u1;
for j:=1 to 72 do
  begin
    betg:=5*j; betr:=betg*pi/180; xm:=R*cos(betr);
    ym:=R*sin(betr); xm1:=x0+Round(xm*mx);
    ym1:=y0-Round(ym*my);
    SetPenColor(clBlack); SetPenWidth(1);
    Circle(xm1,ym1,1); xm2:=xm1; ym2:=ym1;
  end;
SetPenStyle(psSolid);         SetPenWidth(2);
Line(x0,y0,x0-15,270);        //----wheel support
Line(x0,y0,x0+15,270);        SetPenWidth(3);
SetBrushColor(clWhite);       //---wheel hinge
Circle(x0,y0,8);
if (x>29) and (Abs(y)<1) then m:=0;
if(x<0) and (Abs(y)<1) then m:=1;
//-----------------------------Valve-----------------------------
if m=0 then
  begin
    SetPenColor(clWhite);       SetPenWidth(3);
    Line(550,65,630,55);        SetBrushColor(clWhite);
    Circle(550,65,6);           Circle(630,55,6);
    SetPenColor(clBlack);       SetPenWidth(1);
    Line(550,54,630,65);        SetBrushColor(clBlack);
    Circle(550,55,6);           Circle(630,65,6);
```

```pascal
        SetBrushColor(clWhite); Circle(590,60,6);
      end;
    if m=1 then
      begin
        SetPenColor(clWhite);  SetPenWidth(3);
        Line(550,54,630,65);   SetBrushColor(clWhite);
        Circle(550,55,6);      Circle(630,65,6);
        SetPenColor(clBlack);  SetPenWidth(1);
        Line(550,65,630,54);   SetBrushColor(clBlack);
        Circle(550,65,6);      Circle(630,55,6);
        SetBrushColor(clWhite); Circle(590,60,6);
      end;
      SetPenWidth(1);
      Line(502,58,540,58);     Line(508,64,540,64);
      Line(540,58,582,14);     Line(554,54,588,18);
      Line(540,64,580,106);    Line(556,70,586,100);
      Line(556,54,556,70);     Line(640,58,680,58);
      Line(640,64,676,64);     Line(600,14,640,58);
      Line(592,18,623,52);     Line(592,102,624,70);
      Line(600,104,640,64);    Line(624,52,624,70);
      Sleep(100);
      SetPenStyle(psClear);    SetBrushColor(clWhite);
      SetPenColor(clWhite);    SetPenWidth(4);
      Line(x0,y0,x1,y1);        //----wheel
      Circle(x1,y1,8);         Circle(z1,y0,8);
      SetPenColor(clWhite);     SetPenWidth(4);
    Line(x1,y1,z1,y0);          //----crank
    Line(z1,y0,u1,y0);          //----rod
    SetPenWidth(14);
    Line(u1,y0-20,u1,y0+20);  //----piston
    SetPenWidth(1);             SetBrushColor(clWhite);
    if (k mod 10) = 0 then
        writeln('          ',(21-k/10)*2-2);
  end;
end;
//============= Main ========================
var R,L,P: real;
begin
  R:=30; L:=70; P:=80;   StartPos(R,L,P);
  var t1:= new TextABC(100,0,12,
        'Forward-recurrence and rotary movement');
```

```
  var t2:= new TextABC(140,20,11,
  'R = 30 - fly-wheel radius');
  var t3:= new TextABC(140,36,11, 'L = 70 - cranc length');
  var t4:= new TextABC(140,52,11, 'P = 80 - rod length');
  var t5:= new TextABC(140,68,11,
  'T = 5 sec. - rotation period');
  var t6:= new TextABC(250,330,11,
          'Model works 40 sec. It does 200 cycles
          (every cycle 0.1 sec.)');
  var t7:= new TextABC(260,348,11,
     'There are 200 lags (duration 0.1 sec.)
     between the cycles ');
  var t8:= new TextABC(610,0,11,'Input');
  var t9:= new TextABC(605,110,11,'Output');
    Sleep(100);           SetPenColor(clWhite);
    SetPenWidth(1);     SetBrushColor(ClWhite);
    Ellipse(230,90,480,220);
    Rectangle(550,160,678,202);  Movement(R,L,P);
    R:=30; L:=70; P:=80;
    StartPos(R,L,P);
end.
```
Listing 8.3

8.4. Three basketball throws

The educational game program is considered here. It may be useful to confirm knowledge of the elementary physics, elementary algebra and language Pascal ABC. The word "educational" does not mean tedious or tiresome always. The working program challenges to user. Despite of simple game conditions, one may detect that it is not so easy to get high result at once. One will have to do some training throws previously.

The task for a programmer.

Develop the game program that simulates throws of the ball into the basket. The user is given 3 attempts. He gets 2 points in case of direct hit into the ring. One point is set if the ball hits into the basket after rebounding from the backboard.

The throw distance is left out of account. The total result is the sum of points after the third throw.

Game conditions:

1. All trajectories are in the single vertical plane of the picture (fig.50).

2. The Cartesian coordinates of the start point are random values in the intervals: $x = 0 \ldots 12$ meters; $y = 0.8 \ldots 2.0$ meters.

3. The user is to input the initial play data:

- the modulus of the start velocity vector from 6 to 15 m/sec;

- the angle of the vector slope from 25 to 85 grades.

4. The program is to check if these data values are admissible.

Solution.

The program Prog_8_4 (listing 8.4) contains the next subroutines:

- "CoordSys" that draws the rectangle axes, the ring and the back-board;

- "Start" that generates coordinates for the start position of the ball;

- "Parabola" that calculates the trajectiories using the expression from the page 10 and draws the graphic.

```
Program Prog_8_4;
Uses GraphABC, ABCobjects;
  const R=11;  mx=50;  my=80;
  var att, bon, score: integer;

Procedure CoordSys;
va  j, x: integer    var d: string;
begin
//-----------------Sport hall, basket, backboard------------
```

```
SetWindowSize(900, 500);      SetPenStyle(psClear);
SetPenColor(clBlack);         SetPenWidth(1);
Line(100, 30, 870, 30);       Line(100, 440, 870, 440);
Line(100, 30, 100, 440);      SetPenWidth(2);
Line(868, 30, 868, 440);      SetPenWidth(3);
Line(836, 150, 870, 150);     Line(836, 190, 870, 190);
SetBrushColor(ClBlack);       Rectangle(830, 120, 840, 220);
SetBrushColor(ClBlack);       Roundrect(750, 198, 815, 210, 6,
6);
SetPenStyle(psClear);         SetPenColor(clBlack);
SetPenWidth(4);               Line(809, 204, 830, 204);
SetPenWidth(1);               Line(752, 204, 766, 260);
Line(766, 204, 775, 260);     Line(782, 204, 790, 260);
Line(798, 204, 800, 260);     Line(760, 204, 760, 260);
Line(775, 204, 770, 260);     Line(790, 204, 785, 260);
Line(808, 204, 795, 260);
for j := 1 to 5 do
begin
  SetPenWidth(2);   SetPenColor(clBlack);
  Line(100, 440 - 80 * j, 108, 440 - 80 * j);
end;
for j := 1 to 15 do
  begin
    SetPenWidth(2);  SetPenColor(clBlack);
    Line(100 + 50 * j, 440, 100 + 50 * j, 432);
  end;
for j := 0 to 15 do
  begin
    x := 1 * j;  d := IntToStr(x);
  var t := new TextABC(95 + 50 * j, 450, 12, d);
  end;

for j := 1 to 5 do
  begin
    x := 1 * j; d := IntToStr(x);
    var t := new TextABC(85, 432 - 80 * j, 12, d);
  end;
end;

Procedure Start(R: integer;  var xs, ys: real);
var x0, y0: integer;
begin
```

```pascal
    x0 := 100 + Round(xs * mx);  y0 := 440 - Round(ys * my);
    SetPenStyle(psSolid);           SetBrushColor(ClWhite);
    SetPenWidth(4);                 SetPenColor(clBlack);
    Circle(x0, y0, R);
end;

Procedure Parabola(var xs, ys, ang, v: real; att, score: integer);
const g = 9.81;
var k, x1, x2, y1, y2, x0, y0, m, z: integer;
var x, y, a, xc, yc, sh: real;
begin
  SetPenStyle(psClear);           SetBrushColor(ClWhite);
  Rectangle(9,466,890,500);       a := ang * pi / 180;
  x0 := 100 + Round(xs * mx);   y0 := 440 - Round(ys * my);
  x := 0; y := 0; m := 1; bon := 0;
  x0 := 100 + Round(xs * mx);   y0 := 440 - Round(ys * my);
  x := 0;   y := 0; m := 1; bon := 0;
  for k := 1 to 150 do
     begin
       x := x + 0.1;
       y := x * tan(a) - g * x * x / (2 * v * v * cos(a) * cos(a));
       xc := xs + x;   yc := ys + y;
       if yc > 5 then break;
       if (m = 1) and (xc >= 13.0) and (xc < 14.1)
                and  (yc <= 3.1) and (yc > 2.9) then
          begin
            bon := 2; m := 0; break;
          end;
       y1 := y0 - Round(y * my);
       if (xc <= 14.6) then
       x1 := x0 + Round(x * mx);    // direct
       if (xc > 14.5) and (yc > 3.0)
                and (yc < 4.8) then // rebound
          begin
            if xc > 14.5 then
              sh := 640 - xs * 100 else sh := 0;
              z := 920 + Round(sh);
              x1 := z + Round(xs * mx) - Round(x * mx);
          end;
       if (xc > 15.0) and (xc < 16.4) then   // rebound
         begin
           bon := 1;m := 0;
```

```pascal
        end;
    if (xc > 13.9) and (yc < 3.0) then break;
    SetPenWidth(3);   SetPenColor(clBlack);
    SetPenStyle(psDot);
    if (k > 1) and (y > 0) and (y1 > 40) then
      begin
        Line(x1, y1, x2, y2);
      end;
      x2 := x1;y2 := y1; Sleep(20);
  end;
end;
//===================Main=====================
var xs, ys, d, ang, v: real;
var j,z: integer;
var sc: string;
begin
  CoordSys;
  var t1:=new TextABC(300,0,12, 'Three attempts
          to earn 6 points');
  att:=0; score:=0;
  for j:=1 to 3 do
    begin
      ys := 1.0+Random(10)/10; xs := 0.5 + Random(80)/10;
      Start(R, xs, ys); write('  V?');
      var t2:=new TextABC(10,478,12,
            'Input start velocity from  6  to  15  m / sec');
      readln(v);    writeln(' = ',v);
      if (v<6) or (v>15) then
        begin
          writeln(' Inadmissible'); writeln(' value');   break;
        end;
      write('  A?');
      var t3:=new TextABC(310,478,12, 'and now the angle
            of slope  from  25  to  85 grades');
      readln(ang);   writeln(' = ',ang);
      if (ang<25) or (ang>85) then
        begin
          writeln(' Inadmissible'); writeln(' value');   break;
        end;
      Parabola(xs, ys, ang, v, att, score);
      score:=score+bon; writeln('   b = ', bon);
```

```
  writeln('---------------');
  end;
  sc:=IntToStr(score);
  var t3:=new TextABC(700,390,12,
         'Total sum = '+sc);
  SetPenColor(clWhite);   SetBrushColor(clWhite);
  Rectangle(8,470,700,500);
end.
```

Listing 8.4

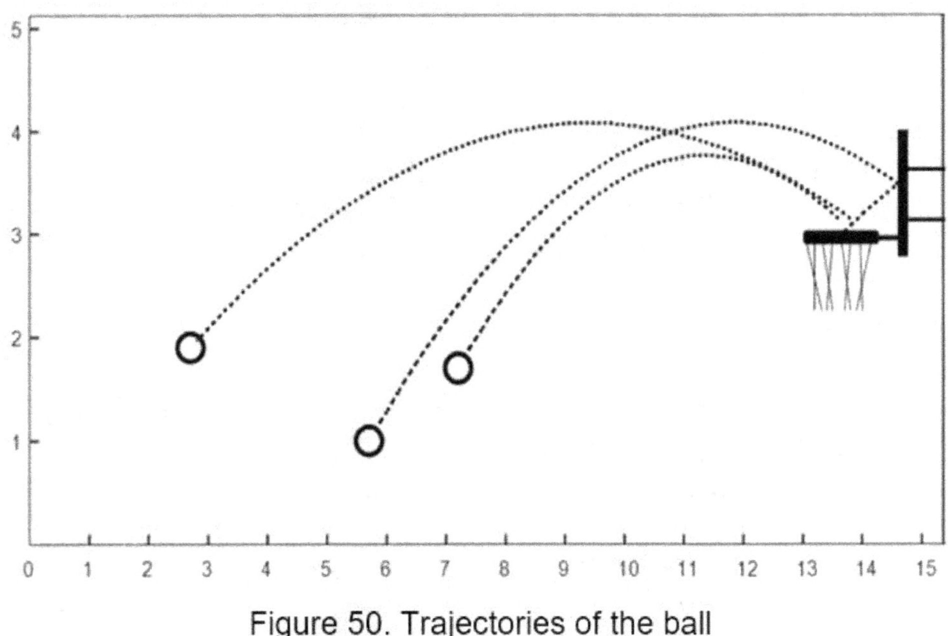

Figure 50. Trajectories of the ball

8.5. Automatic control system

The model of the automatic control system (ACS) is described usually with the differential equation system added by
algebraical and trigonometrical equations. The typical ACS
(fig.51) is intended to maintain the given state of the controlled
object.

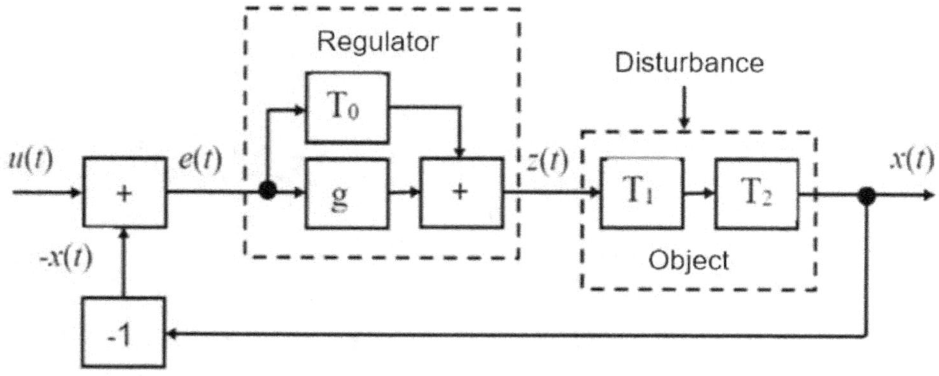

Figure 51. The automatic control system example

The object state is described by the output functioin x(t) which is to repeat the input drive-signal u(t) despite of any external dusturbance. The ACS has the regulator that measures the difference

$$e(t) = u(t) - x(t).$$

The function e(t) is called a control error. The value "-x(t)" is available due to the negative feedback contour. The regulator transforms the error e(t) into the signal z(t) which makes the object to change its state x(t) in so manner that the error e(t) would aspire to zero.

Let us suppose that three differential equations (8.1) describe the object behaviour. The inertialness of object is determined by two constant-time T_1 and T_2.

Once more differential equation is for the regulator that contains an integrator with the constant-time T_0 and a scale link with gain coefficient "g". Thus, we have the equation system in terms of the infinite small values (8.1).

There are three non-homogeneous first-order differential equations and one algebraical equation:

$$\begin{cases} T_1 \dfrac{dy(t)}{dt} + y(t) = z(t); \\[2mm] T_2 \dfrac{dx(t)}{dt} + x(t) = z(t); \\[2mm] gT_0 \dfrac{dz(t)}{dt} + z(t) = T_0 \dfrac{de(t)}{dt}; \\[2mm] u(t) - x(t) = e(t) \end{cases} \qquad (8.1)$$

We take the denotes to prepare the system for a computer solution:

$$C_0 = \frac{T_0}{T}; \qquad C_1 = \frac{T_1}{T}; \qquad C_2 = \frac{T_2}{T};$$

$$A_0 = \frac{1}{C_0}; \qquad A_1 = \frac{1}{1+C_1}; \qquad A_2 = \frac{1}{1+C_2}$$

where T – time quantum step.

Using left approximation of the derivatives (see page 47), we'll change the system (8.1) by equations in the finite differences (8.2):

$$\begin{cases} y(k) = A_1 \cdot [z(k) + C_1 \cdot y(k-1)] \\ x(k) = A_2 \cdot [y(k) + C_2 \cdot x(k-1)] \\ z(k) = (A_0 + g)^* e(k) - g^* e(k-1) + z(k-1) \\ e(k) = u(k) - x(k-1), \end{cases} \qquad (8.2)$$

Solution.

The ACS is simulated by the program Prog_8_5 (listing 8.5). It contains two subroutines:

- "CoordSys" that draws the rectangle axes;

- "ACS" that simulates the control system response upon the standard test signal, estimates the control error and draws the diagram (fig.52).

```pascal
Program Prog_8_5;
Uses GraphABC, ABCobjects;

Procedure CoordSys;
 const x0=100;
 var j, k, xc, yc: integer;  var x,y: real;  var s: string;
 begin
  SetWindowSize(800,460); SetPenStyle(psClear);
  SetPenColor(clBlack);      SetPenWidth(1);
  Line(140, 10, 140, 410);    Line(140, 310, 760, 310);
  Line(140, 10, 760, 10);     Line(140, 410, 760, 410);
  SetPenWidth(3);             SetPenColor(clGray);
  Line(760, 10, 760, 410);    SetPenStyle(psDot);
  Line(140, 110, 750, 110);   // Level y = 1
  SetPenStyle(psSolid);       SetPenWidth(1);
  for k:=1 to 20 do
     begin
        Line(140+30*k, 310, 140+30*k, 318);
     end;
  for k:=0 to 18 do
     begin
        Line(132, 390-20*k, 140, 390-20*k);
     end;
  for k:=2 to 11 do
     begin
        x:=0.5*(k-1); Str(x:2:1,s);
        var t:= new TextABC(72+60*k,320,12,s);
     end;
  for k:=1 to 10 do
     begin
        y:=0.2*(k-1)-0.4; Str(y:2:1,s);
        var t:= new TextABC(104,420-40*k,12,s);
     end;
end;
```

```pascal
Procedure ACS;
const x0=140; y0=310; mx=120; my=200;
var k, x1, x2, y1, y2, z1, z2, e1, e2: integer;
var x, y,  z,e, u: array[0..120] of real;
var t, dt, t0, t1, t2, a0, a1, a2, c0, c1, c2, g, gn: real;
var s0, s1, s2, s3 : string;
begin
  dt:=0.05; t0:=0.2; t1:=1.5; t2:=0.4; g:=9;
  c0:=t0/dt; c1:=t1/dt;  c2:=t2/dt;
  a0:=1/c0; a1:=1/(1+c1); a2:=1/(1+c2);
  gn:=g+a0;
 t:=0; x[0]:=0; y[0]:=0; z[0]:=0; e[0]:=0; u[0]:=0;
 for k:=1 to 100 do
   begin
    u[k]:=1;  e[k]:=u[k]-x[k-1];   z[k]:=gn*e[k]-g*e[k-1]+z[k-1];
    y[k]:=a1*(z[k]+c1*y[k-1]);   x[k]:=a2*(y[k]+c2*x[k-1]);
    t:=t+dt;
    x1:=x0+Round(t*mx);        y1:=y0-Round(x[k]*my);
    e1:=y0-Round(e[k]*my);
    if k>1 then
      begin
        SetPenWidth(4);        SetPenColor(clBlack);
        Line(x1,y1,x2,y2);      SetPenWidth(3);
        SetPenColor(clBlack); Line(x1, e1, x2, e2);
      end;
     x2:=x1; y2:=y1; e2:=e1;
  end;
  SetPenWidth(4);  SetPenColor(clBlack);
  Line(x0,y0,x0+8,y0-10);
  var v1:=new TextABC(450,160,12,
          'Automatic control system');
  Str(t0:2:1,s0);  Str(t1:2:1,s1);   Str(t2:2:1,s2);  Str(g:2:1,s3);
  var v2:=new TextABC(480,190,12,  'Time-constants:');
  var v3:=new TextABC(440,215,12,
          'T0 = '+s0+' T1 = '+s1+'    T2 = '+s2);
  var v4:=new TextABC(490,245,12,  'Gain  g =  '+s3);
  var v5:=new TextABC(230,20,12,'x(t)');
  var v6:=new TextABC(340,260,12,'e(t)');
end;
```

```
//===== Main =====
var s0, s1, s2: string;
begin
  CoordSys; ACS;
end.
```

Listing 8.5

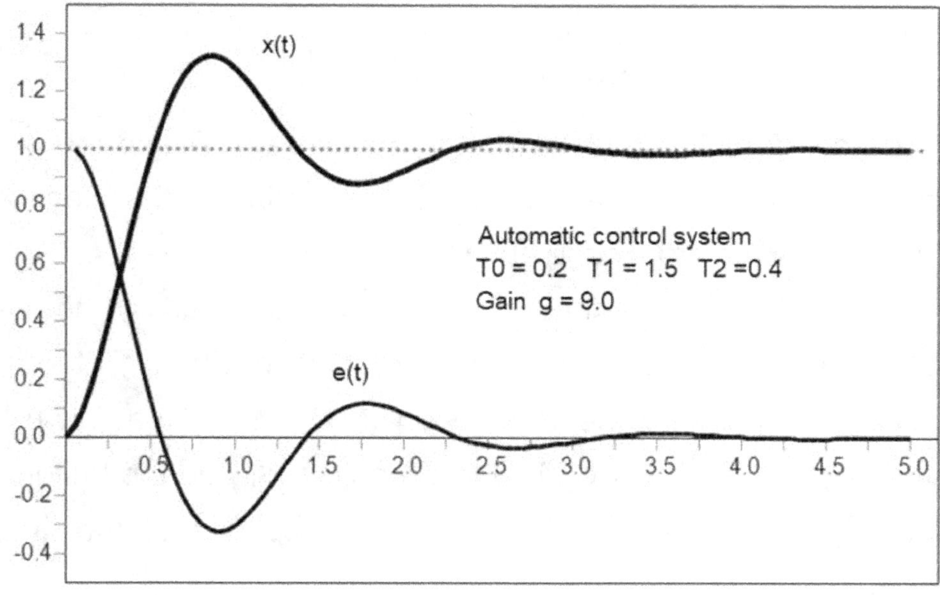

Figure 52. The transition response

Models of the ACS are created using typical links – linear and non-linear. A link is called linear if its coefficients are constant in the mathematical description of the response at the input signal. There are examples: scale, inertial, differential, integral, forced links.

The class of non-linear links involves restrictors, functional transformers (logarithmical, trigonometrical), relay-type, hysteresis-type links and other with behavior being described by non-linear continuos functions, broken functions with finite breaks or by lagging functions.

The linear links are more attractive for those who studies the language Pascal ABC. These links are investigated usually in both spaces – time and frequency. The dynamic model shows the link behavior in real time by means a differential equation but the behaviour may be predicted also by means of the magnitude-frequency and phase-frequency responses (MFR and PFR).

8.6. An inertial link

A quadrupole circuit formed by the resistor R and the capacitor C (fig. 53) is the very wide-spread link in the electrotechnic, radioelectronic and automatic.

The RC-link behavior in time space is desribed by the differential equation (5.1). The equation solution x(k) in finite differences is given by the expression (5.4), where the link parameter b = Δ/T, T = R*C. The link reaction x(k) on the uniform input signal u(k) is called a transition response of the link, where k = 1, 2, 3, …

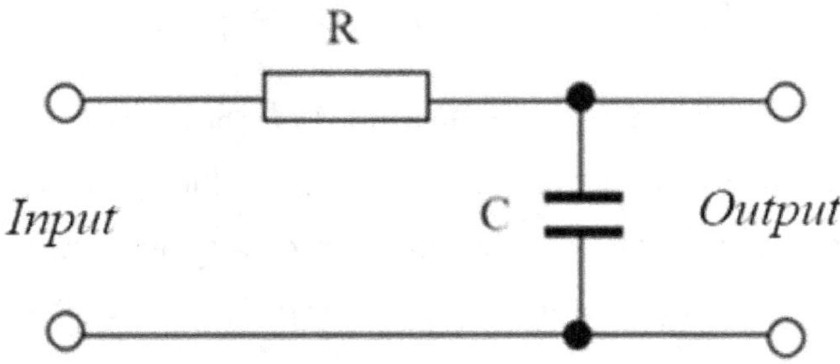

Figure 53. The inertial RC-link

Example 8.1.

Develop the program that calculates and draws the transition curve family of the inertial link with elements C = 0.1 mcF and R = 0.51, 0.82, 1.2 kOm.

Solution.

The program Prog_8_6 (listing 8.6) contains two subrou-tines:

- CoordSys that draws the rectangle axes;

- Equat that calculates and draws the response family (fig.54).

```pascal
Program Prog_8_6;
Uses GraphABC, ABCobjects;

Procedure CoordSys;
var j, k: integer;
var x, xc, yc: real;
var s: string;
begin
  SetWindowSize(800,500);   SetPenStyle(psClear);
  SetPenColor(clBlack);     SetPenWidth(1);
  Line(100,20,750,20);      Line(750,20,750,400);
  Line(100,20,100,400);     Line(100,400,750,400);
  for k:=1 to 30 do
   begin
     Line(100+21*k, 400, 100+21*k,392);
   end;
  for k:=2 to 16 do
   begin
     x:=0.2*(k-1); Str(x:3:1,s);
     var t:=new TextABC(48+42*k,408,12,s);
   end;
  for k:=1 to 11 do
   begin
     Line(100,436-36*k,108,436-36*k);
   end;
  for k:=1 to 11 do
   begin
     yc:=0.1*(k-1); Str(yc:2:1,s);
     var t:=new TextABC(72,426-36*k,12,s);
   end;
for k:=1 to 11 do
   begin
     Line(742,436-36*k,750,436-36*k);
```

```pascal
    end;
  for k:=1 to 11 do
    begin
      yc:=0.1*(k-1);  Str(yc:2:1,s);
      var t:=new TextABC(758,428-36*k,12,s);
    end;
end;

Procedure Equat(cc, rr: real);
const x0=100; y0=400; mx=2100; my=360;
var k, x1, x2, y1, y2, u1, u2, z: integer;
var u, y: array[0..99] of real;
var b, br, t, dt, tt: real;
begin
  tt:=rr*cc; dt:=0.01;  b:=dt/tt; y[0]:=0; t:=0;  y2:=0;
  for k:=1 to 30 do
    begin
      t:=t+dt;                y[k]:=b+(1-b)*y[k-1];
      x1:=x0+Round(t*mx);   y1:=y0-Round(y[k]*my);
      SetPenWidth(3);         SetPenColor(clBlack);
      if k>1 then
        begin
          Line(x1,y1,x2,y2);
        end;
      x2:=x1; y2:=y1;
    end;
  z:=Round(b*370);  Line(x0,y0,x0+22,y0-z);
end;
//===============Main==================
var cc, rr:real;
var s1, s2, s3, s4: string;
begin
  CoordSys;
  var t1:= new TextABC(525,185,12,'Inertial link');
    cc:=0.1;     Str(cc:3:1,s1);
   var t2:= new TextABC(540,210,12,'C = '+s1);
   rr:=0.51; Str(rr:3:2,s2);
   var t3:= new TextABC(180,104,12,'R = '+s2);
   Equat(cc,rr);  rr:=0.82; Str(rr:3:2,s3);
   var t4:= new TextABC(296,94,12,'R = '+s3);
   Equat(cc,rr);   rr:=1.2;  Str(rr:3:2,s4);
   var t5:= new TextABC(400,104,12,'R = '+s4);
```

Equat(cc,rr);
end.

Listing 8.6

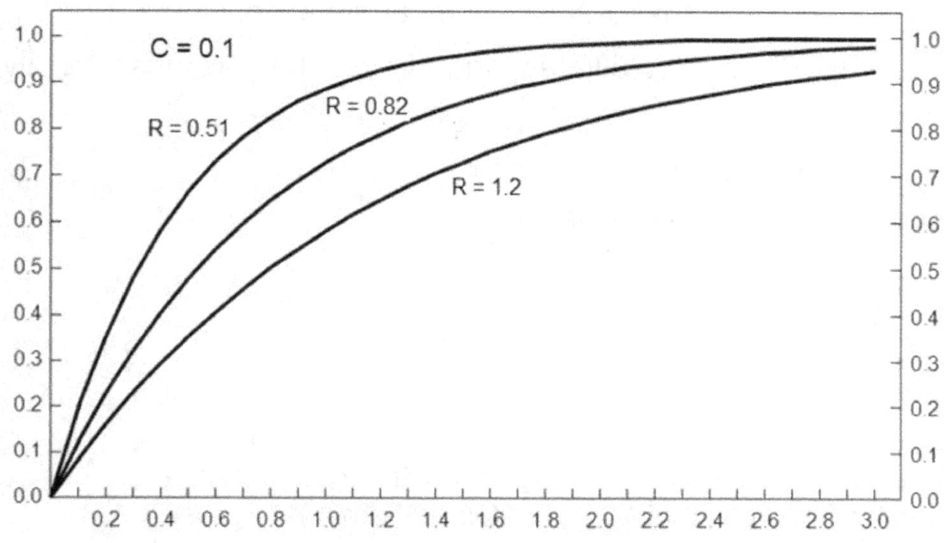

Figure 54. The transition response of the inertial link

The inertial link is investigated in the frequency space as its response at a harmonic input signal in the given frequency diapason with the signal amplitude being constant.

There are the expressions of the MFR and the PFR:

$$A(\omega) = \frac{1}{\sqrt{(\omega T)^2 + 1}};$$

$$\varphi(\omega) = -arctg(\omega T)$$

where $\omega = 2\pi F$ – angular frequency, radian per sec.;

F – harmonic frequency, Hz.

Example 8.2.

Develop the family of the MFR for the inertial filter using the parameters from the Example 8.6.

Solution.

The program Prog_8_7 (listing 8.7) contains two subroutines:

- "CoordSys" that draws the rectangle axes;

- "MFR" that calculates and draws the response family (fig.55).

```
Program Prog_8_7;
Uses GraphABC, ABCobjects;

Procedure CoordSys;
var j, k: integer;
var x, xc, yc: real;
var s: string;
begin
  SetWindowSize(800,500);  SetPenStyle(psClear);
  SetPenColor(clBlack);    SetPenWidth(1);
  Line(100,20,750,20);     Line(750,20,750,400);
  Line(100,20,100,400);    Line(100,400,750,400);
  for k:=1 to 30 do
   begin
     Line(100+21*k,400,100+21*k,392);
   end;
  for k:=2 to 16 do
   begin
     x:=0.2*(k-1); Str(x:3:1,s);
     var t:=new TextABC(48+42*k,408,12,s);
   end;
  for k:=1 to 11 do
   begin
     Line(100,436-36*k,108,436-36*k);
   end;
  for k:=1 to 11 do
   begin
     yc:=0.1*(k-1); Str(yc:2:1,s);
     var t:=new TextABC(72,426-36*k,12,s);
   end;
for k:=1 to 11 do
   begin
```

```pascal
      Line(742,436-36*k, 750 ,436-36*k);
    end;
for k:=1 to 11 do
   begin
     yc:=0.1*(k-1); Str(yc:2:1,s);
      var t:=new TextABC(758, 428-36*k, 12,s);
    end;
end;


Procedure MFR(cc,rr: real);
const pi=3.14158; x0=100; y0=400; mx=21; my=360;
var k, x1, x2, y1, y2, u1, u2, z: integer;
var tt, df, f, cf, am: real;
begin
  tt:=rr*cc;  f:=0;  df:=0.5;
   for k:=1 to 60 do
     begin
       f:=f+df;     cf:=2*pi*f;    am:=1/sqrt(cf*tt*cf*tt+1);
       x1:=x0+Round(f*mx);  y1:=y0-Round(am*my);
       SetPenWidth(3);         SetPenColor(clBlack);
       if k>1 then
         begin
            Line(x1,y1,x2,y2);
         end;
        x2:=x1;  y2:=y1;
     end;
end;


//=============Main===================
var cc, rr:real;
var s1, s2, s3, s4: string;
begin
  CoordSys;
  var t1:= new TextABC(525,185,12,'Inertial link');
  cc:=0.1;  Str(cc:3:1,s1);
  var t2:= new TextABC(540,210,12,'C = '+s1);
  rr:=0.51; Str(rr:3:2,s2);
  var t3:= new TextABC(280,260,12,'R = '+s2);
  MFR(cc,rr); rr:=0.82; Str(rr:3:2,s3);
  var t4:= new TextABC(290,310,12,'R = '+s3);
  MFR(cc,rr); rr:=1.2;  Str(rr:3:2,s4);
  var t5:= new TextABC(160,320,12,'R = '+s4);
```

  MFR(cc,rr);
end.

Listing 8.7

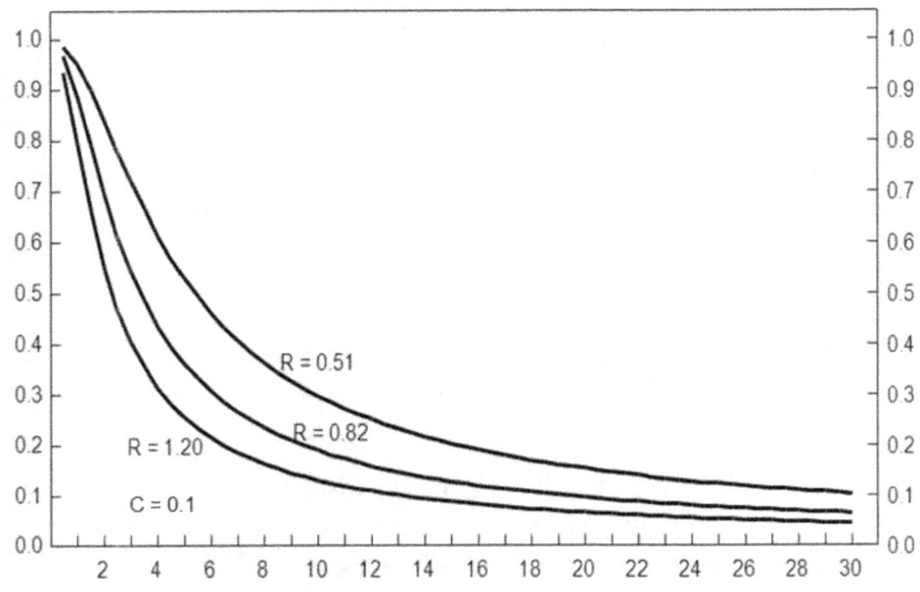

Figure 55. The magnitude-frequency
response of the inertial link

Example 8.3.

Develop the family of the PFR for the inertial filter using
the parameters from the Example 8.6.

Solution.

The program Prog_8_8 (listing 8.8) contains two subroutines:

- "CoordSys" that draws the rectangle axes;

- "MFR" that calculates and draws the response family
(fig.56).

```
Program Prog_8_8;
Uses GraphABC, ABCobjects;
Procedure CoordSys;
var j, k: integer;
var x, xc, yc: real;   var s: string;
```

```pascal
begin
  SetWindowSize(800,500);  SetPenStyle(psClear);
  SetPenColor(clBlack);        SetPenWidth(1);
  Line(100,30,100,390);        Line(100,390,750,390);
  Line(100,30,750,30);         Line(750,30,750,390);
  for k:=1 to 30 do
    begin Line(100+21*k,30,100+21*k,38);
  end;
  for k:=1 to 16 do
    begin
     x:=2*(k-1); Str(x:3:0,s);
     var t:=new TextABC(44+42*k,10,12,s);
    end;
  for k:=1 to 18 do
    begin
     Line(100,30+20*k,108,30+20*k);
    end;
  for k:=2 to 10 do
    begin
     yc:=-10*(k-1); Str(yc:2:0,s);
     var t:=new TextABC(74,40*k-20,12,s);
    end;
  for k:=1 to 18 do
    begin
     Line(742,30+20*k,750,30+20*k);
    end;
  for k:=2 to 10 do
    begin
     yc:=-10*(k-1); Str(yc:2:0,s);
     var t:=new TextABC(758,40*k-20,12,s);
    end;
end;

Procedure PFR(cc,rr: real);
const pi=3.14158;  x0=100;  y0=30;  mx=21;  my=4;
var k, x1, x2, y1, y2, u1, u2, z: integer;
var tt, df, f, cf, p, ph: real;
begin
  tt:=rr*cc;  f:=0;  df:=0.5;
  for k:=1 to 60 do
    begin
     f:=f+df;   cf:=2*pi*f;
```

```pascal
    p:=-arctan(cf*tt);        ph:=p*180/pi;
    x1:=x0+Round(f*mx);  y1:=y0-Round(ph*my);
    SetPenWidth(3);        SetPenColor(clBlack);
    if k>1 then
      begin
        Line(x1,y1,x2,y2);
      end;
      x2:=x1;  y2:=y1;
    end;
    z:=Round(tt*700); Line(x0,y0,x0+10,y0+z);
end;
//================ Main =================
var cc, rr: real;   var s1, s2, s3, s4: string;
begin
  CoordSys;
  var t1:= new TextABC(525,185,12, 'Inertial link');
  cc:=0.1;  Str(cc:3:1,s1);
  var t2:= new TextABC(540,210,12,'C = '+s1);
  rr:=0.51; Str(rr:3:2,s2);
  var t3:= new TextABC(224,260,12,'R = '+s2);
  PFR(cc,rr);
  rr:=0.82; Str(rr:3:2,s3);
  var t4:= new TextABC(224,300,12,s3);
  PFR(cc,rr);
  rr:=1.2;  Str(rr:3:2,s4);
  var t5:= new TextABC(130,320,12,'R = '+s4);
  PFR(cc,rr);
end.
```

Listing 8.8

Conclusions:

The magnitude-frequence response shows:

1. The inertial link is a low-pass filter which is transparent for the low frequency harmonics. The transfer coefficient is maximum if F = 0.

2. The harmonic amplitude decreases twice if F = 3… 5 Hz and falls 10-20 times if F > 30 Hz.

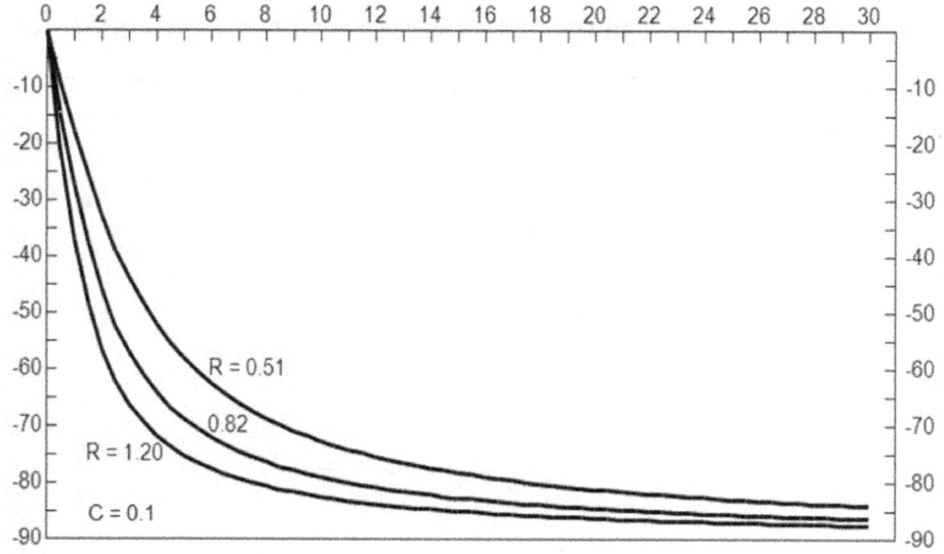

Figure 56. The phase-frequency
response of the inertial link

The phase-frequency response indicates that the link provides the negative phase shift which aspires to -90^0 if F > 5-6 Hz. It means in the time space that a signal lag is several milliseconds in our example.

The inertial link is usually applied as the first-order low-pass filter for the hindrances suppression outside the frequencies spectrum of the useful signal. It works as a smoothing device in the time space.

An eccessive lag in the automatic control system is dangerous so far as it may change the sign of the compensation signal in the feedback contour provoking the control instability. The feedback must always be negative.

8.7. A real differential link

Ingineer-electronics call this link (fig.57) as "dif-circuit". The capacity C serves a complete insulator for the direct current. Due to this fact the link responses to the signal changes only.

The reaction is like a differentiation, but it is not an exact mathematical operation. At this reason the operation is named a "real" differentiation.

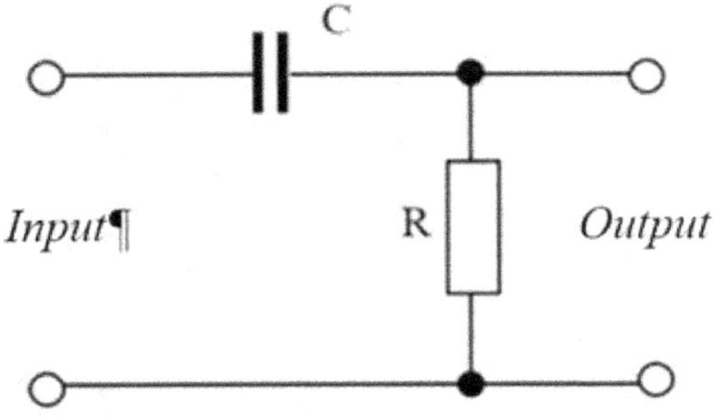

Figure 57. The real differential link

The response x(t) to the input signal u(t) is described by the first-order differential equation:

$$T\frac{dx(t)}{dt} - x(t) = T\frac{du(t)}{dt}, \qquad T = RC$$

The equation solution in finite differences is:

x(k) =u(k) − u(k-1) +(1 − b) * x(k − 1),

b = Δt / T; k = 1, 2, 3, …

The MFR and the PFR of the link are:

$$A(\omega) = \frac{\omega T}{\sqrt{(\omega T)^2 + 1}};$$

$$\varphi(\omega) = \frac{\pi}{2} - arctg(\omega T)$$

where $\quad\omega = 2\pi F\quad$ – angular frequency, radian per sec.;

$\qquad\quad$ F $\qquad\qquad$ – harmonic frequency, Hz.

Example 8.4.

Develop the program that calculates and draws the transition curves family of the dif-circuit with elements

$$C = 0.1\ mcF,\ R = 0.51,\ 0.82,\ 1.2\ kOm.$$

Solution.

The program Prog_8_9 (listing 8.9) contains two subroutines:

- "CoordSys" that draws the rectangle axes;

- "Equat" that calculates and draws the response family (fig.58).

```
Program Prog_8_9;
Uses GraphABC, ABCobjects;

Procedure CoordSys;
var j, k: integer;
var x, xc, yc: real;
var s: string;
begin
  SetWindowSize(800,500);  SetPenStyle(psClear);
  SetPenColor(clBlack);        SetPenWidth(1);
  Line(100,20,750,20);       Line(750,20,750,400);
  Line(100,20,100,400);     Line(100,400,750,400);
  for k:=1 to 30 do
    begin
     Line(100+21*k, 400,100+21*k,392);
    end;
  for k:=2 to 16 do
    begin
     x:=0.2*(k-1); Str(x:3:1,s);
     var t:=new TextABC(48+42*k,408,12,s);
    end;
  for k:=1 to 11 do
```

```pascal
    begin
      Line(100,436-36*k, 108,436-36*k);
    end;
  for k:=1 to 11 do
    begin
      yc:=0.1*(k-1); Str(yc:2:1,s);
      var t:=new TextABC(72,426-36*k,12,s);
    end;
  for k:=1 to 11 do
    begin
      Line(742,436-36*k,750,436-36*k);
    end;
    for k:=1 to 11 do
      begin
        yc:=0.1*(k-1);  Str(yc:2:1,s);
        var t:=new TextABC(758,428-36*k,12,s);
      end;
end;

Procedure Equat(cc, rr: real);
const x0=100; y0=400;  mx=2100; my=360;
var k, x1, x2, y1, y2, u1, u2, z, z1, z2, z3: integer;
var u, y: array[0..99] of real;
var b, br, t, dt, tt: real;
begin
  tt:=rr*cc; dt:=0.01;   b:=dt/tt;
  y[0]:=0;  y[1]:=1; t:=0; x2:=0; y2:=0;
  for k:=2 to 32 do
    begin
      t:=t+dt;              y[k]:=(1-b)*y[k-1];
      x1:=x0+Round(t*mx);  y1:=y0-Round(y[k]*my);
      SetPenWidth(3);
      SetPenColor(clBlack);
      if k>2 then
        begin
          Line(x1,y1,x2,y2);
        end;
      x2:=x1; y2:=y1;
    end;
    z:=42+Round(b*350);  Line(x0,40,x0+22,z);
end;
```

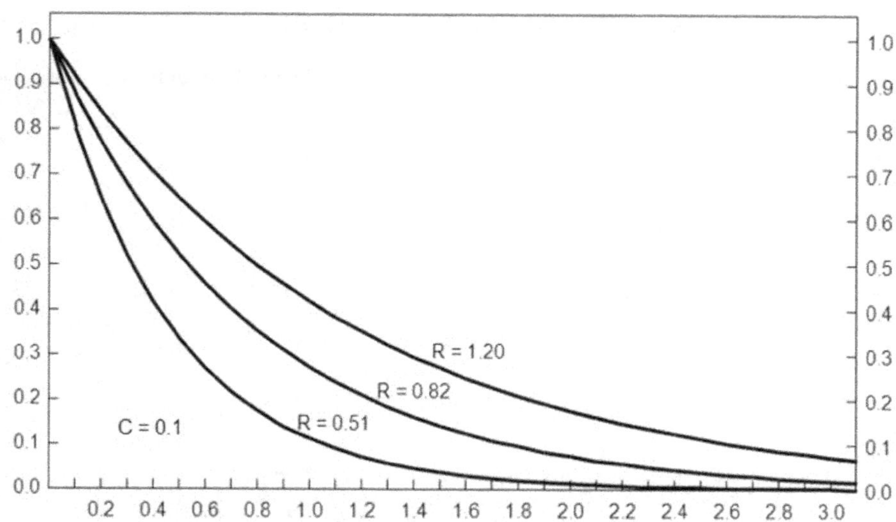

Figure 58. The transition curves
of the real differenital link

```
//==========Main===============
var cc, rr: real;  var s1, s2, s3, s4: string;
begin
  CoordSys;
  var t1:= new TextABC(530,165,12,'Dif-circuit');
  cc:=0.1;  Str(cc:3:1,s1);
  var t2:= new TextABC(540,200,12,'C = '+s1);
  rr:=0.51; Str(rr:3:2,s2);
  var t3:= new TextABC(300,338,12,'R = '+s2);
  Equat(cc,rr);  rr:=0.82; Str(rr:3:2,s3);
  var t4:= new TextABC(364,312,12,'R = '+s3);
  Equat(cc,rr);  rr:=1.2;  Str(rr:3:2,s4);
  var t5:= new TextABC(408,280,12,'R = '+s4);  Equat(cc,rr);
end.
```

Listing 8.9

Example 8.5.

Develop the program that calculates and draws the MFR family of the dif-circuit with the same elements.

Solution.

The program Prog_8_10 (listing 8.10) contains two sub-routines:

- "CoordSys" that draws the rectangle axes;

- "MFR" that calculates and draws the response family (fig.59).

```
Program Prog_8_10;
Uses GraphABC, ABCobjects;

Procedure CoordSys;
var j, k: integer;
var x, xc, yc: real;
var s: string;
begin
  SetWindowSize(800,500);  SetPenStyle(psClear);
  SetPenColor(clBlack);       SetPenWidth(1);
  Line(100,20,750,20);        Line(750,20,750,400);
  Line(100,20,100,400);       Line(100,400,750,400);
  for k:=1 to 30 do
    begin
      Line(100+21*k, 400, 100+21*k,392);
    end;
  for k:=2 to 16 do
    begin
      x:=2*(k-1);
      Str(x:2:0,s);
      var t:=new TextABC(48+42*k,408,12,s);
    end;
  for k:=1 to 11 do
    begin
      Line(100,436-36*k, 108, 436-36*k);
    end;
  for k:=1 to 11 do
    begin
      yc:=0.1*(k-1);   Str(yc:2:1,s);
      var t:=new TextABC(72,426-36*k,12,s);
    end;
  for k:=1 to 11 do
    begin
      Line(742,436-36*k, 750, 436-36*k);
    end;
  for k:=1 to 11 do
    begin
```

```pascal
      yc:=0.1*(k-1); Str(yc:2:1,s);
      var t:=new TextABC(758, 428-36*k, 12, s);
    end;
end;

Procedure MFR(cc,rr: real);
const pi=3.14158; x0=100;  y0=400; mx=30; my=360;
var  k, x1, x2, y1, y2, u1, u2, z: integer;
var tt, df, f, cf, am: real;
begin
  tt:=rr*cc;  f:=0;  df:=0.5;
  for k:=1 to 42 do
    begin
      f:=f+df;    cf:=2*pi*f;      am:=cf*tt/sqrt(cf*tt*cf*tt+1);
      x1:=x0+Round(f*mx);  y1:=y0-Round(am*my);
      SetPenWidth(3);          SetPenColor(clBlack);
      if k>1 then
        begin
          Line(x1,y1,x2,y2);
        end;
        x2:=x1;  y2:=y1;
    end;
    z:=Round(tt*(1-tt*0.5)*1150);  Line(x0,y0,x0+15,y0-z);
end;
//===============Main============================
var cc, rr:real;
var s1, s2, s3, s4: string;
begin
  CoordSys;
  var t1:= new TextABC(515,185,12, 'Dif-circuit'); cc:=0.1;
  Str(cc:3:1,s1);
  var t2:= new TextABC(525,210,12, 'C = '+s1);   rr:=0.51;
  Str(rr:3:2,s2);
  var t3:= new TextABC(140,50,12, 'R = '+s2);   MFR(cc,rr);
  rr:=0.82; Str(rr:3:2,s3);
  var t4:= new TextABC(200,90,12,s3);
  MFR(cc,rr);   rr:=1.2;    Str(rr:3:2,s4);
  var t5:= new TextABC(218,120,12,'R = '+s4);
  MFR(cc,rr);
end.
```

Listing 8.10

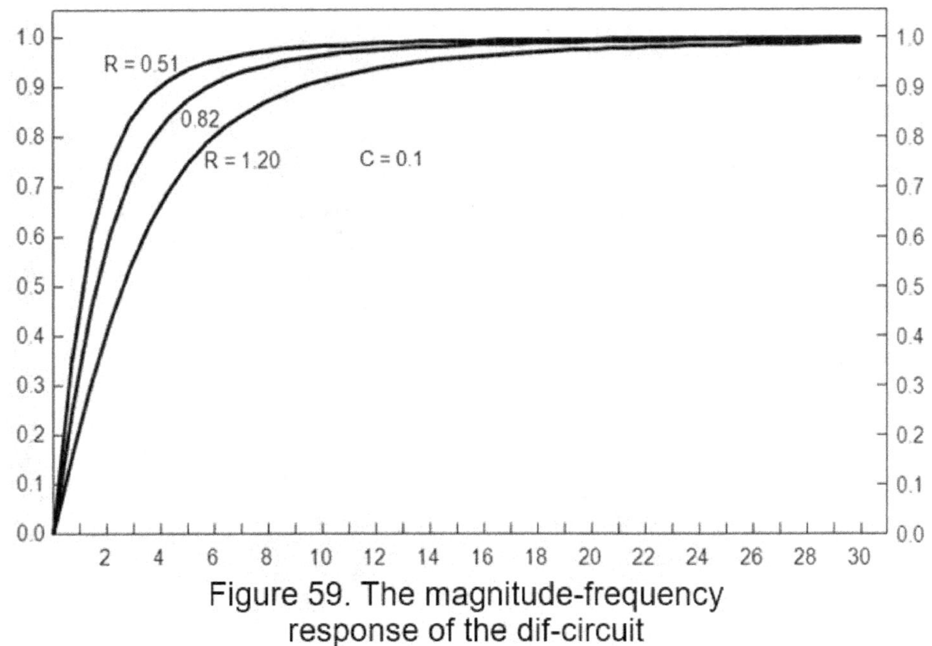

Figure 59. The magnitude-frequency
response of the dif-circuit

Example 8.6.

Develop the program that calculates and draws the PFR family of the dif-circuit with the same elements.

Solution.

The program Prog_8_11 (listing 8.11) contains two sub-routines:

- "CoordSys" that draws the rectangle axes;

- "PFR" that calculates and draws the response family (fig.60).

```
Program Prog_8_11;
Uses GraphABC, ABCobjects;

Procedure CoordSys;
var j, k: integer;
var x, xc, yc: real;  var s: string;
begin
  SetWindowSize(800,500);  SetPenStyle(psClear);
  SetPenColor(clBlack);       SetPenWidth(1);
  Line(100,20,750,20);        Line(750,20,750,400);
```

```pascal
    Line(100,20,100,400);        Line(100,400,750,400);
     for k:=1 to 30 do
       begin
          Line(100+21*k, 400,100+21*k,392);
       end;
  for k:=2 to 16 do
     begin
       x:=2*(k-1); Str(x:2:0,s);
       var t:=new TextABC(48+42*k,408,12,s);
     end;
   for k:=1 to 9 do
     begin
       Line(100,400-41*k, 108, 400-41*k);
     end;
   for k:=1 to 9 do
     begin
       yc:=90-10*(k-1); Str(yc:2:0,s);
       var t:=new TextABC(80,41*k-20,12,s);
     end;
  end;

Procedure PFD(cc,rr: real);
const pi=3.14158; x0=100; y0=390; mx=30; my=4;
var  k, x1, x2, y1, y2, u1, u2, z: integer;
var tt, df, f, cf, p, ph: real;
begin
  tt:=rr*cc;  f:=0;  df:=0.3;
   for k:=1 to 70 do
      begin
        f:=f+df;                 cf:=2*pi*f;
        p:=pi/2-arctan(cf*tt);     ph:=p*180/pi;
        x1:=x0+Round(f*mx);     y1:=y0-Round(ph*my);
        SetPenWidth(3);          SetPenColor(clBlack);
        if k>1 then
          begin
             Line(x1,y1,x2,y2);
          end;
        x2:=x1;  y2:=y1;
      end;
   z:=Round(tt*480);   Line(x0,30,x0+10,30+z);
end;
```

```
//================Main==================
var cc, rr: real;
var s1, s2, s3, s4: string;
begin
  CoordSys;
  var t1:= new TextABC(500,185,12,'Dif-circuit');
  cc:=0.1;  Str(cc:3:1,s1);
  var t2:= new TextABC(510,210,12,'C = '+s1);
  rr:=0.51; Str(rr:3:2,s2);
  var t3:= new TextABC(280,260,12,'R = '+s2);
  PFD(cc,rr);
  rr:=0.82; Str(rr:3:2,s3);
  var t4:= new TextABC(240,284,12,s3);
  PFD(cc,rr);
  rr:=1.2;  Str(rr:3:2,s4);
  var t5:= new TextABC(150,310,12,'R = '+s4);
  PFD(cc,rr);
end.
```

Listing 8.11

Conclusions:

The magnitude-frequence response shows:

1. The real differential link is a high-pass filter which is transparent for the high frequency harmonics. The transfer coefficient is zero if F = 0.

2. The harmonic amplitude increases if F > 0 and the transfer coefficient aspires to level 1 if F > 5 … 6 Hz.

3. The phase shift is positive and aspires to 90^0 when F → 0. The shift aspires to 0 when F → ∞.

The dif-circuit is applied for the solution of the next tasks:

- suppression of the infra-low oscillations;

- phase correction the feedback signal in the ACS, i.e. compensation of a lag.

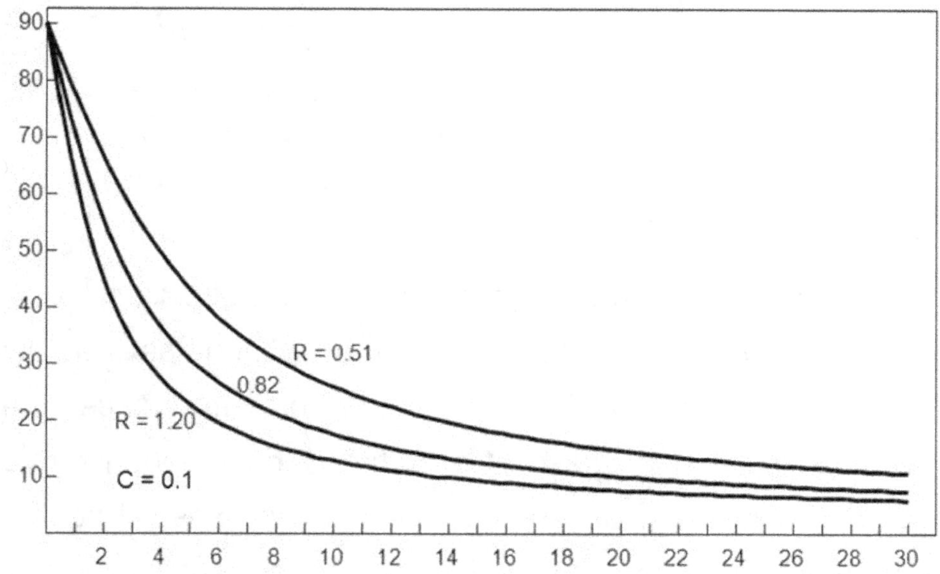

Figure 60. The phase-frequency response of the dif-circuit

8.8. A resonant link

Resonant phenomenons are often used in the radio-electronic and automatic systems. Resonant circuit provides the signal selection or the signal generation at the given frequency. There are many examples of dangerous resonances in buildings, bridges, ships, aeroplanes et cetera.

We'll consider the resonant quadrupole circuit (fig. 61) formed with the inductance L and the capacitor C. An assumption is made that both elements have ideal reactive conductance. Its real energy loses are simulated with the resistor R.

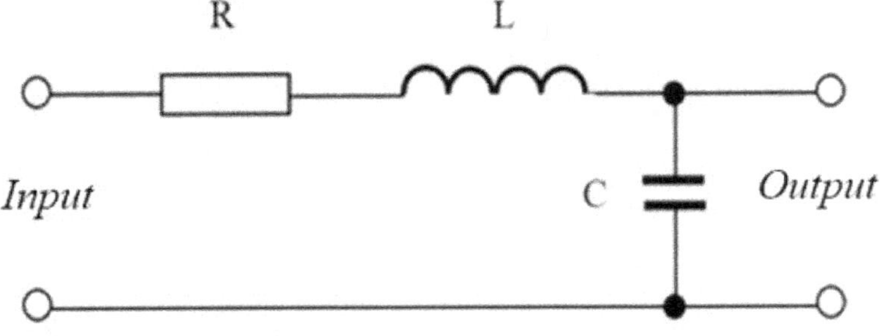

Figure 61. The resonant circuit

This quadrupole works as a divider of an alternating voltage. It is evident that a signal is transfered without loses at low frequenses if active load R_a >> R. At very high frequences the quadrupole is like a low-pass filter, so far as the impedance of L arises and capacity C becomes a short curcuit. Special behaviour of the circuit is observed at the angular frequency ω_0:

$$\omega_0 C = \frac{1}{\omega_0 L}$$

It is just a resonant frequency of the oscillating link. The frequency value is determined by parameters **L** and C.

A circuit response at the input unit pulse is described by the second-order differential equation (5.5). Dynamical properties are determined by pair of constant-times:

$$T_1 = \sqrt{LC}; \quad T_2 = RC$$

Solution of the equation may be exact (5.8) or approximated (5.11). The solution may occur convergent or divergent depending on the values T_1 and T_2. In one's turn the convergent solution may be or not be optimum as to a swing damping.

Here we investigate frequency responces only:

$$A(\omega) = \frac{1}{\sqrt{(1 - \omega^2 T_1^2)^2 + \omega^2 T_2^2}}$$

$$\varphi(\omega) = -arctg\,\frac{\omega T_2}{1 - \omega^2 T_1^2}$$

Example 8.7.

Develop the program that calculates and draws the MFR family of the resonant circuit with time-constant $T_1 = 20$ ms and $T_2 = 10, 14, 20$ ms.

Solution.

The program Prog_8_12 (listing 8.12) contains two sub-routines:

- "CoordSys" that draws the rectangle axes;

- "MFR" that calculates and draws the response family (fig.62).

```
Program Prog_8_12;
Uses GraphABC, ABCobjects;

Procedure CoordSys;
var j, k: integer;
var x, xc, yc: real;
var s: string;
begin
 SetWindowSize(770,500);  SetPenStyle(psClear);
 SetPenColor(clBlack);    SetPenWidth(1);
 Line(100,26,720,24);     Line(720,26,720,420);
 Line(100,26,100,420);    Line(100,420,720,420);
 for k:=1 to 20 do
  begin
   Line(100+30*k,420,100+30*k,412);
  end;
```

```
      for k:=2 to 11 do
        begin
          x:=2*(k-1); Str(x:2:0,s);
          var t:=new TextABC(32+60*k,428,12,s);
        end;
       for k:=1 to 12 do
        begin
          yc:=0.2*(k-1); Str(yc:2:1,s);
          var t:=new TextABC(72,446-34*k,12,s);
        end;
    end;

    Procedure MFR(tt1,tt2: real);
    const pi=3.14158; x0=100; y0=440;  mx=30;  my=190;
    var k, x1, x2, y1, y2: integer;
    var a, b, f, df, cf, am: real;
    begin
      f:=0; df:=0.1;
       for k:=1 to 200 do
        begin
          f:=f+df;    cf:=2*Pi*f;
          a:=cf*tt1;  b:=cf*tt2;
          am:=1/sqrt((1-a*a)*(1-a*a)+b*b);
          x1:=x0+Round(f*mx);
          y1:=y0-Round(am*my);
          SetPenWidth(3); SetPenColor(clBlack);
          if k>1 then
            begin
              Line(x1,y1,x2,y2);
            end;
           x2:=x1;  y2:=y1;
        end;
    end;
    //======Main========
    var tt1, tt2:real;
    var s1, s2, s3, s4: string;
    begin
      CoordSys;
      var t1:= new TextABC(490,125,12, 'Resonant link');
      tt1:=0.02;  Str(tt1:3:2,s1);
      var t2:= new TextABC(506,145,12,'T1 = '+s1);
      tt2:=0.01; Str(tt2:3:2,s2);
```

```
var t3:= new TextABC(230,50,12,'T2 = '+s2);
MFR(tt1,tt2);  tt2:=0.014;
Str(tt2:4:3,s3);
var t4:= new TextABC(286,128,12,'T2 = '+s3);
MFR(tt1,tt2);  tt2:=0.02;
Str(tt2:3:2,s4);
var t5:= new TextABC(270,200,12,'T2 = '+s4);
MFR(tt1,tt2);
end.
```

Listing 8.12

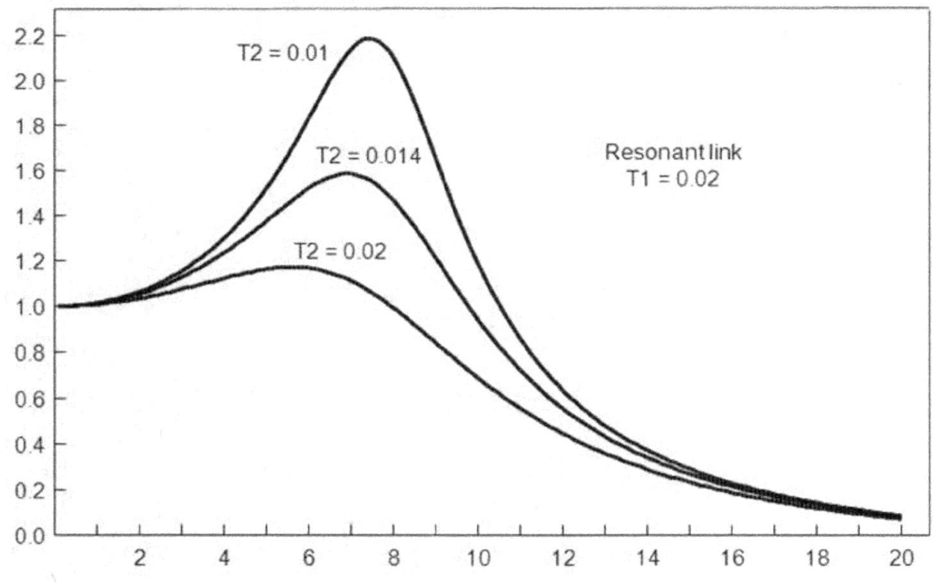

Figure 62. The MFR of the resonant link

Example 8.8.

Develop the program that calculates and draws the PFR family of the resonant circuit with the same time-constants.

Solution.

The program Prog_8_13 (listing 8.13) contains two sub-routines:

- "CoordSys" that draws the rectangle axes;

- "PFR" that calculates and draws the response family (fig.63).

```pascal
Program Prog_8_13;
Uses GraphABC, ABCobjects;

Procedure CoordSys;
var j, k: integer;
var x, xc, yc: real;
var s: string;
begin
  SetWindowSize(770,500);  SetPenStyle(psClear);
  SetPenColor(clBlack);        SetPenWidth(1);
  Line(100,26,720,24);         Line(720,26,720,430);
  Line(100,26,100,430);        Line(100,430,720,430);
  Line(100,230,720,230);
   for k:=1 to 20 do
     begin  Line(100+30*k, 230, 100+30*k, 223);
   end;
       for k:=1 to 11 do
           begin
             x:=2*(k-1); Str(x:3:0,s);
             if(x>1) and (x<8) then
                 begin
                   var t:=new TextABC(28+60*k, 204, 12, s);
                 end;
             if x>8 then
                 begin
                   var t:=new TextABC(28+60*k, 235, 12, s);
                 end;
         end;
    for k:=1 to 20 do
        begin
           Line(100, 30+20*k, 108, 30+20*k);
        end;
    for k:=2 to 11 do
        begin
           yc:=-110+20*(k-1); Str(yc:2:0,s);
           var t:=new TextABC(72,480-40*k,12,s);
        end;
end;

Procedure PFR(tt1,tt2: real);
const pi=3.14158; x0=100; y0=230; mx=30; my=2;
```

```pascal
var k, x1, x2, y1, y2, u1, u2, z: integer;
var a, b, f, df, cf, p, ph: real;

begin
  f:=0; df:=0.005;
  for k:=1 to 4000 do
    begin
      f:=f+df;    cf:=2*Pi*f;      a:=cf*tt1;  b:=cf*tt2;
      p:=-arctan(b/(1-a*a));   ph:=p*180/pi;
      x1:=x0+Round(f*mx);   y1:=y0-Round(ph*my);
      SetPenWidth(3);         SetPenColor(clBlack);
      if k>1 then
          begin
            Line(x1,y1,x2,y2);
          end;
          x2:=x1;  y2:=y1;
    end;
end;
//==================== Main ====================
var tt1, tt2:real;
var s1, s2, s3, s4: string;
begin
  CoordSys;
  var t1:= new TextABC(490,325,12, 'Resonant link');
  tt1:=0.02;  Str(tt1:3:2,s1);
  var t2:= new TextABC(506,345,12,'T1 = '+s1);
  tt2:=0.01; Str(tt2:3:2,s2);
  var t3:= new TextABC(396,170,12,'T2 = '+s2);
  PFR(tt1,tt2);
  tt2:=0.014; Str(tt2:4:3,s3);
  var t4:= new TextABC(570,128,12,'T2 = '+s3);
  PFR(tt1,tt2);
  tt2:=0.02;  Str(tt2:3:2,s4);
  var t5:= new TextABC(430,100,12,'T2 = '+s4);
  PFR(tt1,tt2);
  SetPenWidth(1); SetPenColor(clBlack);
  Line(490,160,570,140);
end.
```

Listing 8.13

Conclusions:

1. The stability of the resonant link is to be investigated so far as the solution of the differential equation may be both convergent and divergent depending on time-constant values T_1, T_2.

2. The convergent solution is optimum if $T_1 = T_2$. In this case the transition curve has the single aperiodic swing with magnitude 30% and the magnitude-frequency response has a rise near 20%.

3. The resonant frequency is worthy special attention so far as the phase sign of the output signal jumps here from -90^0 to 90^0.

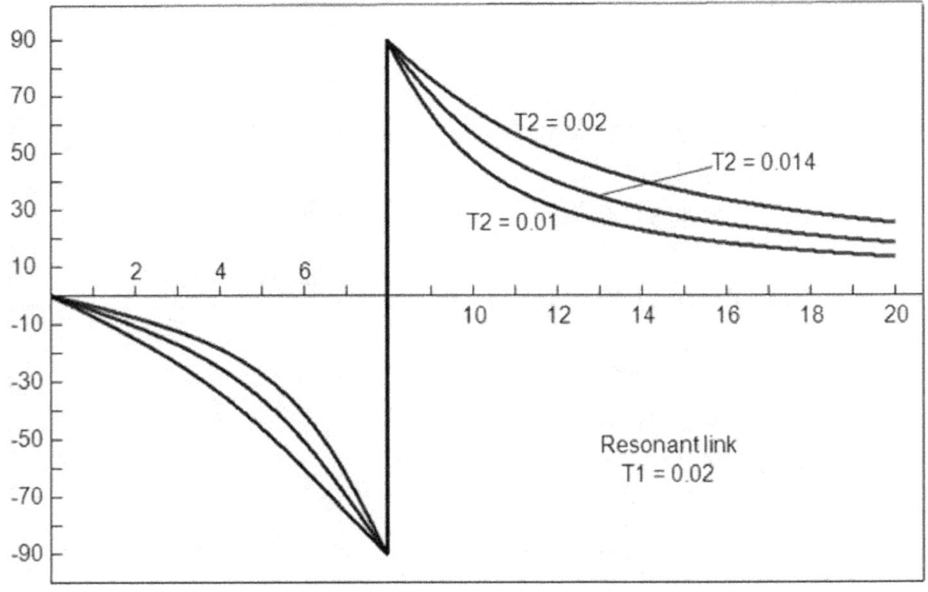

Figure 63. The PFR of the resonant link

8.9. An integrator and its applications

The integrating link is described by the equation:

$$x(t) = \frac{1}{T} \int_0^t u(t)dt \; ; \quad \text{where T} - \text{time-constant.}$$

Sometimes the inertial link is used for a rough simulation of an integrator. It is admissible during small time interval and if the inertialness has enough big value. Indeed, in this case the transition curve of the inertial link may be approximated by a straight line which is similar a solution of an integral equation. There is much more exact realization of an integrator in analog computer technology. Is is an operational amplifier with an input resistor R and a capacitor C in the feedback circuit (fig.64).

The operational amplifier is called differential since it has two inputs the voltage difference being gained. The gain coefficient is near 10^5 if the feedback is disconnected. The amplifier becomes a scaling link when a resistor is in the feedback circuit. In this case the transfer coefficient is

$$K = - R_2/R_1,$$

where R_1 – resistor in the input circuit;

               R_2 – resistor in the feedback circuit.

Usually $R_2 > R_1$.

The amplifier is called a sign invertor if $R_1 = R_2$.

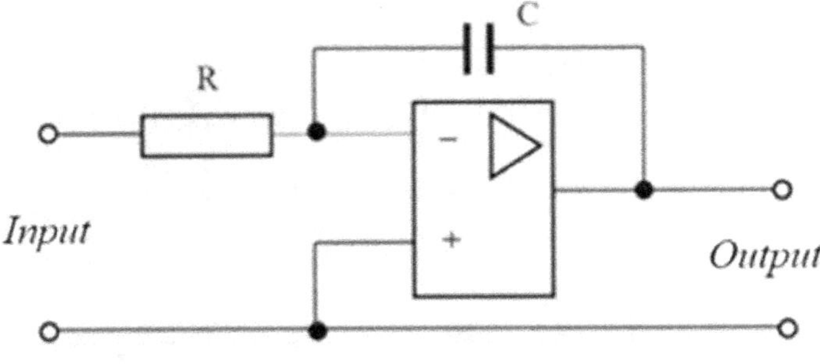

Figure 64. The analog integrator

Integrating link is out of a programmer interest since its transition function is a straight line being an integral of a constant value in the linear diapason near ±10V restricted by the power supply voltage. The MFR has a primitive form as well:

$$A(\omega) = \frac{1}{\omega T}, \quad T{=}RC$$

A phase shift is equal to -90^0 and does not depend on a frequency (in the low passband of course).

Hence, the integrator is not a problem for a programmer. Nevertheless, an integrator is a remarkable link due to its function in different applications. Let us see only two examples.

Example 8.9. An integrator in the automatic.

Perhaps, it would be interesting to know for uninitiated in the automatic control theory that the automatic regulation has two versions – statical and astatical. The regulators in old steam-engines were typical statical so far as if only small error was necessary to form a control command. Since the 20th century practically all automates became astatical with the null control error owing to application of the integrating links. Long ago before invention of an electronic integrator, the piston in the cylinder realized the integrator functioin – it moves with a constant linear velocity depending on the work gas volume and pressure. Then electrical motor was used as an integrator – its roller rotates with a constant angular velocity depending on the power supply voltage.

Example 8.10. An integrator in the radioelectronic.

For the purpose to demonstrate possibilities of the integrating link (and the language Pascal ABC too) we'll develop the program which simulates the pulse amplitude selector. This device is to find a short-time pulse signal in the mixture with the "white" noise. The selector automatically regulates the amplitude threshold due to the integrator in so manner that the noise splashes can overcome the threshold with a very low probability. Such device is called a selector with a low level of false alarm [4].

The selector (fig.65) contains two comparators K_1 and K_2, a pulse generator G, RS-trigger, a volatage divider M and the integrator.

Let us consider the selection algorithm. There is a mixture of the useful pulse signal with the white noise at the device input. The threshold voltage at the input of the comparator K_2 cannot be overcame by many of the noise splashes. Such voltage is regulated by the integrator depending on the variable noise level at the input of the comparator K_1. The trigger state depends on two pulse flows at its inputs S and R. If the state "1" prevalences in time, the output integrator voltage rises as well as the threshold voltage does. Otherwise the threshold voltage decreases if the state "0" prevalences. The variable threshold provides the maintenance of the false alarm frequence at the given level.

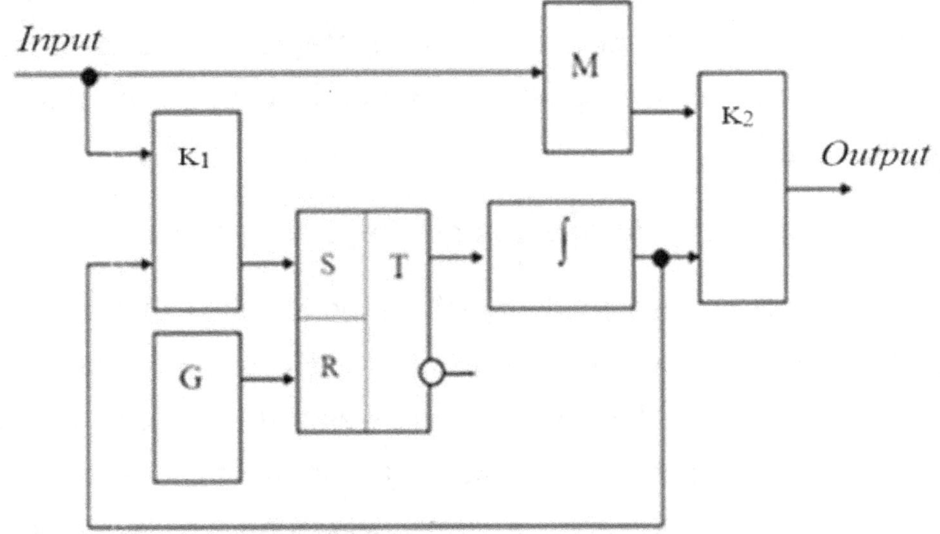

Figure 65. The amplitude selector

The program Prog_8_10 (listing 8.14 and fig.66) is model-ling this algorithm.

```
Program Prog_8_14;
Uses GraphABC, ABCobjects;

procedure CoordSys;
var j, k: integer;
var d: string;
begin
  SetWindowSize(860, 550); SetPenStyle(psClear);
  SetPenColor(clBlack);      SetPenWidth(2);
  Line(70, 140, 830, 140);   Line(70, 180, 830, 180);
  Line(70, 220, 830, 220);   Line(70, 300, 830, 300);
  Line(70, 340, 830, 340);   Line(70, 450, 830, 450);
  var t1 := new TextABC(80, 16, 12, 'T h r e s h o l d');
  var t2 := new TextABC(10, 70, 12, 'Mixture');
  var t3 := new TextABC(10, 90, 12, 'signal');
  var t4 := new TextABC(10, 110, 12, '+noise');
  var t5 := new TextABC(26, 170, 12, 'Set');
  var t6:=new TextABC(8,210,12,'Reset');
  var t7:=new TextABC(4,250,12,'Trigger');
  var t8:=new TextABC(4,290,12,'Signal');
```

```pascal
    var t9:=new TextABC(4,330,12,'Trouble');
    var t10:=new TextABC(4,420,12,'Integral');
 end;

Procedure Noise;
  const x0=60; y0=140; mx=15; my=10; mz=6;
  var j, k, x1, x2, y1, y2, z1, z2, s, r, z, sz: integer;
  var trab, tr1, tr0, trig, trsh: integer;
  var x, t, dt, y, ns, ds, lev, dlev, tresh: real;
  begin
    t:=0; dt:=0.125; sz:=12;
    SetPenColor(clBlack);
    SetPenStyle(psSolid);
    lev:= 9;
    for k:=1 to 400 do
       begin
         t:=t+dt;
         if k=77 then ns:=12 else ns:=10*Random;
         x1:=x0+Round(t*mx);
         y1:=y0-Round(ns*my);
         if k=77 then SetPenWidth(3)
                 else SetPenWidth(1);
         if k>1 then Line(x1,y0,x1,y1);     // Noise
         if (k mod 8)=0 then
           begin
             tr0:=1;   SetPenWidth(3);
             Line(x1,220,x1,205); trig:=0;
             SetPenWidth(2);
             Circle(x1,260,2);   sz:=sz-1;
           end;
         dlev:=sz*0.00003;    lev:=lev+dlev;
         trsh:=y0-Round(lev*my);
         SetPenColor(clGray);
         Circle(x1,trsh-8,1);           // Treshold
         SetPenColor(clBlack);
         if ns>lev then
           begin
             tr1:=1;   SetPenWidth(3);
             Line(x1,180,x1,165);  // Set-Comparator_1
             trig:=1;
             SetPenWidth(2);
             Circle(x1,245,2);       // Trigger_1
```

```pascal
          sz:=sz+1;
       end;
     If k=77 then  Line(x1,300,x1,285);  // Signal
     if (ns>lev+0.85) and (k>77) then  // Comparator_2
       begin
          trab:=1; Line(x1,340,x1,325);      // Trouble
       end;
     z1:=Round(sz*mz);
     SetPenWidth(4);
     if k>1 then Line(x1,450-z1,x2,450-z2);
     x2:=x1; z2:=z1;
    end;
   end;
//========Main===
begin
   CoordSys; Noise;
end.
```

Listing 8.14

The subroutine "Noise" simulates the noise using the library function Random. The noise splashes are unipolar. There is the single signal pulse. The voltage thresold is more then the most part of the noise splashes.

Line "Set" shows the output pulses of the comparator K_1. Line "Reset" contains the generator pulses. Line "Trigger" demonstrates the direct output of the trigger.

There is the single false trouble in the fig.66. In practice several noise splashes can overcome the threshold and it is admissable because the amplitude selection is only the first stage of the noise suppression.

Further other signal properties are used to delete the false alarms, for instance knowing the exact signal period and using time selection method.

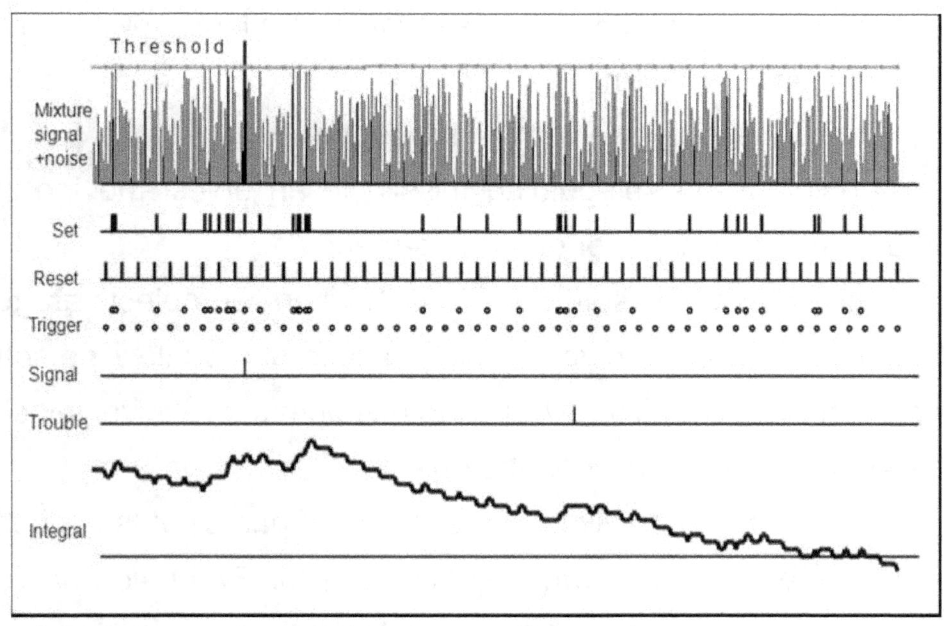

Figure 66. The diagram of the pulse signal selection

Chapter 9. SOLUTION OF APPROACH PROBLEM

9.1. Statement of the problem

To draw together two moving objects is a dynamical task too. It is solved in real time and embraces a wide range of the vehicle control problems. There are several examples:

- space vehicle docking at near-earth orbit;
- aircraft refueling in air;
- ship manoeuvreing near sea-wall or in a lock;
- making up a train;
- tower assemblage using a helicopter;
- guiding and self-guiding missile to an object moving in the air, on sea surface, or on earth surface.

The approach is soft (non-conflict) when two objects are to be connected but it is rigid (conflict) in case of a battle. The soft approach is reached by reduction to null both the relative velocity and the distance between the objects.

In case of the conflict approach one of participants aspires to avoid meeting. The game situation arises. Every side has a chance to win depending on several conditions including the start positions, the real technical parameters, and the play skill. Both participants are destroyed in result of the conflict meeting.

The most complex approach problem is studied in this chapter, namely the dynamical modelling of the aim and the guided missile trajectories. Several model simplifications are admitted in the model. Both participants are considered as mathematical points but some material restrictions are applied to their trajectories, and before all being engendered by the inertialness of both objects.

Firstly, their trajectories are smooth functions with admissable curvature radius owing to the finite mass and due to the big movement velocity. Secondly, the missile manoeuvrability is restricted by the possibility of the rudder empennage as well as by the body solidity.

On these reasons the trajectory curvature radius, demanded by the approach algorithm, cannot be less a value which is permissable by the missile inertial properties. The centripetal acceleration cannot be more then the value w = 9.81*n, where n is an admissable overload. The value "n" is a measure of the control quality of the transversal velocity vector projection. So, not every cinematic trajectory may be realized by the missile planer.

Remark.

The approch control is being made in two space planes. We'll model the trajectories in the vertical plane only where a straight line connects two mathematical points at every time moment. This line is called a sight line despite of being set by not optical devices only. Depending on the environment, a sight line may be built with lazers, photo-cameras, TV-cameras, radars, sonars.

9.2. Direct pursuit

The direct pursuit can be met it the wild nature at the top of a food chain. Carnivore beast of prey pursuits its sacrifice ingeneously being sure of own superiority in speed and endurance exclusively.

Remark.

We do not consider here group hunt methods which are used by wolves and dolphins. We take in account the situations "one-to-one" only.

Let us estimate the technical practicability of the direct pursuit method. We'll suppose that the active object A has technical means for self-guiding to passive (or escaping) object B. These means allow to build the sight line continuously. It is supposed also there are not any information at the object A concerning possible manoeuvres of the object B. In other words the object A has no possibility to control own velocity vector to forestall the aim trajectory. The control system goal is to provide the coincidence the velocity vector A with the sight line at every moment of approaching (fig.67). The object A is to move from the point "a" to the point "c" in a small interval of time.

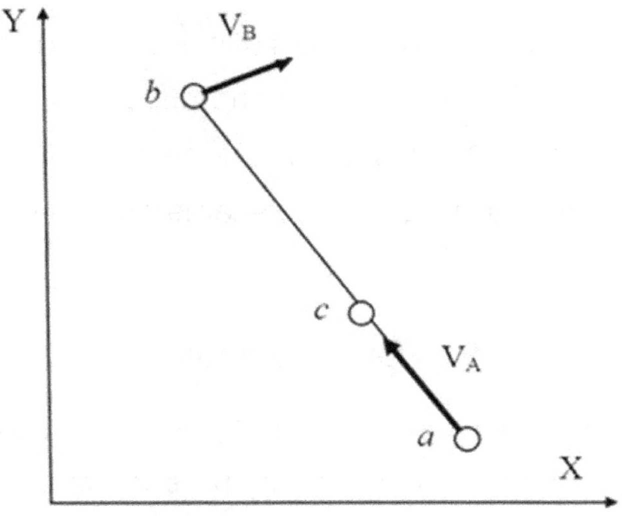

Figure 67. The scheme of the direct pursuit

Preliminary remarks.

1. The short-distance battle is modelled at horizontal distance less than 4 km and the aim's height less than 2 km. The aim B starts to move on the axes OY at height 1500…1800 m. The object A has start point on the axes OX being chosen by the user in the interval 0 … 4000 m. It is supposed that device sensibility on the object A is enough to build the sight line during all approach time, so the object A maintains the continious visual contact with the object B.

2. In all examples below the measurement units are used: meters for distances, seconds for current time and meter per second for linear velocity.

Example 9.1.

Build the approach trajectories of objects A and B (fig.68) under the next conditions:

- start coordinates: xb = 0; yb = 1800; ya = 0; xa is choosed by the user;

- velocity vectors: vb = 100; direction is parallel to axes OX; va = 130.

It is necessary to display the data:

- position of both objects and the sight line in every 0.5 sec.;

- the current distance between objects beginning from 400 m;

- the current overload of the object A.

There is the mathematical description of the direct pursuit trajectory using denotes from fig.69:

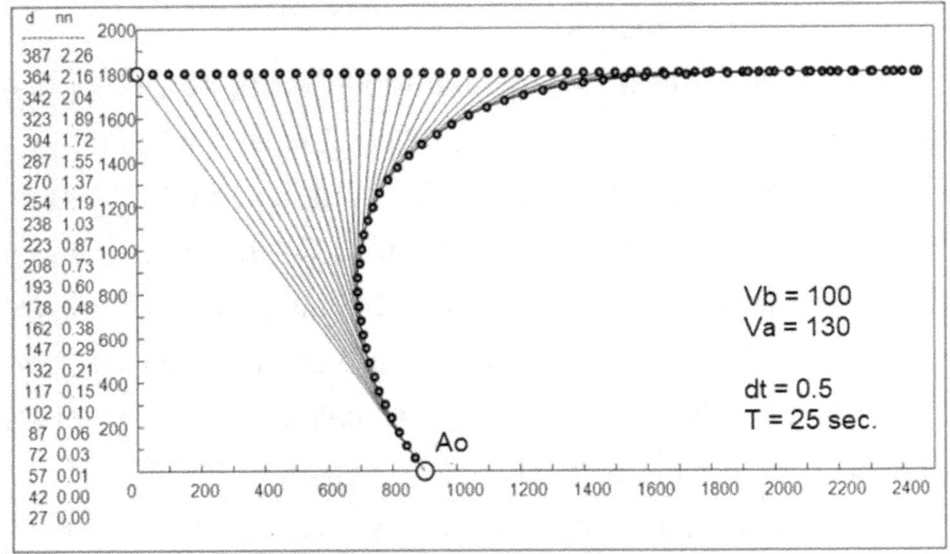

Figure 68. The traejctory of approach
with a non-manoeuvreing aim

1. The ojects move during every interval of time dt at distances:

$$sa = va*dt,$$

$$sb = vb*dt.$$

2. If the sight line is inclined at left from the vertical line, cyclical calculations are being fulfiled by the next formulae:

$$e(k) = \arctan([yb(k\text{-}1) / (xa(k\text{-}1) - (xb(k\text{-}1))];$$

$$\Delta x = sa*\cos[e(k)]; \Delta y = sa*\sin[e(k)];$$

$$xa(k) = xa(k\text{-}1) - \Delta x; ya(k) = ya(k\text{-}1) + \Delta y$$

3. If the sight line position takes position where the next conditions

$$xb(k\text{-}2) < xa(k\text{-}2) \text{ and } xb(k\text{-}1) > xa(k\text{-}1)$$

are true, then the coordinates of the object A are computed by formulae:

$$xa(k) = xa(k-1) + sb/5.5;$$

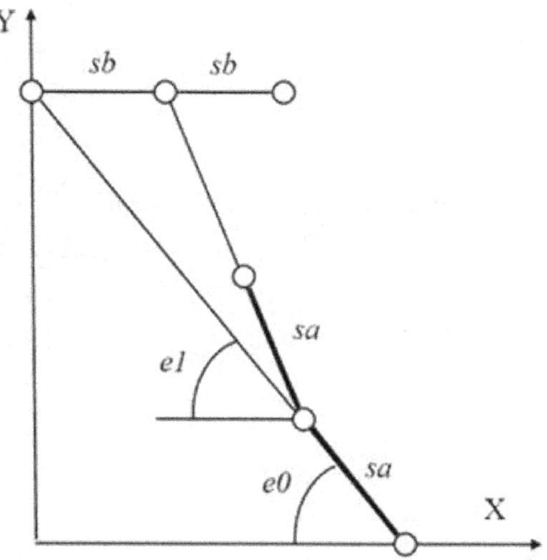

Figure 69. The scheme for computing
of the pursuit trajectory

$$ya(k) = ya(k-1) + sa.$$

4. If the sight line is inclined at right from the vertical line, cyclical calculations are being fulfiled by the formulae:

$$e(k-1) = arctan([yb(k-1) - (ya(k-1) / (xb(k-1) - (xa(k-1))];$$

$$\Delta x = sa*cos[e(k)];$$

$$\Delta y = sa*sin[e(k)];$$

$$xa(k) = xa(k-1) + \Delta x;$$

$$ya(k) = ya(k-1) + \Delta y.$$

The solution is made by the program Prog_9_1 (listing 9.1) containing two subroutines:

- "CoordSys" to draw the cartesian axis;

- "Pursuit" to receive initial data, calculate trajectories and overload, to draw the graphic, that is shown at fig.68.

```pascal
Program Prog_9_1;
Uses ABCobjects, GraphABC;

Procedure CoordSys;
var k, x, y: integer;
var d: string;
begin
  SetWindowSize(900,520);   SetPenStyle(psClear);
  SetPenWidth(1);           SetPenColor(clBlack);
  Line(120,20,870,20);      Line(120,20,120,420);
  Line(120,420,870,420);    Line(870,20,870,420);
  for k:=1 to 25 do
    begin
      Line(120+30*k, 420,120+30*k,414);
    end;
  for k:=1 to 13 do
    begin
      x:=200*(k-1); d:=IntToStr(x);
      var t:=new TextABC(50+60*k,428,12,d);
    end;
  for k:=1 to 20 do
    begin
      Line(120,20*k,126,20*k);
    end;
  for k:=1 to 10 do
    begin
      y:=200+200*(k-1); d:=IntToStr(y);
      var t:=new TextABC(84,412-40*k,12,d);
    end;
end;

Procedure Pursuit(vb,va: real; xs: integer);
const pi=3.14158; x0=120; y0=420;
mx=0.3; my=0.2; g=9.81;
var j, k, n, M, x, y, u, v: integer;
var t, dt, dx, dy, xx, dd, de, be, ro, nn: real;
var dxb, dyb, ph, w, a, b,v, g, sb, sa: real;
var xb, yb, xa, ya, e, d: array[0..220] of real;
var tot, p, f, z, miss: real;
```

```pascal
var tb, ta, tp, tt, nw, dm, rd: string;
Rectangle(10,480,850,510);
writeln('  d   nn');        writeln('  ------------');
  dt:=0.5;    sb:=vb*dt;      sa:=va*dt;
  var t01:=new TextABC(200,460,12, 'Direct pursuit');
  Str(vb,tb);  Str(va,ta);
  var t02:=new TextABC(400,460,12,  'Vb = '+tb+
                          '   Va = '+ta);

  xb[0]:=0; yb[0]:=1800;       //--- Start B
  xa[0]:=xs; ya[0]:=0;         //--- Start A
  x:=x0+Round(xb[0]*mx);   y:=y0-Round(yb[0]*my);
  SetPenStyle(psSolid);        SetPenWidth(2);
  Circle(x,y,6);               //--- Start B
  u:=x0+Round(xa[0]*mx);   v:=y0-Round(ya[0]*my);
  SetPenStyle(psSolid);        Circle(u,v,8);    //--- Start A
  u:=x0+Round(xa[0]*mx);   v:=y0-Round(ya[0]*my);
  var t2:=new TextABC(u+10,400,12,'Ao');
  SetPenColor(clBlack);        SetPenWidth(1);   Line(x,y,u,v);
  Sleep(100);                   k:=1;
  xb[1]:=xb[0]+sb;             yb[1]:=yb[0];
  x:=x0+Round(xb[1]*mx);   y:=y0-Round(yb[1]*my);
  SetPenColor(clBlack);        SetPenWidth(3);
  SetPenStyle(psSolid);        Circle(x,y,3);
  e[1]:=arctan(yb[0]/xs);    dx:=sa*cos(e[1]);
  dy:=sa*sin(e[1]);           xa[1]:=xs-dx;
  ya[1]:=dy;
  u:=x0+Round(xa[1]*mx);   v:=y0-Round(ya[1]*my);
  SetPenStyle(psSolid);        SetPenColor(clBlack);
  Circle(u,v,3);               SetPenColor(clBlack);
  SetPenWidth(1);             Line(x,y,u,v);
  tot:=dt;
    //------------------------------ Pursuit cycle
    k:=1;    d[k]:=yb[k];
    while d[k]>20 do
      begin
        k:=k+1;
        //------------Trajectory B
        xb[k]:=xb[k-1]+sb;        yb[k]:=yb[k-1];
        x:=x0+Round(xb[k]*mx); y:=y0-Round(yb[k]*my);
        SetPenColor(clBlack);    SetPenWidth(3);
        SetPenStyle(psSolid);    Circle(x,y,3);
```

```pascal
//--------------Trajectory A, --------------------Condition 1
    if xb[k-1]<xa[k-1] then
    begin
        e[k]:=arctan(yb[k-1]/(xa[k-1]-xb[k-1]));
        dx:=sa*cos(e[k]);              dy:=sa*sin(e[k]);
        xa[k]:=xa[k-1]-dx;             ya[k]:=ya[k-1]+dy;
        u:=x0+Round(xa[k]*mx);  v:=y0-Round(ya[k]*my);
        SetPenStyle(psSolid);      SetPenColor(clBlack);
        Circle(u,v,3);                  SetPenWidth(1);
        SetPenColor(clBlack);      Line(x,y,u,v);
    end;
//------------------------------------Condition 2
  if (xb[k-2]<xa[k-2]) and (xb[k-1]>xa[k-1]) then
  begin
      xa[k]:=xa[k-1]+sb/5.5;      ya[k]:=ya[k-1]+sa;
      u:=x0+Round(xa[k]*mx);  v:=y0-Round(ya[k]*my);
      SetPenStyle(psSolid);      SetPenColor(clBlack);
      Circle(u,v,3);                  SetPenColor(clBlack);
      SetPenWidth(1);              Line(x,y,u,v);
  end;
//------------------------------------Condition 3
  if (xb[k-2]>xa[k-2]) and (xb[k-1]>xa[k-1]) then
  begin
   e[k]:=arctan((yb[k-1]-ya[k-1])/(xb[k-1]-xa[k-1]));
   dx:=sa*cos(e[k]);              dy:=sa*sin(e[k]);
   xa[k]:=xa[k-1]+dx;             ya[k]:=ya[k-1]+dy;
   u:=x0+Round(xa[k]*mx);   v:=y0-Round(ya[k]*my);
   SetPenStyle(psSolid);      SetPenColor(clBlack);
   Circle(u,v,3);                  SetPenColor(clBlack);
   SetPenWidth(1);              Line(x,y,u,v);
  end;
  tot:=tot+dt*k;  dx:=xb[k]-xa[k];  dy:=yb[k]-ya[k];
  d[k]:=sqrt(dx*dx+dy*dy);       //---left d[k] m
  if d[k]<20 then break;
//---------------------------------------------------- Miss
  if (xs > 2400) and (ya[k]>yb[k]-sa-230)
                and (d[k]>150) then
    begin
        miss:=d[k]; Str(miss:3:0,dm);
        var tm:=new TextABC(750,300,12, 'Miss '+dm+' m');
        break;
```

```pascal
    end;
  if k>120 then break;
//-------------------------------------------------Overload
    if (d[k]<400) and (xb[k-1]>xa[k-1]) then
     begin
       e[k-1]:=arctan((yb[k-1]-ya[k-1])/(xb[k-1]-xa[k-1]));
       e[k]:=arctan((yb[k]-ya[k])/(xb[k]-xa[k]));
       de:=e[k-1]-e[k];
       be:=(pi-de)*0.5;
       ro:=sa/(2*cos(be));
       nn:=va*va/(g*ro);
       writeln('   ',d[k]:3:0,' ',nn:3:2);
       if nn>10 then
       begin
         Str(nn:3:1,nw);
         var tw1:=new TextABC(500,340,12,
                  'Acceleration = '+nw+' g');
         var tw2:=new TextABC(500,358,12,
                  'Rudder is ummovable');
         Str(d[k]:3:0,rd);
         var tw3:=new TextABC(500,376,12,
                  'Left-over distance '+rd+' м');
         break;
        end;
      end;
   Sleep(300);
end;
  Str(dt:2:1,tt);  tot:=(1+k)*dt;   Str(tot:3:1,tp);
  var tap:=new TextABC(600,460,12,
          'dt = '+tt+'    T  = '+tp+' sec.');
end;
//=====================Main===================
var vb,va: real;  var xs: integer;
begin
  CoordSys;  vb:=100; va:=130;
  var t1:=new TextABC(155,490,11, 'Input the start point');
  var t2:=new TextABC(290,490,11, 'of the object  "A"
          at the axes  OX  - number  from  100  to  4800 m');
  readln(xs);   Pursuit(vb,va,xs);
end.
```

Listing 9.1

The cinematic trajectory quality is estimated by the maximum control overload. The overload is usualy less than 10g if a missile is intended for a neighbouring battle, for instance, the "Stinger" (USA). Missiles of this type are small: the calibre is less than 100 mm, the length is less than 2 m. The value n = 10g is used as the technical practicability criteria of an approach trajectory in all next examples. We'll consider that the rudder becames unmovable further if n > 10g. It means the control loss. In this case the trajectory of the object A is out of modelling, and a miss is registered.

Now let us see how the current curvature radius and overload are to be computed. We'll use denotes shown in fig.70.

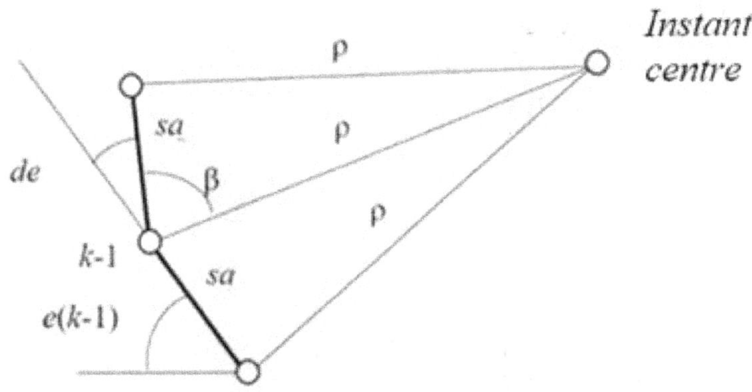

Figure 70. Curvature radius of a trajectory

$$tg[e(k-1)] = \frac{ya(k-1)-ya(k-2)}{xa(k-2)-xa(k-1)};$$

$$tg[e(k)] = \frac{ya(k)-ya(k-1)}{xa(k-1)-x(k)};$$

$$\Delta e = e(k) - e(k-1);$$

$$\beta = \pi - de ; \qquad p = \frac{sa}{2\cos(\beta)};$$

The instant control overload is $\quad n = \dfrac{va^2}{g*\rho};$

There is the example of an inadmissible overload (fig.71).

One ought not to forget that object B is also ready to fulfil own task. He is sure to be attacked. We can model his possible tactics. Most likely he would make the object A to do some mistake, forcing inadmissible actions. All fighter pilots are taught to fulfil an anti-missile manouvre. The simplest of them is "a hill" with lowering near the earth aim. The count is made on the eccessive curvature of the missile trajectory and an inadmissible overload to be the reason of the control loss.

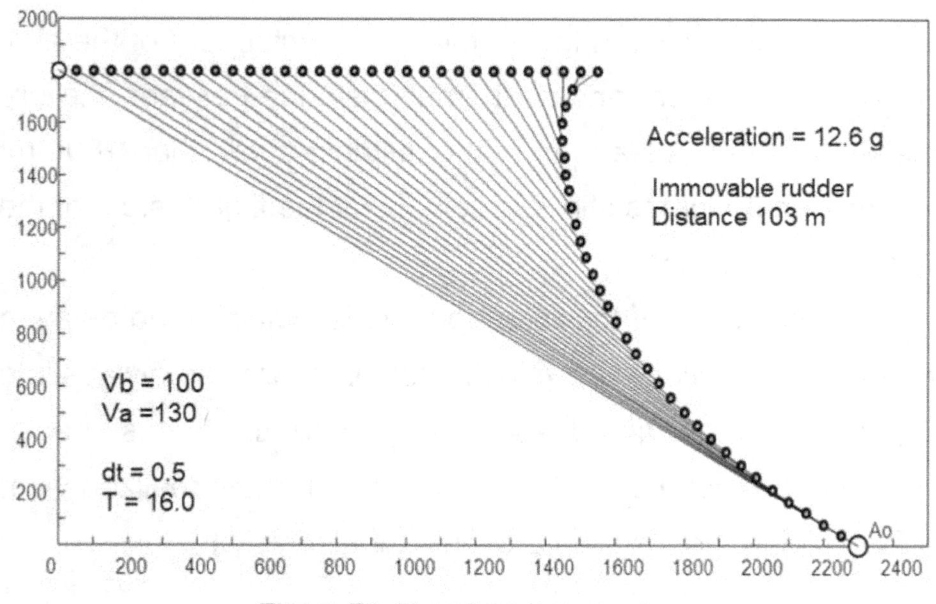

Figure 71. How the miss occures

The program is developed to simulate a manouvre of the object B. It is distinct from the previous program by the single fragment concerning the object B only. The program run result is shown at fig.72. The pilot skill gives an evident advantage.

The object A is needed to restrict essentially the moment and place of the start.

Remark.

The message "Immovable rudder" (fig.72) means that the possibi-lity to enlarge the transversal acceleration is exausted. Further the movement of the object A is out of control.

Conclusions.

1. The direct pursuit method provides the approach with admissible overload if the attack is made into the back semi-sphere of the object B as a rule. It is disadvantageous tactic since the object B has the time to invade into the protected terri-tory.

2. If an attack is made into the front semisphere of the aim, the overload problem arises at small distance between the ob-jects when object A starts from the distance more than 1800 m. The problem becomes still less solvable if the flight height of the object B is being lower.

The drawback of the direct pursuit is owing to the delay in the percecutor actions. Its velocity vector is directed always into the point where the aim does not stay more, the aim is already has gone. In other words, the trajectory control is realized with-out forestalling, hence the percecutor is always late.

9.3. Proportional approach

The "rational" percecutor directs his velocity vector strictly along the sight line only when the avoiding object is not manou-vreing. In another cases he tries to predict the manouvre of the object B.

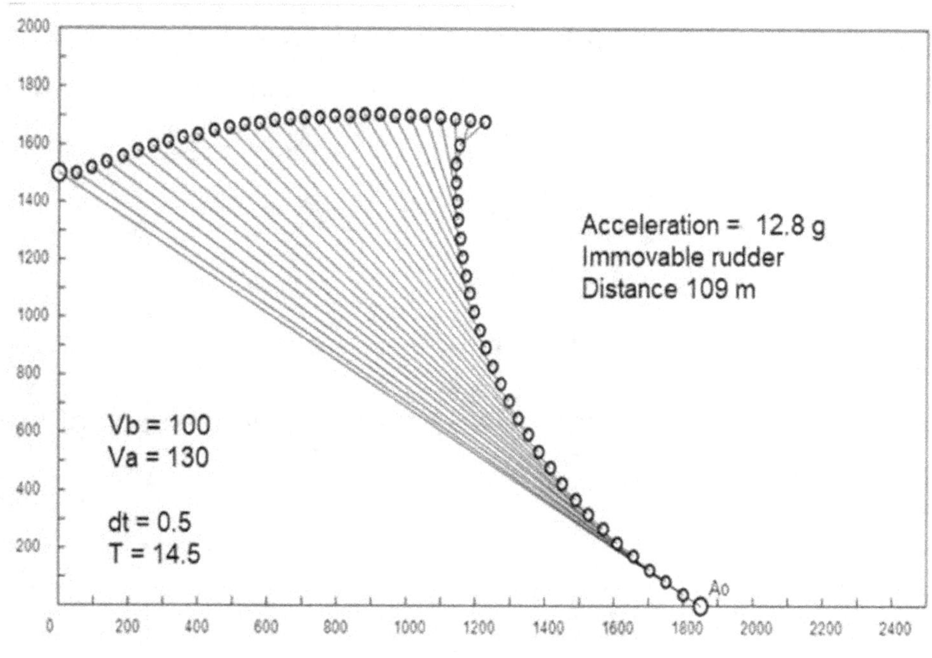

Figure 72. Unfeasible approach

The percecutor is to decline the vector from the sight line depending on the behaviour of the aim. There is a real technical possibility to forestall the aim manouvre. The giroscope is a device capable to measure the angular rotating speed of the sight line.

Remark.

There are two types of giroscope – positional and corrected. The first one is used to build the board support coordinate system which is immovable relatively the stars. The corrected giroscope is the very device which is building the sight line continuosly despite of all space evolutions of the missile body on the trajectory.

The angular rotating speed of the sight line is calculated by the angle position increment Δe of this line in a discrete time interval. This information is enough to determine the sign and value of the missile velocity vector increment. In accordance with a linear extrapolation theory (see the chapter 7 and [4]), the rotation speed increment $\Delta\Theta$ of the missile vector is to be larged proportionally to Δe:

$$\frac{d\Theta}{dt} = C\frac{de}{dt}$$

where C – a navigational constant.

The angle forestalling: $q(t) = \Theta(t) - e(t)$. Joint solution of the equations gives the expression:

$$\frac{de}{dt} = \frac{1}{C-1} * \frac{dq}{dt}$$

Expression $C \to \infty$ is the condition of an ideal forestalling. The approach trajectory is alike to the trajectory of the direct pursuit if $C \to 1$.

The first step in the model of "proportional" approach is to be done without forestalling, i.e. along the sight line because the previous iformation is absent (fig.73). Further the angle out-stripping may be calculated.

There are formulae of the approach algorithm:

1. If xb(k-1) < xa(k-1), then

$$e(k-2) = arctg\frac{yb(k-2)}{xa(k-2)};$$

$$e(k-1) = arctg\frac{yb(k-1)}{xa(k-1)-xb(k-1)};$$

$\Delta e = e(k-1) - e(k-2); \quad \Delta v = C*\Delta e;$

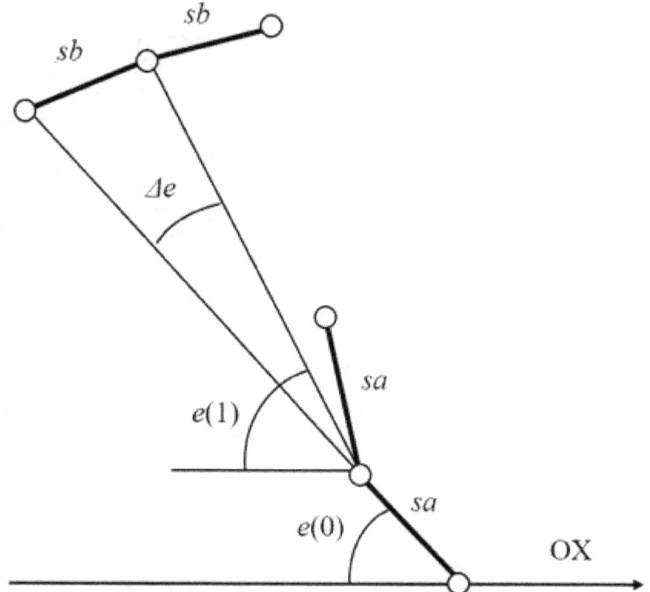

Figure 73. The scheme of the proportional approach

$av(k) = e(k-2) + \Delta v;$

$\Delta xa = sa*cos[av(k)];$

$xa(k) = xa(k-1) - \Delta xa;$

$\Delta ya = sa*sin[av(k)];$

$ya(k) = ya(k-1) + \Delta ya.$

2. If xb(k-2) < xa(k-2) и xb(k-1)>xa(k-1), then

$\Delta e = \pi / 2 - e(k-1);$

$\Delta v = C*\Delta e;$

$av(k) = e(k-1) + \Delta v;$

$xa(k) = xa(k-1);$

$ya(k) = ya(k-1).$

3. Если xb(k-2) > xa(k-2) и xb(k-1)>xa(k-1), то

$$e(k-2) = arctg\frac{yb(k-2)-ya(k-2)}{xb(k-2)-xa(k-2)};$$

$$e(k-1) = arctg\frac{yb(k-1)-ya(k-1)}{xb(k-1)-xa(k-1)};$$

Δe = e(k-1) − e(k-2);

Δv = C*Δe; av(k) = e(k-2) + Δv;

xa = sa*cos[av(k)];

xa(k) = xa(k-1) + Δxa;

Δya = sa*sin[av(k)];

ya(k) = ya(k-1) + Δya.

These formulae are used in the program Prog_9_2 (listing 9.2).

The approach with a non-manouvreing object is shown in fig.74.

```
Program Prog_9_2;
Uses ABCobjects, GraphABC;

Procedure CoordSys;
var k, x, y: integer;
var d: string;
begin
  SetWindowSize(900,520);  SetPenStyle(psClear);
  SetPenWidth(1);          SetPenColor(clBlack);
  Line(120,20,870,20);      Line(120,20,120,420);
  Line(120,420,870,420);    Line(870,20,870,420);
  for k:=1 to 25 do
   begin
    Line(120+30*k,420,120+30*k,414);
   end;
```

```pascal
  for k:=1 to 13 do
   begin
     x:=200*(k-1); d:=IntToStr(x);
     var t:=new TextABC(50+60*k,428,12,d);
   end;
  for k:=1 to 20 do
   begin
     Line(120,20*k,126,20*k);
   end;
  for k:=1 to 10 do
    begin
      y:=200+200*(k-1); d:=IntToStr(y);
      var t:=new TextABC(84,412-40*k,12,d);
    end;
end;

Procedure Approach(vb,va: real; xs: integer);
const pi=3.14158; x0=120; y0=420;
mx=0.3; my=0.2; g=9.81;
var j, k, x, y, u, v: integer;
var t, dt, dx, dy, xx, dd, dxa, dya, ph, f, a, b, sb, sa: real;
var xa, xb, ya, yb, d, e, av: array[0..920] of real;
var de,dv, p, pf, tot, w, z, dxb, dyb, be, nn, ro: real;
var tb, ta, tapp, ddt, nw, rd: string;
begin
  Rectangle(10,480,850,510);
  writeln('   d       nn');  writeln('  -------------');
  dt:=0.5; sb:=vb*dt;  sa:=va*dt;
  var t01:=new TextABC(250,450,12,
          'Proportional approach');
  Str(vb,tb);   Str(va,ta);

  var t02:=new TextABC(500,450,12,
  'Vb = '+tb+'   Va = '+ta);
  xb[0]:=0;    yb[0]:=1800;    //--- Start B
  xa[0]:=xs; ya[0]:=0;         //--- Start A
  x:=x0+Round(xb[0]*mx);    y:=y0-Round(yb[0]*my);
  SetPenStyle(psSolid);        SetPenWidth(2);
  SetPenColor(clBlack);        Circle(x,y,6);        //--- Start B
  var t1:=new TextABC(130,36,12,' ');
  u:=x0+Round(xa[0]*mx);   v:=y0-Round(ya[0]*my);
  var t2:=new TextABC(u+8,400,12,'Ao');
```

```pascal
Circle(u,v,6);                          //--- Start A
Sleep(100);              SetPenColor(clBlack);
SetPenWidth(1);          Line(x,y,u,v);       //---Sight line 0
dx:=xa[0]-xb[0];         dy:=yb[0]-ya[0];
d[0]:=sqrt(dx*dx+dy*dy);                 //--Start distance
k:=1;
//-------------------Trajectory B-----------------
xb[k]:=xb[k-1]+sb;          yb[k]:=yb[k-1];
x:=x0+Round(xb[k]*mx);      y:=y0-Round(yb[k]*my);
SetPenColor(clBlack);       SetPenWidth(2);
SetPenStyle(psSolid);       Circle(x,y,4);
av[0]:=arctan(yb[0]/xa[0]); xa:=sa*cos(av[0]);
dya:=sa*sin(av[0]);         xa[1]:=xa[0]-dxa;
ya[1]:=ya[0]+dya;           u:=x0+Round(xa[1]*mx);
v:=y0-Round(ya[1]*my);      SetPenStyle(psSolid);
SetPenColor(clBlack);       Circle(u,v,4);
x:=x0+Round(xb[1]*mx);      y:=y0-Round(yb[1]*my);
SetPenWidth(1);             Line(x,y,u,v);
dx:=xa[1]-xb[1];            dy:=yb[1]-ya[1];
d[1]:=sqrt(dx*dx+dy*dy);    tot:=dt;
//============ Approach cycle==========
k:=1; z:=1.5; f:=-0.5;
xb[1]:=sb;  yb[1]:=yb[0]; j:=0;
p:=7/(1+0.0075*xs);              //--- Coefficient p
While  d[k]>30 do
  begin
   k:=k+1;
//---------------------Trajectory B-------------------
   xb[k]:=xb[k-1]+sb;          yb[k]:=yb[k-1];
   x:=x0+Round(xb[k]*mx);   y:=y0-Round(yb[k]*my);
   SetPenColor(clBlack);       SetPenWidth(2);
   SetPenStyle(psSolid);       Circle(x,y,4);
//----------------------------Trajectory A----------------------
   if xb[k-1]<xa[k-1] then    //--start of condition. 1
     begin
       e[k-2]:=arctan(yb[k-2]/xa[k-2]);
       e[k-1]:=arctan(yb[k-1]/(xa[k-1]-xb[k-1]));
       de:=e[k-1]-e[k-2];          dv:=de*p;
       av[k]:=e[k-2]+dv;           dxa:=sa*cos(av[k]);
       dya:=sa*sin(av[k]);         xa[k]:=xa[k-1]-dxa;
       ya[k]:=ya[k-1]+dya;         u:=x0+Round(xa[k]*mx);
```

```pascal
      v:=y0-Round(ya[k]*my); SetPenStyle(psSolid);
      SetPenColor(clBlack);    Circle(u,v,4);
      SetPenColor(clBlack);    SetPenWidth(1);
      Line(x,y,u,v);
    end;
  //----------------------------------------start of condition 2
  if (xb[k-2]<xa[k-2]) and (xb[k-1]>xa[k-1]) then
    begin
      de:=pi/2-e[k-1]; dv:=de*p; av[k]:=e[k-1]+dv;
      xa[k]:=xa[k-1];              ya[k]:=ya[k-1];
      u:=x0+Round(xa[k]*mx);
      v:=y0-Round(ya[k]*my);
      SetPenStyle(psSolid);    SetPenColor(clBlack);
      Circle(u,v,4);              SetPenColor(clBlack);
      SetPenWidth(1);            Line(x,y,u,v);
    end;
  //----------------------------------------start of condition 3
    if (xb[k-2]>xa[k-2]) and (xb[k-1]>xa[k-1]) then
    begin
      e[k-2]:=arctan((yb[k-2]-ya[k-2])/(xb[k-2]-xa[k-2]));
      e[k-1]:=arctan((yb[k-1]-ya[k-1])/(xb[k-1]-xa[k-1]));
      de:=e[k-1]-e[k-2];        dv:=de*p;
      av[k]:=e[k-2]+dv;         dxa:=sa*cos(av[k]);
      dya:=sa*sin(av[k]);       xa[k]:=xa[k-1]+dxa;
      ya[k]:=ya[k-1]+dya;       u:=x0+Round(xa[k]*mx);
      v:=y0-Round(ya[k]*my); SetPenStyle(psSolid);
      SetPenColor(clBlack);    Circle(u,v,4);
      SetPenColor(clBlack);    SetPenWidth(1);
      Line(x,y,u,v);
    end;
  dx:=xa[k-1]-xb[k-1];  dy:=yb[k-1]-ya[k-1];
d[k]:=sqrt(dx*dx+dy*dy);
    if d[k]<130 then break;
    if (xa[k]>3000) and (ya[k-1]>yb[k-1]) then
    begin
    var mis1:=new TextABC(680,250,12,'The miss!');
      break;
    end;
  if (y<20) or (x>970) then
    begin
```

```pascal
  var mis2:=new TextABC(600,50,12,
              'Object A has exausted its resourse');
    break;
  end;
//----------------------Rudder overload----------------------
  if (d[k]<600) and (xb[k-1]>xa[k-1]) then
    begin
    e[k-1]:=arctan((yb[k-1]-ya[k-1])/(xb[k-1]-xa[k-1]));
    e[k]:=arctan((yb[k]-ya[k])/(xb[k]-xa[k]));
    de:=e[k-1]-e[k];          be:=(pi-de)*0.5;
    ro:=sa/(2*cos(be));    nn:=va*va/(g*ro);
    writeln('   ',d[k]:3:0,' ',nn:3:2);
    if nn>10 then
    begin
      Str(nn:3:1,nw);
      var tw1:=new TextABC(720,190,12,
      'Acceleration = '+nw+' g');
      var tw2:=new TextABC(720,208,12,
      'Unmovable rudder');
      Str(d[k]:3:0,rd);
      var tw3:=new TextABC(720,226,12,
      'Left '+rd+' m'); break;
    end;
  end;
  Sleep(30);
 end;                      //--------end of cycle----
  tot:=tot+dt*k;  Str(dt,ddt);     Str(tot,tapp);
  var tap:=new TextABC(680,450,11,
  'dt = '+ddt+'  T = '+tapp+' sec.');
end;
//===============Main=============
begin
  var vb, va: real;
  var xs: integer;
  CoordSys;
  vb:=100; va:=130;
  var t1:=new TextABC(115,490,12, 'Input the start point');
  var t2:=new TextABC(290,490,12, 'of the object "A"
        on the axes  OX  - number  from  100  to  2400 m');
  readln(xs);
```

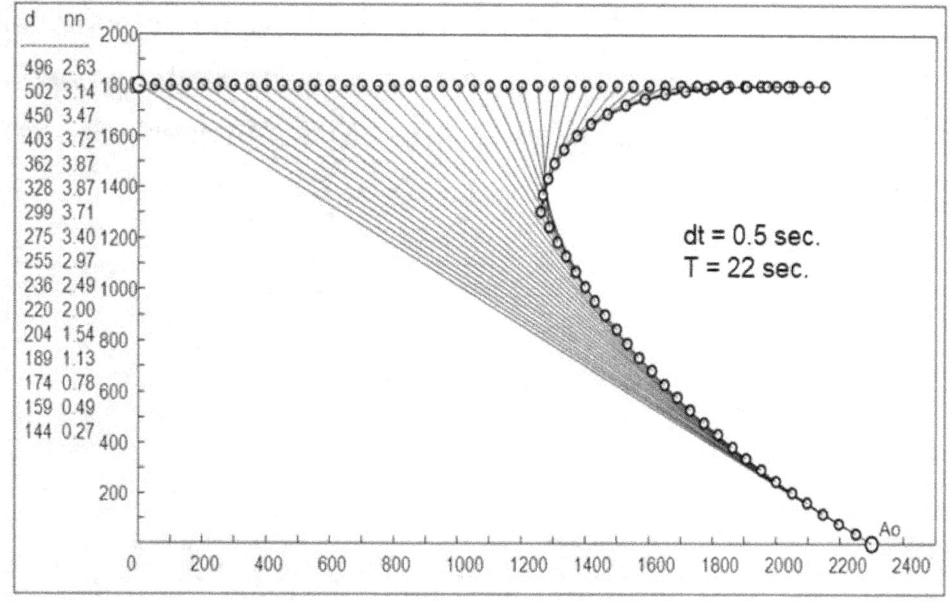

Figure 74. The proportional approach trajectory

```
if xs<50 then
   begin
      Rectangle(10,480,850,510);
      var t3:=new TextABC(15,490,12, 'Input number > 50');
   end;
 if xs>40 then Approach(vb,va,xs);
end.
```

Listing 9.2

Conclusion.

The proportional approach has been realized successfully with the same conditions where the direct approach occurred uneffective because of an admissable overload (see fig.71 and fig.72). The forestalling guidance allowed to reduce the maximum overload to value n = 3.87 in the trajectory point where the left distance was 329 m.

The proportional approach is also effective despite of the aim is making the anti-missile manoeuvre "a hill" (fig.75). The overload value occurred n = 5.21 at the very distorted trajectory segment, i.e. twice less than the admittable value. Proprtional approach is the basic guiding method realised in the weapon class "Man-Portable Air Defence" ("Red Eye", "Stinger", "Mistrale", "SATCP" and etc.).

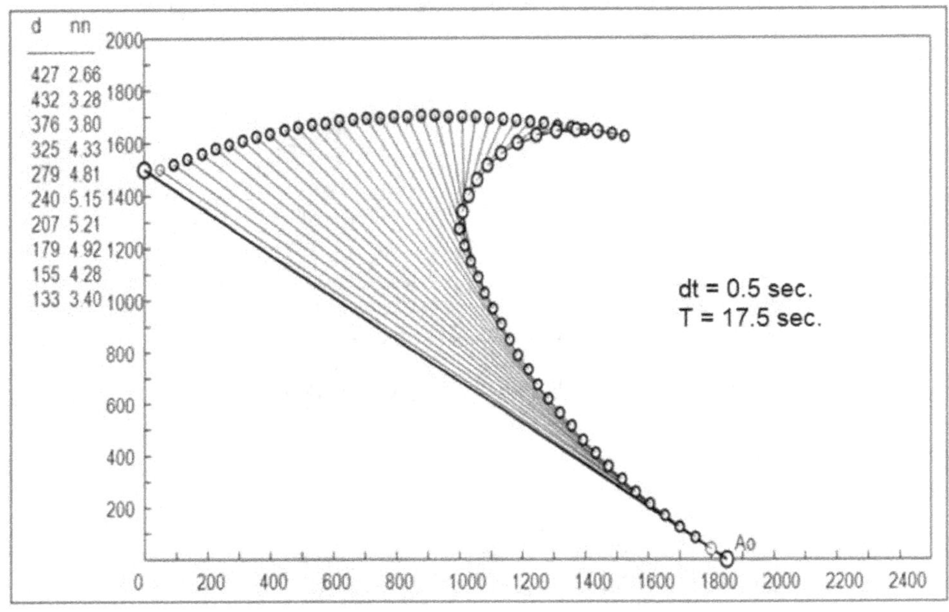

Figure 75. The proportional approach with the manoeuvreing aim

9.4. Beam approach

Two mathematical points take part in the models of the direct pursuit and the proportional approach. Such aproach methods are called "two-point" ones.

The meeting problem is solved more effective in many aspects if the third participant occures. It is to observe the aim movement and to provide he missile guiding.

Remark.

The tactical advantages and shortages of guide-missile systems are not considered in this monograph. Alike problems are out of the book item. The mathematical and algorithmical analysis of the approach problems one may find here. So, we'll consider the conflict approach of two objects A and B, the third object C being involved on the side of the object A. It is supposed the object C has the information about the trajectory of the aim and has also technical means to correct object A trajectory. In other words, the object C is a command post.

The sight line joins the points B and C. This line can be build with the optical devices or the radar placed in the point C. These apparatuses measure the angular deflection the object A from the sight line. The command post computer forms the control signals which are transmitted to the object A. The signals are used by the object A to minimize the deflections and to hold the point A on the sight line continuously. Thus, three points are always on the sight line (fig.76).

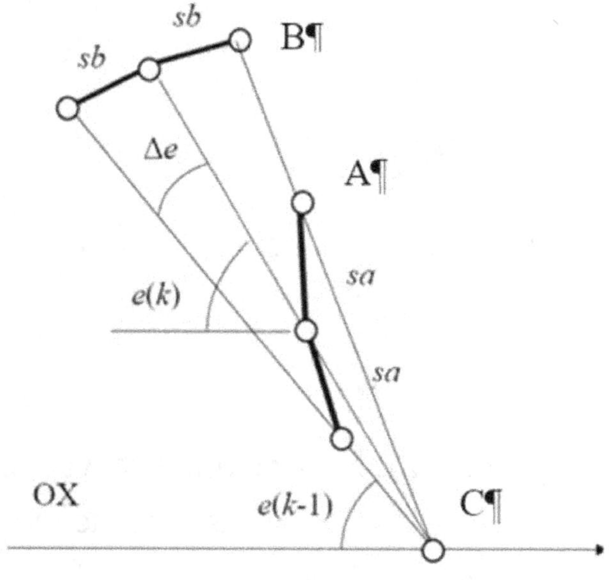

Figure 76. Scheme of three-point approach

The sight line is build with the narrow beam of a device in optical or radio range of wave length. In this reason the three-point approach is called a beam method also. The space position of the sight line is described by the expresson:

$$e(k) = arctg \frac{yb(k)}{xc - xb(k)}.$$

There are formulae for calculating of the object A coordinates:

1. If xb(k) < xc, then:

$\Delta e = e(k) - e(k\text{-}1)$;

$z1 = s(k\text{-}1)^*\cos(\Delta e)$;

$q = s(k\text{-}1)^*\sin(\Delta e)$;

$z2 = \sqrt{sa^2 - q^2}$;

$s(k) = z1 + z2$;

$xa(k) = xc - s(k)^*\cos[e(k)]$;

$ya(k) = s(k)^*\sin[e(k)]$,

$$db = \sqrt{yb^2 + (xc - xb)^2};$$

$$da = \sqrt{ya^2 + (xc - xb)^2};$$

$$d(k) = db - da.$$

where da, db q, s, z1, z2 – intermediate variables;

d – distance left to the aim.

2. if k > 1 and e(k) < 0 then

$e(k) = - e(k)$; $\Delta e = e(k\text{-}1) - e(k)$;

$z1 = s(k\text{-}1)^*\cos(\Delta e)$; $q = s(k\text{-}1)^*\sin(\Delta e)$;

$$z2 = \sqrt{sa^2 - q^2}$$

s(k) = z1 + z2;mm

xa(k) = xc − s(k)*cos[e(k)];

ya(k) = s(k)*sin[e(k)];

$$da = \sqrt{ya^2 + (xa - xc)^2};$$

$$db = \sqrt{yb^2 + (xb - xc)^2};$$

d(k) = db - da

There are the formulae for the calculating of the current curvature of the trajectory, hence for the control overload. At first it is necessary to know the incline of the object A velocity vector:

$$av(k - 1) = arcg\,\frac{ya(k-1)-ya(k-2)}{xa(k-2)-xa(k-1)};$$

$$av(k) = arctg\,\frac{ya(k)-ya(k-1)}{xa(k-1)-xa(k)};$$

1. If 1 < k <28 и xc < 1400, then

If av(k) < 0.5π, then Δe = av(k) − av(k-1);

If Δe < 0, then b Δe = π + Δe;

$$\beta = 0.5(\pi - \Delta e);$$

$$\rho = \frac{sa}{2cos\beta};$$

$$n = \frac{va^2}{9.81 * \rho};$$

The approach is considered over when d(k) < 10.

The object A starts from the point C. To get comparable results, the point C is set at the distance 2200 m from support point of the Cartesian coordinates (fig.77).

2. If k > 11 and xc ≥ 1400, then

$$\Delta e = av(k) - e(k\text{-}1);$$

If $\Delta e < 0$, then $\Delta e = \pi + \Delta e$;

$$\beta = 0.5(\pi - \Delta e);$$

$$\rho = \frac{sa}{2cos\beta};$$

$$n = \frac{va^2}{9.81 * \rho};$$

The model is realized by the program Prog_9_3 (listing 9.3).

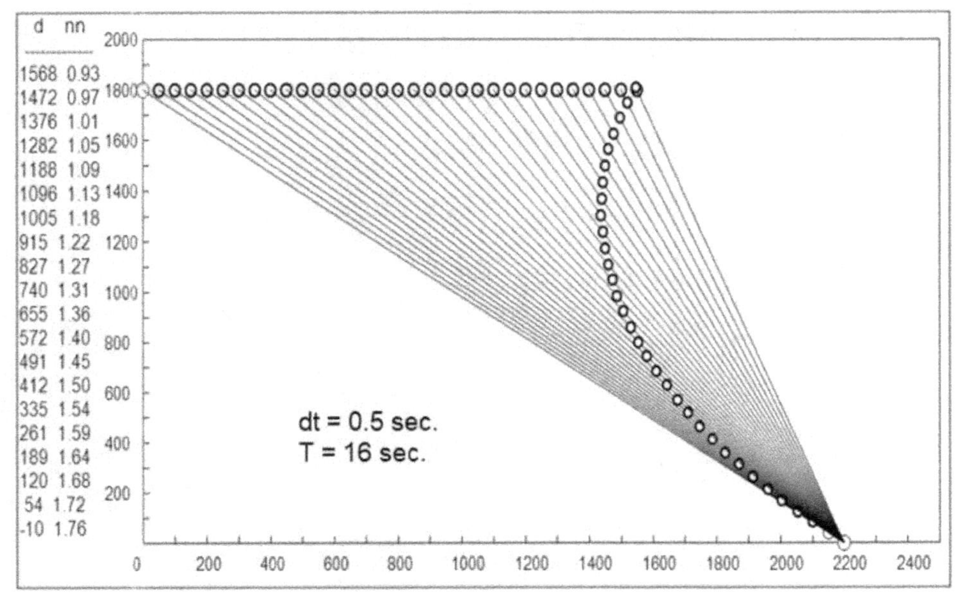

Figure 77. The beam approach method. Variant 1.

```pascal
Program Prog_9_3;
Uses ABCobjects, GraphABC;

Procedure CoordSys;
var k, x, y: integer;
var d: string;
begin
  SetWindowSize(900,520);  SetPenStyle(psClear);
  SetPenWidth(1);              SetPenColor(clBlack);
  Line(120,20,870,20);      Line(120,20,120,420);
  Line(120,420,870,420);  Line(870,20,870,420);
  for k:=1 to 25 do
   begin
    Line(120+30*k,420,120+30*k,414);
   end;
  for k:=1 to 13 do
   begin
    x:=200*(k-1);
    d:=IntToStr(x);
    var t:=new TextABC(50+60*k,428,11,d);
   end;
  for k:=1 to 20 do
   begin
    Line(120,20*k,126,20*k);
   end;
  for k:=1 to 10 do
   begin
    y:=200+200*(k-1); d:=IntToStr(y);
    var t:=new TextABC(84,412-40*k,11,d);
   end;
end;

Procedure Approach(vb, va, xc: real);
const pi=3.14158;   x0=120;   y0=420;
mx=0.3;   my=0.2;   g=9.81;
var j, k, n, M, x, y, u, v, uc, vc: integer;
var t, dt, dx, dy, dd, yc, to, dxa, dya, ph, w, a, b, sb,sa: real;
var xb, yb, xa, ya, av, e, s, d: array[0..200] of real;
var de, dav, z1, z2, q, bm, be, da, db, ro, nn, xmin, ymin: real;
var tb, ta, tapp, ddt, nw, rd: string;
begin
  Rectangle(10,480,850,510);
```

```pascal
writeln(' d   nn');  writeln(' -------------');
dt:=0.5;  sb:=vb*dt;   sa:=va*dt;
//---------------------------------------------------
xb[0]:=0; yb[0]:=1800; //--- Start B
yc:=0;                      //--- Start A
Sleep(30);
x:=x0+Round(xb[0]*mx);  y:=y0-Round(yb[0]*my);
SetPenStyle(psSolid);    SetPenWidth(1);
SetPenColor(clBlack);
Circle(x,y,6);                //--- Start B
u:=x0+Round(xc*mx);     v:=y0;
Circle(u,v,6);                //--- Point C
SetPenColor(clBlack);     Line(x,y,u,v);
//------------------------------- Step k = 0
e[0]:=arctan(yb[0]/xc);     xa[0]:=xc-sa*cos(e[0]);
ya[0]:=sa*sin(e[0]);        u:=x0+Round(xa[0]*mx);
v:=y0-Round(ya[0]*my);   SetPenStyle(psSolid);
SetPenWidth(1);            SetPenColor(clBlack);
Circle(u,v,5);                dx:=xc; dy:=yb[0];   s[0]:=sa;
d[k]:=sqrt(dx*dx+dy*dy)-sa;
Sleep(300);   k:=0; d[k]:=yb[0];
while d[k] > 0 do
 begin
   k:=k+1;
//-------------------Trajectory B-------------------
   xb[k]:=xb[k-1]+sb;          yb[k]:=yb[k-1];
   x:=x0+Round(xb[k]*mx);   y:=y0-Round(yb[k]*my);
   SetPenColor(clBlack);      SetPenWidth(2);
   SetPenStyle(psSolid);      Circle(x,y,5);
   uc:=x0+Round(xc*mx);     vc:=y0;
   SetPenColor(clBlack);      SetPenWidth(1);
   Line(x,y,uc,vc);
//-----------------------Trajectory A------------------------
   e[k]:=arctan(yb[k]/(xc-xb[k]));
//----------------------------------------------- Condition 1
   if xb[k]<=xc then
     begin
       de:=e[k]-e[k-1];          z1:=s[k-1]*cos(de);
       q:=s[k-1]*sin(de);        z2:=sqrt(sa*sa-q*q);
       s[k]:=z1+z2;              xa[k]:=xc-s[k]*cos(e[k]);
       ya[k]:=s[k]*sin(e[k]);
```

```pascal
    da:=sqrt(ya[k]*ya[k]+(xc-xa[k])*(xc-xa[k]));
    db:=sqrt(yb[k]*yb[k]+(xc-xb[k])*(xc-xb[k]));
    d[k]:=db-da;               u:=x0+Round(xa[k]*mx);
    v:=y0-Round(ya[k]*my); SetPenStyle(psSolid);
    SetPenWidth(2);            SetPenColor(clBlack);
    Circle(u,v,4);             tot:=tot+dt;
  end;
//----------------------------------------------------- Condition 2
  if (k>1) and (e[k]<0) then
   begin
    e[k]:=-e[k];               de:=e[k-1]-e[k];
    z1:=s[k-1]*cos(de);        q:=s[k-1]*sin(de);
    z2:=sqrt(sa*sa-q*q);       s[k]:=z1+z2;
    xa[k]:=xc+s[k]*cos(e[k]);  ya[k]:=s[k]*sin(e[k]);
    da:=sqrt(ya[k]*ya[k]+(xa[k]-xc)*(xa[k]-xc));
    db:=sqrt(yb[k]*yb[k]+(xb[k]-xc)*(xb[k]-xc));
    d[k]:=db-da;
    u:=x0+Round(xa[k]*mx);  v:=y0-Round(ya[k]*my);
    SetPenStyle(psSolid);      SetPenColor(clRed);
    Circle(u,v,5);             tot:=tot+dt;
  end;
//-------------------------------------------------Rudder overload
  if (k>1) and (k<28) and (xc<1400) then
    begin
    av[k-1]:=arctan((ya[k-1]-ya[k-2])/(xa[k-2]-xa[k-1]));
    av[k]:=arctan((ya[k]-ya[k-1])/(xa[k-1]-xa[k]));
    if av[k]<pi/2 then
      begin
        de:=av[k]-av[k-1];
        if de<0 then de:=pi+de;
        be:=(pi-de)/2;         ro:=sa/(2*cos(be));
        nn:=va*va/(g*ro);  writeln(' ',d[k]:3:0,'  ',nn:3:2);
      end;
    end;
   if (k>11) and (xc>=1400) then
     begin
       av[k-1]:=arctan((ya[k-1]-ya[k-2])/(xa[k-2]-xa[k-1]));
       av[k]:=arctan((ya[k]-ya[k-1])/(xa[k-1]-xa[k]));
       if av[k]<pi/2 then
       begin
         de:=av[k]-av[k-1];
```

```pascal
            if de<0 then de:=pi+de;
            be:=(pi-de)/2;  ro:=sa/(2*cos(be));
            nn:=va*va/(g*ro);
            writeln(' ',d[k]:3:0,' ',nn:3:2);
          end;
        end;
    if d[k]<50 then break;
    if k>200 then break;
    Sleep(300);
  end;
  tot:=tot+dt; Str(tot,tt); Str(dt,tdt);
  var t5:=new TextABC(630,460,12,
        'dt = '+tdt+'    T = '+tt+' sec.');
end;
//=====================Main=================
begin
  var vb, va, xc: real;
  var ta, tb: string;
  CoordSys;
  var t1:=new TextABC(190,460,12,
        'Beam approach method');
  vb:=100; va:=130;  Str(vb,tb);  Str(va,ta);
  var t2:=new TextABC(390,460,12, Vb = '+tb+ ',
        Va = '+ta+' m / sec., ');
  var t3:=new TextABC(15,490,12,
        'Input the position  of the command post "C" ');
  var t4:=new TextABC(440,490,12,
        'on the axes OX   number from  10  to  2400 m');
  readln(xc);
  Approach(vb,va,xc);
end.
```

Listing 9.3

The beam method solves the approach problem successively despite of the anti-missile manouevre (fig.78). The maximum overload is not more 1.35g. It is the excellent result!

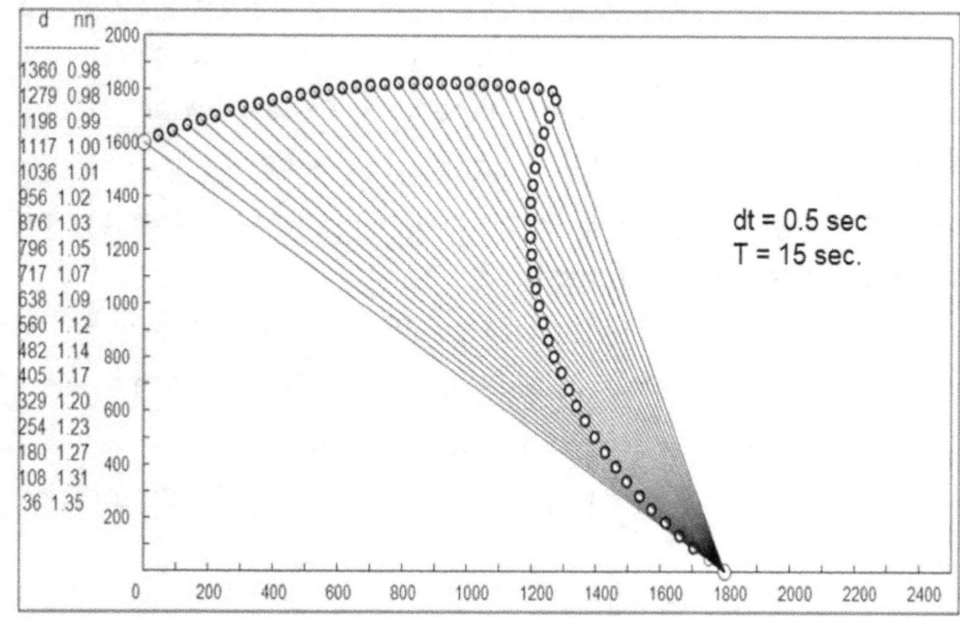

Figure 78. The beam approach method. Variant 2

Conclusion.

The three-point guiding is realized with the cinematic over-load which is 2-3 times less than in the case of proportional self-guiding.

This advantage is achieved by means of the complex expensive equipment. Once more shortage is vulnarability of the command post. It must be defended.

Nevertheless, three-point method is realised in anti-tank and air defence guided weapon of NATO ("Milan", "Tow", "Javelin", "Blowpipe", "RBS-70" and etc.).

Chapter 10. ECONOMIC MODELS

10.1. Market equilibrium

It is known from economical manuals that a commodity price accumulates cost of production, fiscal extra charge, transport expences, overhead expenses. At last, when a commodity occures at a market, demand and supply complete the price forming. So, a business conflict is over peacefully between a seller and a purchaser.

Interests of both sides are contrary. The seller wants to rise the price, the purchaser aspires to buy cheaply. A market is considered free if the next conditions are satisfied:

- the market is not regulated by authority directly or indirectly;

- the criminal pressure is absent;

- there is no price collusion between monopolistic suppliers.

If it so, the commodity price depends on the demand and the suggestion only. Sellers stimulate the demand by lowering the price if the commodity is in plenty. In so manner they can force sale and avoid the loss of perishable commodities. If the commodity is in shortage, sellers do not miss the opportunity to rise the price. When above mentioned mechanism works, a dynamical balance of demand and supply takes place in some time and the price is stabilized at a definite level.

We'll consider the model of the market equilibrium using the next variables:

q   – commodity quantity at the market;

pd   – demand price;

ps   – suggest price.

World practice promts the tight relation between quantity and price but this relation is sooner qualitative. It may be set by means of statistical observation. The exact mathematical dependences do not exist. They are replaced by different approximations and trend lines.

Let us suppose the approximate functions are found:

$$pd(q) = 32 - 0.27q; \qquad\qquad (10.1)$$

$$ps(q) = 1.1 + 0.9*exp(0.032q). \qquad (10.2)$$

Alike equations are usually used in economical manuals for building web-diagrams which illustrate the price-forming process. The reciprocal task occures also – how to find the argumetnt value if the function is known:

$$q = (32 - pd) / 0.27; \qquad\qquad (10.3)$$

$$q = 31.25*ln[(ps - 1.1) / 0.9]. \qquad (10.4)$$

The program Prog_10_1 (listing 10.1) calculates using the formulae (10.1) ... (10.4) and draws the graphic (fig.79).

The rectangle spiral is a market price trajectory. It starts in the arbitrary point q1 near the zero. The verical line is drawn up to the function "Demand price". The horizontal line is built at the level price = 25 right to the curve "Suggest price". The cross point has coordinate q2. New vertical line is drawn down to the function "Demand price". Then the horizontal line goes left to the curve "Suggest price" in the point q3. So, the first cycle of the spiral is over.

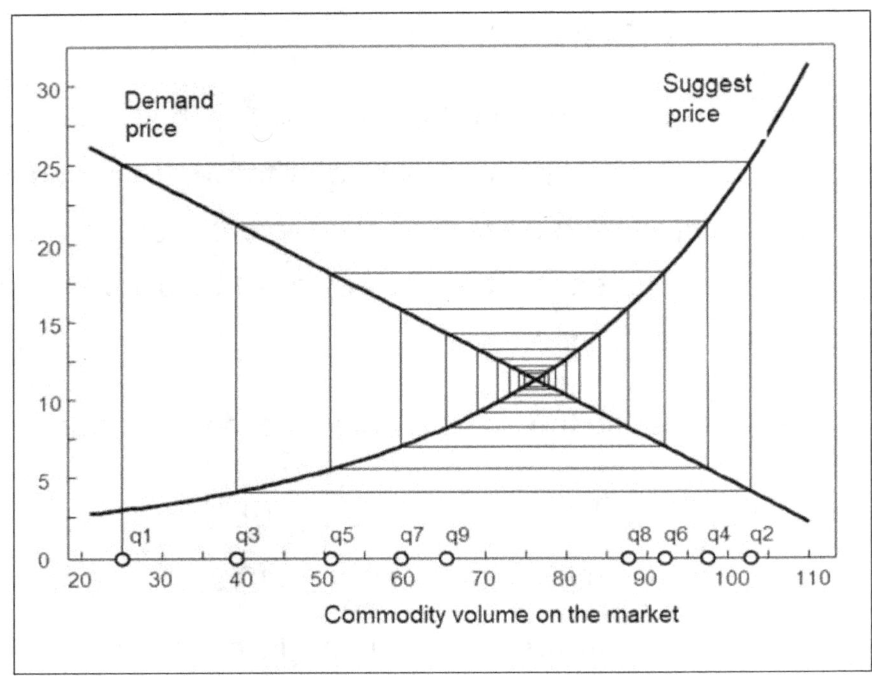

Figure 79. The market equilibrium model. Variant 1

New cycle begins in the point q3. Cyclic calculations are continued till one may distinguish horizontal lines of the spiral on the diagram. Modelling may be stopped at the 8th cycle in our example. Process is proved to be quick convergent. The convergence depends on the form of functions (10.1) … (10.4).

```
Program Prog_10_1;
Uses ABCobjects, GraphABC;

Procedure CoordSys;
var  k, x, y: integer;
var  d: string;
begin
  SetWindowSize(700, 520);  SetPenStyle(psClear);
  SetPenWidth(1);           SetPenColor(clBlack);
  Line(100, 22, 670, 22);   Line(100, 22, 100, 440);
  Line(100, 440, 670, 440); Line(670, 22, 670, 440);
  for k := 1 to 19 do
  begin
    Line(80 + 30 * k, 440, 80 + 30 * k, 434);
  end;
```

```pascal
    for k := 2 to 11 do
      begin
        x := 10 * k;     d := IntToStr(x);
        var t := new TextABC(60 * k - 20, 448, 12, d);
      end;
    for k := 1 to 13 do
      begin
        Line(100, 438 - 32 * k, 106, 438 - 32 * k);
      end;
    for k := 1 to 7 do
     begin
        y := 5 * (k - 1);
        d := IntToStr(y);
        var t := new TextABC(78, 496 - 64 * k, 12, d);
     end;
end;


Procedure Web;
const  x0=-10; y0 = 440; mx = 6.0;  my = 12.8;
var  j, k, n, st, xs, x1, x2, y1, y2, z1, z2: integer;
var  u, x, y, z, q, pd, ps : array[1..50] of integer;
var d, s: real;
var sq: array[1..50] of string;
begin
   SetPenWidth(3); SetPenColor(clBlack);  n:=20;
    for k := 1 to 90 do
      begin
        n:=n+1;  d := 32 - 0.27 * n;
        s := 1.1 + 0.9 * exp(0.032 * n);
        x1 := x0 + Round(n * mx);
        y1 := y0 - Round(d * my);
        z1 := y0 - Round(s * my);
        if k > 1 then
          begin
            Line(x1, y1, x2, y2); Line(x1, z1, x2, z2);
          end;
        x2 := x1;y2 := y1;z2 := z1;
      end;
SetPenWidth(2);           SetBrushColor(clWhite);
SetPenStyle(psSolid);     Circle(x1,y0,5);
var t1:=new TextABC(x1+6,412,12,'q1');
```

```pascal
    Circle(x2, y0, 5);
    var t2:=new TextABC(x2,412,12,'q2');
    y1:=y0-Round(ps[2]*my);   SetPenWidth(1);
    Line(x1,y1,x2,y1);            pd[2]:=32-0.27*q[2];
    x1:=x0+Round(q[2]*mx);   y1:=y0-Round(pd[2]*my);
    y2:=y0-Round(ps[2]*my);   Line(x1,y1,x1,y2);
    for j:= 3 to 20 do
      begin
        SetPenWidth(1); ps[j]:=pd[j-1];
        q[j]:=31.25*ln((ps[j]-1.1)/0.9);
        x1:=x0+Round(q[j-1]*mx);      x2:=x0+Round(q[j]*mx);
        y1:=y0-Round(ps[j]*my);
        if j>2 then Line(x1,y1,x2,y1); pd[j]:=32-0.27*q[j];
        x1:=x0+Round(q[j]*mx);         y1:=y0-Round(pd[j]*my);
        y2:=y0-Round(ps[j]*my);
        if j>2 then  Line(x1,y1,x1,y2);
      end;
    for k:=3 to 9 do
      begin
        SetPenStyle(psSolid);        SetPenWidth(2);
        SetPenColor(clBlack);        SetBrushColor(clWhite);
        x1:=x0+Round(q[k]*mx);   Circle(x1, y0, 5);
        sq[k]:=IntToStr(k);
        var t:= new TextABC(x1, 412, 12, 'q'+sq[k]);
      end;
//===================== Main =============
begin
  CoordSys; Web;
  var t1 := new TextABC(130, 60, 12, 'Demand');
  var t2 := new TextABC(130, 80, 12, 'price');
  var t3 := new TextABC(564, 60, 12, 'Suggest');
  var t4 := new TextABC(580, 80, 12, 'price');
  var t5 := new TextABC(300, 470, 12,
          'Commodity volume at the market');
  var t6 := new TextABC(300, 40, 12,
          'Market equlibrium');
end.
```

Listing 10.1

There are two ways to build the diagram of the price oscillation:

- using the array of the web model points;

- solving the equation system (10.1) and (10.2).

We prefer the second way. Firstly, we'll find an expected market price (fig.80) using the numerical method from the chapter 2.

This diagram is drawn by the program Prog_10_2 which uses the subroutine "CoordSys" from the Prog_10_1. On this reason listing 10.2 contains the subroutine "Equat" only.

```
Program Prog_10_2;
Uses ABCobjects, GraphABC;

Procedure Equat;
const  x0 = -10;  y0 = 440;
mx = 6;  my = 12.8;  xs = 140;
var k, q, qr, x1, x2, y1, y2, z1, z2,m: integer;
var xw, yw, zw, pd, ps, pr, dp: real;
var u, x, y, z: array[1..30] of integer;
begin
  SetPenWidth(3); SetPenColor(clBlack);
   q:=20; m:=0;
   for k:=1 to 90 do
   begin
     q:=q+1; pd:=32-0.27*q;     ps:=1.1+0.9*exp(0.032*q);
     dp:=abs(pd-ps);
     if dp<1 then begin m:=1;  qr:=q;   end;
     x1:=x0+Round(q*mx); y1:=y0-Round(pd*my);
     z1:=y0-Round(ps*my);
     if k>1 then
       begin
         Line(x1,y1,x2,y2); Line(x1,z1,x2,z2);
       end;
     x2:=x1; y2:=y1; z2:=z1;
   end;
   u[1]:=25;   u[3]:=39;    u[5]:=51;    u[7]:=59; u[9]:=65;
   u[2]:=103; u[4]:=97;   u[6]:=92;    u[8]:=88; u[10]:=84;
   u[11]:=69; u[12]:=82;  u[13]:=71;   u[14]:=80;
   SetPenWidth(1);    SetPenColor(clBlack);
```

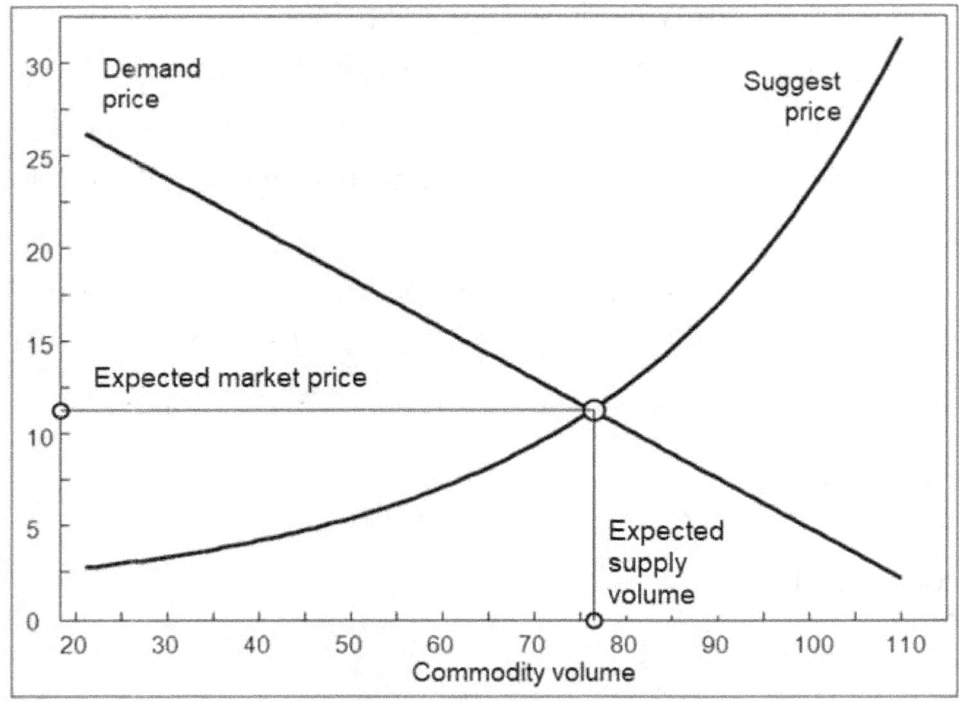

Figure 80. Equatiion system solution

```
  for k:=2 to 14 do
    begin
      xw:=u[k]; ps:=1.1+0.9*exp(0.032*xw);
      zw:=ps;    pd:=32-0.27*xw;
      yw:=pd;    x[k]:=x0+Round(xw*mx);
      y[k]:=y0-Round(zw*my);  z[k]:=y0-Round(yw*my);
    end;
  SetPenStyle(psSolid);      SetPenWidth(2);
  SetPenColor(clBlack);      SetBrushColor(clWhite);
  pr:=32-0.27*qr;            writeln(qr,' ',pr:3:1);
  x1:=x0-3+Round(qr*mx);  y1:=y0-3-Round(pr*my);
  Circle(x1,y1,7);           Circle(100,y1,5);
  Circle(x1,y0,5);           SetPenWidth(1);
  Line(100,y1,x1,y1);        Line(x1,y0,x1,y1);
  var t1 := new TextABC(120, y1-20, 12,
            'Expected market price');
    var t2 := new TextABC(460, y0-70, 12,  'Expected');
    var t3 := new TextABC(460, y0-50, 12,  'supply');
    var t4 := new TextABC(460, y0-30, 12,  'volume');
end;
```

```
//================= Main =======================
begin
  CoordSys;   Equat;
  var t1 := new TextABC(130, 60, 12, 'Demand');
  var t2 := new TextABC(130, 80, 12, 'price');
  var t3 := new TextABC(560, 60, 12, 'Suggest');
  var t4 := new TextABC(580, 80, 12, 'price');
  var t5 := new TextABC(300, 470, 12, 'Commodity volume');
  var t6 := new TextABC(270, 40, 12, 'Market equilibrium');
end.
```
Listing 10.2

The market price oscillations in time (fig.81) are modelled by the Program 10.3 (listing 10.3).

```
Program Prog_10_3;
Uses ABCobjects, GraphABC;

Procedure CoordSys;
  const x0=100;
  var j, k, xc, yc: integer;
  var x, y: real;
  var s: string;
  begin
    SetWindowSize(800,480);  SetPenStyle(psClear);
    SetPenColor(clBlack);       SetPenWidth(1);
    Line(x0, 15, x0, 390);      Line(x0, 390, 775, 390);
    Line(x0, 15, 775, 15);      Line(775, 15, 775, 390);
    SetPenStyle(psSolid);       SetPenWidth(1);
    for k:=1 to 27 do
      begin
        Line(x0+25*k, 390, x0+25*k, 382);
      end;
    for k:=2 to 14 do
      begin
        x:=2*(k-1); Str(x:2:0,s);
      var t:= new TextABC(42+50*k,398,12,s);
      end;
    for k:=0 to 18 do
      begin
        Line(x0, 390-25*k, x0+8, 390-25*k);
```

```
    end;
  for k:=1 to 9 do
   begin
    y:=4*(k-1); Str(y:2:0,s);
    var t:= new TextABC(x0-24,430-50*k,12,s);
   end;
end;
```

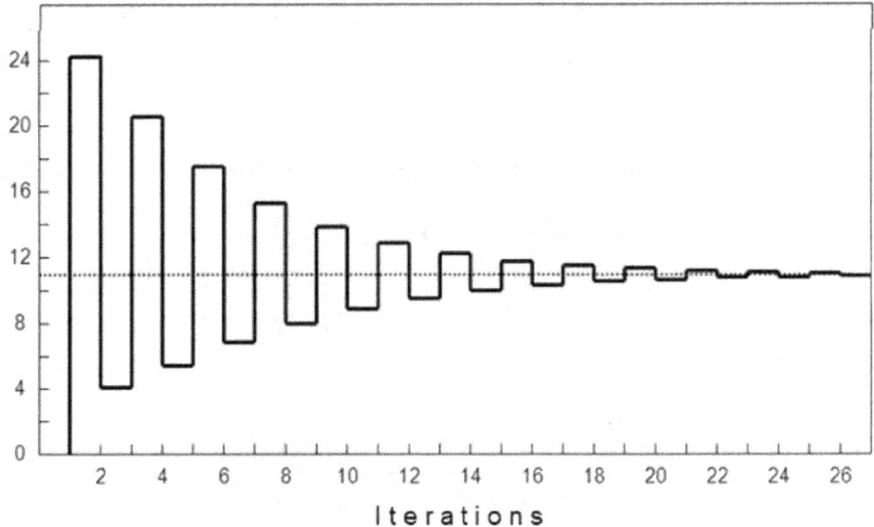

Figure 81. The market price oscillations. Variant 1.

```
Procedure Price;
const x0 = 100;  y0 = 390;
mx = 25;  my = 12;
var j, k ,qr, x1, x2, y1, y2, z1, z2, st: integer;
var x, y, z: real;
var q, pd, ps: array[1..50] of real;
var s1, s2: string;
begin
  SetPenStyle(psDot);   SetPenWidth(2);
  SetPenColor(clBlack);
  y1:=y0-3-Round(11.2*my);
  Line(x0, y1, 760, y1)
  SetPenStyle(psSolid); SetPenWidth(3);
//--------------------------------------------------1
  SetPenColor(clBlack); st:=1;
  q[1]:=25.0; pd[1]:=32-0.27*q[1];
  ps[1]:=1.1+0.9*exp(0.032*q[1]);
```

```pascal
  x1:=x0+Round(st*mx);
  y1:=y0-Round(pd[1]*my);
  z1:=y0-Round(ps[1]*my); Line(x1,y0,x1,y1);
  Line(x1,y1,x1+26,y1);
//------------------------------------------------ 2 - 27
  for j:= 2 to 27 do
    begin
    ps[j]:=pd[j-1];   q[j]:= 31.25*ln((ps[j]-1.1)/0.9);
    pd[j]:=32-0.27*q[j];
    x1:=x0+Round((j-1)*mx);     x2:=x0+Round(j*mx);
    y1:=y0-Round(pd[(j-1)]*my); y2:=y0-Round(pd[j]*my);
    z1:=y0-Round(ps[j-1]*my);   z2:=y0-Round(ps[j]*my);
  if j>2 then
      begin
        Line(x1,y1,x2,y1); Line(x1,y1,x1,z1);
      end;
    end;
end;
//===================== Main ===================
begin
  CoordSys; Price;
  var t1 := new TextABC(360, 40, 12,
  'Market price equilibrium');
  var t2 := new TextABC(390, 420, 12, 'I t e r a t i o n s');
end.
```

Listing 10.3

It is useful to see the market price oscillations using other functions of the demand and the suggestion. For instance, the exponential dependances are possible in both functions pd(q) and ps(q):

$$pd(q) = 40*\exp(-0.018q); \qquad (10.5)$$
$$ps(q) = 1 + 0.8*\exp(0.032q). \qquad (10.6)$$
$$20 \leq q \leq 100$$

We'll find the solution of equations (10.5) and (10.6) relatively the variable q:

$$q = -\frac{500}{9} * \ln\left(\frac{pd}{40}\right);$$
(10.7)

Expressions (10.7) and (10.8) are inputed into the program, developed on the base of the prototype Prog_10_1. Renovated web-diagram is shown in fig.82 and fig.83.

Conclusion.

The non-linear concave functions of the demand and the suggestion accelerate the market price regulation. The cycle quantity of the web spiral is shortened al least twice.

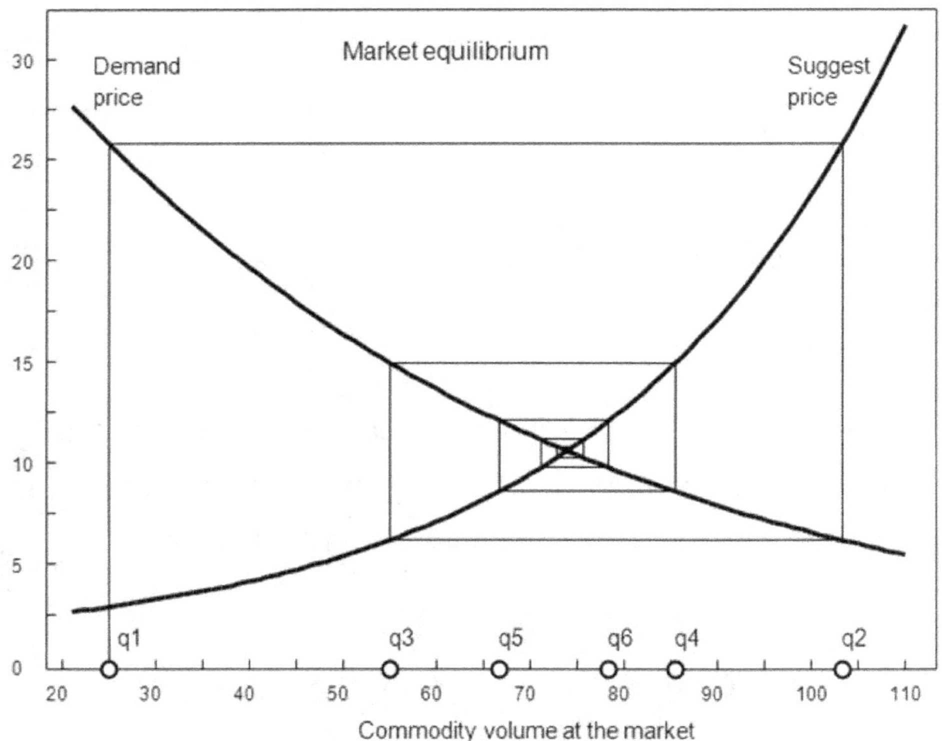

Figure 82. The market equilibrium model. Variant 2

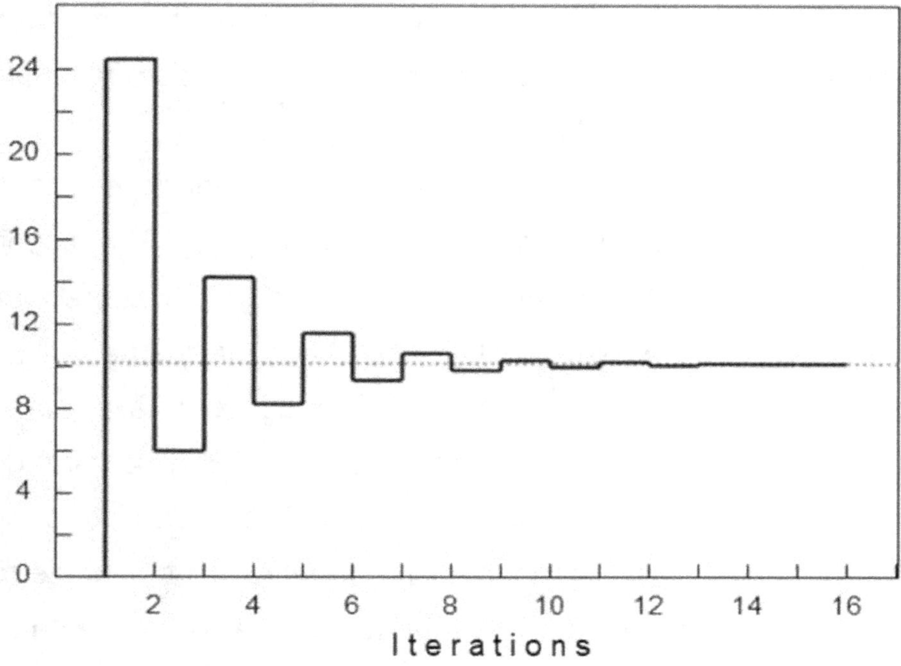

Figure 83. The market price oscillations. Variant 2

10.2. Macroeconomic modelling

Microeconomical modelling is considered in the book [3] using the example of a small business where concrete, grout and building mortat are produced. The model envelopes processes of the material, industrial and financial control. Let us see now macroeconomical model that describes the growth and slump processes of the national gross revenue.

The central place in the economic theory are taken by the dynamic items, namely – the long-term market equlibrium and the growth of the national income which is a main base for well-being of the people. The economic growth is provided by attraction of the investments, the labour resources, also by the scientific and technical progress that promotes the efficiency growth of all resources. On this reason the economical models consider the very important components – the investment, the manpower resources and the productivity of labour. The growth indicators are the joint product output and its important part – public consumption funds.

Two conceptions are known in the modern economic growth theory – Keynesian and neoclassical [2]. The economists Evsey D. Domar and Roy F. Harrod developed the Keynesian model where the Leontyev product function is applied. It is supposed that an excessive suggestion is at the labour market. It involves a constant price level. The capital departure is absent. The capital capacity and the norm of economies are constant. The product issue depends on the single resource only, i.e. on the capital. It is admitted also for the analysis simplification that the investments are transformed into the issue growth at once, so the lag is absent. The technical progress is out of account: the labour productivity is constant.

Despite of the significant simplifications, the model gives possibility to observe the most essential events in the macroeconomic, namely the dependence of the agregate national income on the investments and the public consumption funds.

The income is a sum of the investment and the consumption:

$$y(t) = invest(t) + consume(t) \qquad (10.9)$$

It is supposed that the income growth velocity y(t) is proportional to the investments, so the coefficient "b" of the capital capacity increment is constant:

$$\frac{dy(t)}{dt} = \frac{1}{b} invest(t) \qquad (10.10)$$

The constant $1/b$ is called an increase capital output. The economic state is described by the non-homogeneous differential first-order equation:

$$-b\frac{dy(t)}{dt} + y(t) = consume(t) \qquad (10.11)$$

We'll reduce the expression (10.11) into the finite difference equation. Two operations are needed for this purpose:

- to substitute tne continues time t by the discrete values, t $\rightarrow$ kT, where T – a quantum period;

- to apply the left-sided approximation of the first derivative.

In so manner the finite difference equation is got:

$$-\frac{b}{T}\big[y(k) - y(k-1)\big] + y(k-1) = consume(k) \qquad (10.12)$$

We'll solve it relatively the current income:

$$y(k) = \frac{T}{b}\left[\left(\frac{b}{T}+1\right) * y(k-1) - consume(k)\right] \qquad (10.13)$$

We have got the recursive equation. Solution of alike equations has been considered in the chapter 5.

Further the modelling is shown in two variants when:

- the consumption level is constant;

- the consumption level is growing exponentially.

Variant 1.

Develop the program where the model has parameters:

T = 1; b = 10: y(0) = 100.

The program Prog_10_4 (listing 10.4) solves the equation (10.13) using the consumption parameters $c1 = 40$, $c2 = 60$, $c3 = 75$. The economic growth is shown in fig.84.

Remark.

The model uses relative units of the time and financial measurements.

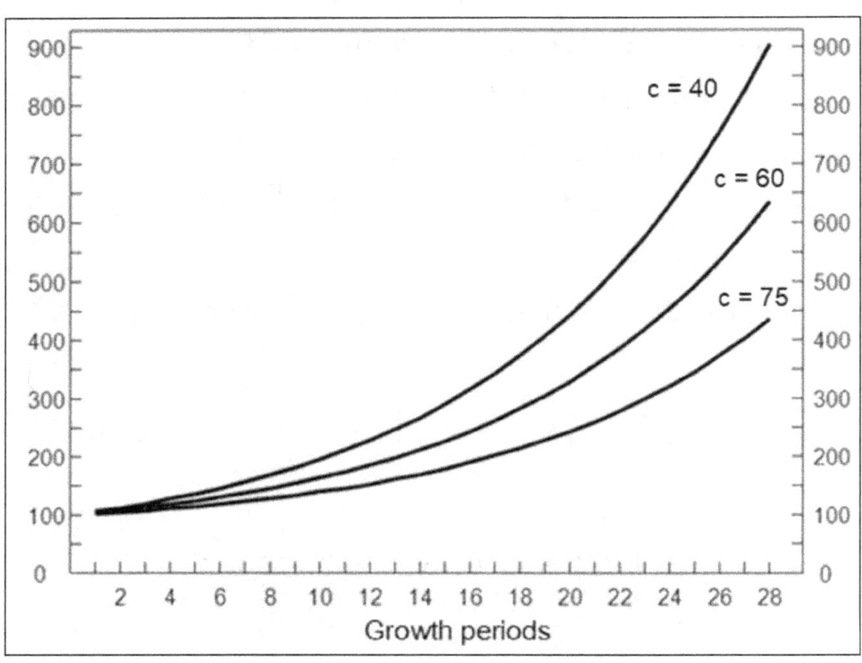

Figure 84. The beginning growth of the gross national income

```
Program Prog_10_4;
Uses ABCobjects, GraphABC;

Procedure CoordSys;
  const x0=72;
  var j, k, xc, yc: integer;
  var x, y: real;
  var s: string;
  begin
    SetWindowSize(700,460);  SetPenStyle(psClear);
    SetPenColor(clBlack);     SetPenWidth(1);
    Line(x0, 0, x0, 390);     Line(x0, 390, 600, 390);
    Line(x0, 0, 600, 0);      Line(600, 0, 600, 390);
```

```pascal
    SetPenStyle(psSolid);        SetPenWidth(1);
   for k:=1 to 28 do
     begin
       Line(x0+18*k, 390, x0+18*k, 382);
     end;
   for k:=2 to 15 do
     begin
       x:=2*(k-1); Str(x:2:0,s);
       var t:= new TextABC(28+36*k,398,12,s);
     end;
    for k:=0 to 18 do
     begin
       Line(x0, 390-21*k, x0+8, 390-21*k);
     end;
    for k:=1 to 10 do
     begin
      y:=100*(k-1); Str(y:2:0,s);
      var t:= new TextABC(x0-30,424-42*k,12,s);
     end;
    for k:=0 to 18 do
     begin
       Line(600, 390-21*k, 592, 390-21*k);
     end;
    for k:=1 to 10 do
     begin
      y:=100*(k-1); Str(y:2:0,s);
      var t:= new TextABC(608,424-42*k,12,s);
     end;
 end;

Procedure Nat_Income(c: real);
const x0 = 72; y0 = 390;
mx = 18; my = 0.42;
var j, k, x1, x2, y1, y2: integer;
var dx, x, T, b, p, q: real;
var y: array[0..300] of real;
var s1, s2: string;
begin
 SetPenStyle(psSolid); SetPenWidth(3);
 SetPenColor(clBlack);
 T:=1; b:=10; y[0]:=100;
 for k:= 1 to 28 do
```

```pascal
    begin
    y[k]:=(T/b)*((b/T+1)*y[k-1]-c);
    x1:=x0+Round(k*mx);   y1:=y0-Round(y[k]*my);
    if k>1 then
      begin
        Line(x1,y1,x2,y2);
      end;
    x2:=x1;  y2:=y1;
    end;
end;
//================= Main =======================
begin
var s1, s2, s3: string;   var c: real;
  CoordSys; c:=40;   Str(c:2:0,s1);
  var t5 := new TextABC(514, 20, 12,  'c = '+s1);
  Nat_Income(c);   c:=60;   Str(c:2:0,s2);
  var t6 := new TextABC(520, 120, 12,  'c = '+s2);
  Nat_Income(c);   c:=75;   Str(c:2:0,s3);
  var t7 := new TextABC(520, 200, 12,  'c = '+s3);
  Nat_Income(c);
  var t1 := new TextABC(200, 30, 12,
 'The gross national revenue');
  var t2 := new TextABC(358, 120, 12, 'Three levels');
  var t3 := new TextABC(360, 140, 12,  'of consumption:');
  var t4 := new TextABC(280, 420, 12,  'Time periods');
end.
```

Listing 10.4

Conclusion.

The modelling results are in accordance with the theory of
the beginning national income growth when the consumption
level is constant. Indeed, the solution of equation (10.11) may
be the exponent only. Naturally, the income growth becomes
lower if the consumption rises.

Variant 2.

The modelling program Prog_10_5 (listing 10.5) is intended for the prediction of consequences beeing result of an unlimited consumption growth which is not provided by the opportunities of the national economic. This program is distinct from its prototype Prog_10_4 with another hypotesis: the consumption is supposed to rise:

$$consume(t) = 20*exp(s*t)$$

This supposition makes the equation (10.11) non-homogeneous. Its solution by the classical methods is difficult enough but the numerical analysis methods allows to reduce the equation to the expression (10.13) and to solve the problem.

The program Prog_10_5 finds the equation solution with three values of the consumption growth degree (fig. 85)

$$s1 = 0.09; \; s2 = 0.105; \; s3 = 0.12$$

```
Program Prog_10_5;
Uses ABCobjects, GraphABC;

Procedure CoordSys;
  const x0=72;
  var j, k: integer;
  var x, y: real;
  var s: string;
  begin
    SetWindowSize(800,460);   SetPenStyle(psClear);
    SetPenColor(clBlack);     SetPenWidth(1);
    Line(x0, 0, x0, 390);     Line(x0, 390, 670, 390);
    Line(x0, 0, 670, 0);      Line(670, 0, 670, 390);
    SetPenStyle(psSolid);     SetPenWidth(1);
    for k:=2 to 17 do
      begin
        x:=2*(k-1); Str(x:2:0,s);
```

```pascal
    var t:= new TextABC(28+36*k,398,12,s);
  end;
  for k:=0 to 18 do
    begin
      Line(662, 390-21*k, 670, 390-21*k);
    end;
   for k:=1 to 10 do
      begin
        y:=100*(k-1); Str(y:2:0,s);
        var t:= new TextABC(678,424-42*k,12,s);
      end;
end;

Procedure Nat_Income(s: real);
const x0 = 72;  y0 = 390;  mx = 18; my = 0.5;
var j, k, x1, x2, y1, y2, N: integer;
var dx, x, T, b, p, q: real;
var y, z: array[0..300] of real;
begin
  SetPenStyle(psSolid);   SetPenWidth(3);
  SetPenColor(clBlack);   T:=1; b:=10;
  y[0]:=100; z[0]:=20;
  if s=0.09 then N:=29;
  if s=0.105 then N:=33;
  if s=0.12 then N:=31;
  for k:= 1 to N do
    begin
      z[k]:=20*exp(s*k);
      y[k]:=(T/b)*((b/T+1)*y[k-1]-z[k]);
      x1:=x0+Round(k*mx);
      y1:=y0-Round(y[k]*my);
     if k>1 then
      begin
          Line(x1,y1,x2,y2);
      end;
    x2:=x1;  y2:=y1;
  end;
end;
```

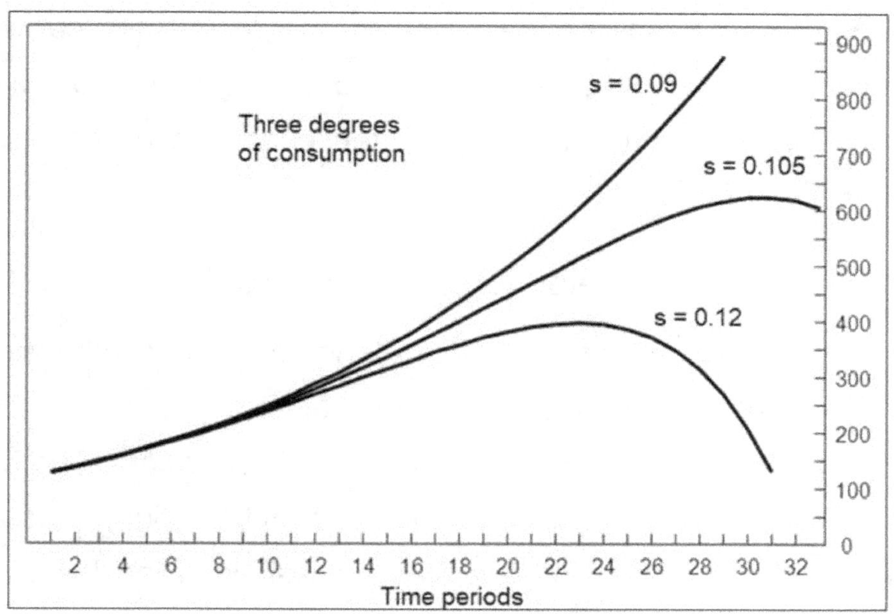

Figure 85. Variants of the income decrease

```
//================= Main =================
begin
  var s1, s2, s3: string;
  var s: real;
  CoordSys;
  s:=0.09;   Str(s:3:2,s1);
  var t5 := new TextABC(514, 30, 12, 's = '+s1);
  Nat_Income(s);
  s:=0.105;   Str(s:4:3,s2);
  var t6 := new TextABC(570, 100, 12, 's = '+s2);
  Nat_Income(s);
  s:=0.12;  Str(s:3:2,s3);
  var t7 := new TextABC(560, 220, 12, 's = '+s3);
  Nat_Income(s);
  var t1 := new TextABC(180, 30, 12, 'Gross income revenue');
  var t2 := new TextABC(338, 120, 12, 'Tree degrees');
  var t3 := new TextABC(340, 140, 12, 'of consumptiion:');
  var t4 := new TextABC(320, 420, 12, 'Time periods');
end.
```

Listing 10.5

Conclusion.

Excessive temps of the consumpton growth can lead to the catastrophic consequences. The model shows the collapse inevitability in case of the short-sighted economic politic. It is the question of time only.

10.3. Detection of the centroid

A cluster is a great number of homogeneous elements joined by some general attributes. The cluster analysis uses the mathematical statistic methods and the decision theory. The element attribute arrays are forming the multidimentional space where clusters may be detected and classified using the distance between the elements. Every element belongs to one of the clusters.

How do the analysts form the clusters? The market specialists study intently the people groups that may be the potential consumers of their commodities and services. It is necessary to begin the exactly directed (addressed) effective reclaim.

The commercial banks have own clusters. The bank allows credit after careful analysis of the client solvency. The set of documents confirms the solvency. This client attributes set is to minimize the bank risk of the failure to return the credit.

The educational cluster investigation is begun usually with modelling the points array on the plane. A distance between two points is calculated as an Euclid metric:

$$d = \sqrt{(x1 - x2)^2 + (y1 - y2)^2}$$

The cluster centroid is a point having minimum average distance to all elememts of this cluster. There are two methods to find a centroid:

- to calculate the arithmetical average of the array coordinates;

- to minimize the quadratic mean of the element deflections from the centroid.

The program Prog_10_6 (listing 10.6 and fig.86) demonstrates the first way.

Remark.

The array of 24 points on the plane is modelled previously in the application Microsoft Excel.

```
Program Prog_10_6;
Uses GraphABC, ABCobjects;

Procedure CoordSys;
 const x0=100;
 var j, k: integer;    var xc, yc: integer;   var s: string;
 begin
  SetWindowSize(660,500);  SetPenStyle(psClear);
  SetPenColor(clBlack);       SetPenWidth(2);
  Line(140, 50, 140, 440);     Line(140, 440, 620, 440);
  SetPenWidth(1);               Line(140, 50, 620, 50);
  Line(620, 50, 620, 440);
  for k:=1 to 16 do
     begin
       Line(140+30*k, 440, 140+30*k, 448);
     end;
  for k:=1 to 13 do
     begin
       Line(140, 440-30*k, 150, 440-30*k);
     end;
  for k:=1 to 12 do
     begin
       xc:=2*(k-1); s:=IntToStr(xc);
       var t:= new TextABC(76+60*k,452,12,s);
     end;
  for k:=2 to 7 do
     begin
       yc:=4*(k-1); s:=IntToStr(yc);
```

```pascal
      var t:= new TextABC(114,490-60*k,12,s);
    end;
end;

Procedure Cluster;
const x0=140; y0=440; mx=30; my=15; me=30;
var j, k, m, x1, y1, x2, y2, xa, ya: integer;
var x, y: array[1..24] of real;
var ax, ay, sx, sy, sq, dx, dy, cx, cy: real;

begin
  x[1]:= 4.05;      x[2]:= 4.25;      x[3]:= 4.85;      x[4]:= 6.45;
  x[5]:= 2.88;      x[6]:= 7.33;      x[7]:= 9.16;      x[8]:= 8.12;
  x[9]:= 4.97;      x[10]:= 5.93;     x[11]:=9.43;      x[12]:= 11.31;
  x[13]:= 10.85;  x[14]:= 10.76;  x[15]:= 7.31;      x[16]:= 11.84;
  x[17]:= 11.99;  x[18]:= 10.92;  x[19]:= 11.19;  x[20]:= 13.21;
  x[21]:= 13.12;  x[22]:= 13.74;  x[23]:= 9.01;      x[24]:= 6.90;

  y[1]:= 16.02;     y[2]:= 22.92;     y[3]:= 18.91;     y[4]:= 21.45;
  y[5]:= 8.08;      y[6]:= 8.72;      y[7]:= 19.06;     y[8]:= 8.64;
  y[9]:= 14.18;     y[10]:=7.90;      y[11]:= 9.03;     y[12]:= 24.97;
  y[13]:= 17.40;    y[14]:= 6.87;     y[15]:= 12.47;    y[16]:= 21.36;
  y[17]:= 8.63;     y[18]:= 14.24;    y[19]:= 11.94;    y[20]:= 3.87;
  y[21]:= 19.80;    y[22]:= 9.74;     y[23]:= 5.72;     y[24]:= 16.17;

  for k:=1 to 24 do
    begin
      x1:=x0+Round(x[k]*mx);  y1:=y0-Round(y[k]*my);
      SetPenStyle(psSolid);   SetPenWidth(2);
      SetPenColor(clBlack);   Circle(x1,y1, 5);
    end;
  sx:=0; sy:=0;
  for k:=1 to 24 do
    begin
      sx:=sx+x[k];            sy:=sy+y[k];
      ax:=sx/24;             x1:=x0+Round(ax*mx);
      ay:=sy/24;             y1:=y0-Round(ay*my);
      Line(x1-40,y1,x1+40,y1);
      Line(x1,y1+40,x1,y1-40);
      SetPenWidth(4);     Circle(x1,y1,8);
      writeln(' ');   writeln('x = ',ax:3:2);  writeln('y = ',ay:3:2);
    end;
  end;
```

```
//======== Main ======
begin
   CoordSys;    Cluster;
end;
```

Lissting 10.6

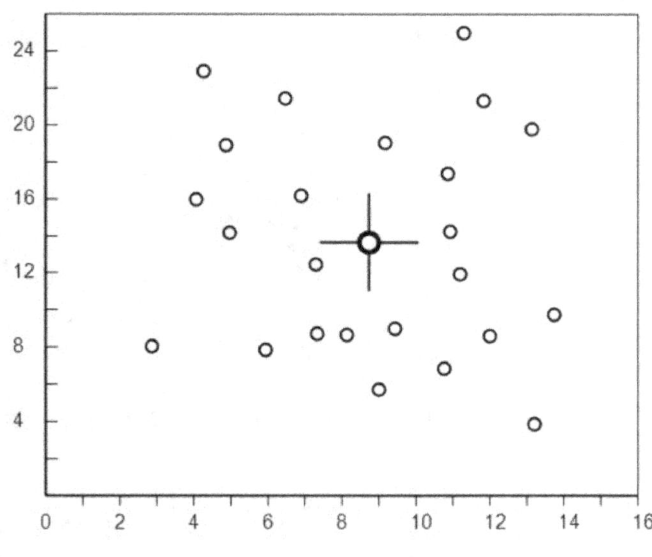

Figure 86. Cluster centroid

The program Prog_10_6 was suplemented by the procedures of coordinate automatical selection firstly along the vertical line (x = 8.73, fig.87) then along the horizontal line (y = 13.67, fig.88). The completion of the centroid selection is shown at the figure 89.

It is worthy to see how the program optimizes the centroid coordinates. For the purpose the program Prog_10_7 (listing 10.7) has been developed. The fig.90 shows how the quadratic deflection average is changing when it is not stopped in the optimum position. The extremum is blunt enough, hence the selection accuracy may be restricted with the second decimal digit of the Euclid metric.

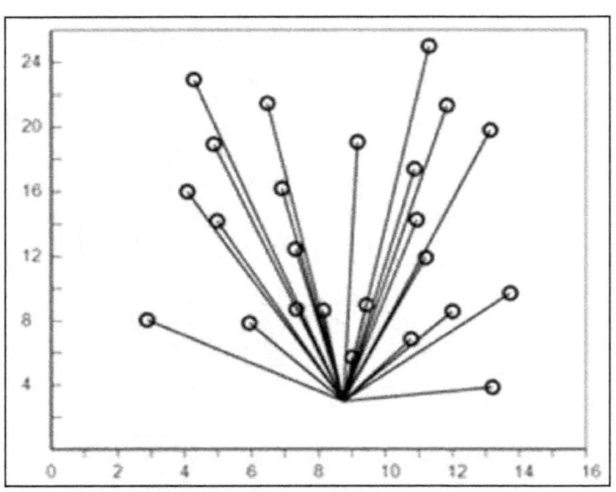

Figure 87. The beginning of centroid selection at the axes Y

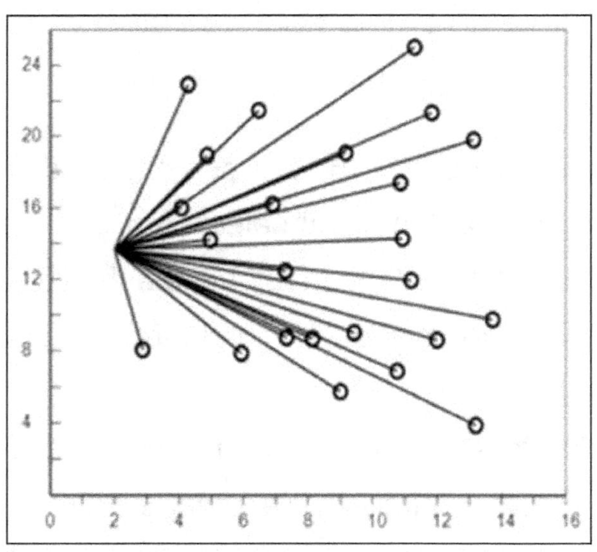

Figure 88. The beginning of centroid selection at the axes X

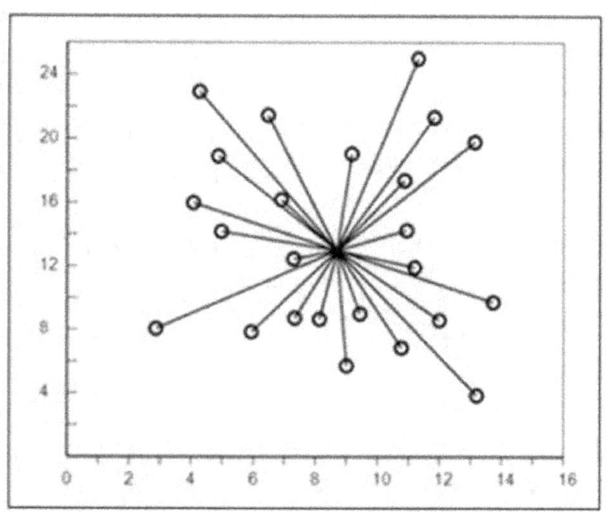

Figure 89. The completion of the centroid selection

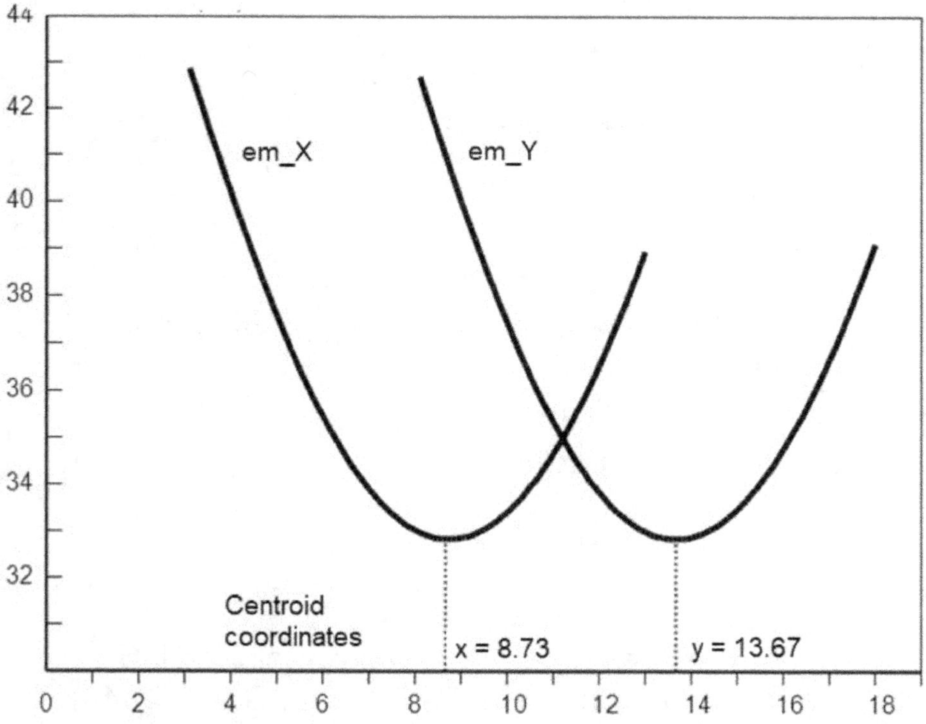

Figure 90. Euclid metric trajectories

Program Prog_10_7;
Uses GraphABC, ABCobjects;

Procedure CoordSys;
  const x0=100;

```
    var j, k, xc, yc: integer;
    var s: string;
    begin
      SetWindowSize(660,500);  SetPenStyle(psClear);
      SetPenColor(clBlack);       SetPenWidth(2);
      Line(140, 50, 140, 440);
      Line(140, 440, 620, 440);  SetPenWidth(1);
      Line(140, 50, 620, 50);     Line(620, 50, 620, 440);
      for k:=1 to 16 do
          begin
             Line(140+30*k, 440, 140+30*k, 448);
          end;
      for k:=1 to 13 do
          begin
             Line(140, 440-30*k, 150, 440-30*k);
          end;
       for k:=1 to 12 do
          begin
             xc:=2*(k-1);   s:=IntToStr(xc);
             var t:= new TextABC(76+60*k,452,12,s);
          end;
       for k:=2 to 7 do
          begin
             yc:=4*(k-1);       s:=IntToStr(yc);
             var t:= new TextABC(114,490-60*k,12,s);
          end;
    end;

Procedure CoordEuclid;
  const x0=100;
  var j, k, xc, yc: integer;
  var s: string;
  begin
    SetWindowSize(760,500);   SetPenColor(clBlack);
    SetPenWidth(2);                Line(140, 20, 140, 440);
    Line(140, 440, 710, 440);   SetPenWidth(1);
    Line(140, 20, 710, 20);     Line(710, 20, 710, 440);
    for k:=1 to 19 do
        begin
           Line(140+30*k, 440, 140+30*k, 448);
        end;
     for k:=1 to 10 do
```

```
    begin
      xc:=2*(k-1);  s:=IntToStr(xc);
     var t:= new TextABC(76+60*k,452,12,s);
     end;
  for k:=1 to 14 do
     begin
       Line(140, 440-30*k, 150, 440-30*k);
     end;
  for k:=2 to 8 do
    begin
       yc:=30 +2*(k-1); s:=IntToStr(yc);
      var t:= new TextABC(114,490-60*k,12,s);
     end;
end;

Procedure Cluster;
const x0=140; y0=440; mx=30; my=15; me=30;
var j, k, m: integer;
var x1, y1, x2, y2, xa, ya: integer;
var x, y: array[1..24] of real;
var ax, ay, sx, sy, sq, em1, em2, dem, dx, dy, cx, cy: real;
begin
  x[1]:= 4.05;     x[2]:= 4.25;     x[3]:= 4.85;     x[4]:= 6.45;
  x[5]:= 2.88;     x[6]:= 7.33;     x[7]:= 9.16;     x[8]:= 8.12;
  x[9]:= 4.97;     x[10]:= 5.93;    x[11]:=9.43;     x[12]:= 11.31;
  x[13]:= 10.85;  x[14]:= 10.76;   x[15]:= 7.31;    x[16]:= 11.84;
  x[17]:= 11.99;  x[18]:= 10.92;   x[19]:= 11.19;   x[20]:= 13.21;
  x[21]:= 13.12;  x[22]:= 13.74;   x[23]:= 9.01;    x[24]:= 6.90;

  y[1]:= 16.02;    y[2]:= 22.92;    y[3]:= 18.91;    y[4]:= 21.45;
  y[5]:= 8.08;     y[6]:= 8.72;     y[7]:= 19.06;    y[8]:= 8.64;
  y[9]:= 14.18;    y[10]:=7.90;     y[11]:= 9.03;    y[12]:= 24.97;
  y[13]:= 17.40;  y[14]:= 6.87;    y[15]:= 12.47;   y[16]:= 21.36;
  y[17]:= 8.63;    y[18]:= 14.24;  y[19]:= 11.94;   y[20]:= 3.87;
  y[21]:= 19.80;  y[22]:= 9.74;    y[23]:= 5.72;    y[24]:= 16.17;

  for k:=1 to 24 do      // Array display
    begin
       x1:=x0+Round(x[k]*mx);   y1:=y0-Round(y[k]*my);
       SetPenStyle(psSolid);    SetPenWidth(2);
       PenColor(clBlack);
     end;                       // Ariyhmetic average
  sx:=0; sy:=0;
```

```pascal
    for k:=1 to 24 do
      begin
        ay:=sy/24;     y1:=y0-Round(ay*my);
        sx:=sx+x[k];   sy:=sy+y[k];
      end;
    ax:=sx/24;                x1:=x0+Round(ax*mx);
    SetPenWidth(4);        writeln(' ');
    writeln('x = ',ax:3:2);     writeln('y = ',ay:3:2);
    y1:=y0-Round((em1-30)*me);
      if j>1 then
        begin
          Line(x1,y1, x2, y2);
        end;
      em2:=em1;     x2:=x1; y2:=y1;
    end;
//---------------------Centroid Y-------------------
  ax:=8.73;  ay:=8.0;  dem:=-1;
  SetPenWidth(4);
  SetPenColor(clBlack);
  for j:=1 to 100 do
   begin
     sq:=0;  ay:=ay+0.1;
     for k:=1 to 24 do
       begin
         dx:=x[k]-ax; dy:=y[k]-ay;
         sq:=sq+dx*dx+dy*dy;
       end;
     em1:=sqrt(sq);   dem:=em1-em2;
     x1:=x0+Round(ay*mx);
     y1:=y0-Round((em1-30)*me);
     if j>1 then
       begin
         Line(x1,y1, x2, y2);
       end;
     em2:=em1; x2:=x1; y2:=y1;
   end;
end;
//==================== Main ====================
begin
 CoordEuclid;   Cluster;
 var t1:=new TextABC(140,0,12, 'Euclid metric (em)');
```

```
var t2:=new TextABC(280, 395,12, 'Centroid start');
var t3:=new TextABC(290, 414,12,  'coordinates');
var t4:=new TextABC(410,414,12,'x = 8.73');
var t5:=new TextABC(560,414,12,'y = 13.67');
var t6:=new TextABC(260,100,12,'em_X');
var t7:=new TextABC(410,100,12,'em_Y');
  SetPenWidth(2);   SetPenColor(clBlack);
  SetPenStyle(psDot);
  Line(400,354,400,440);   Line(550,354,550,440);
end.
```

Listing 10.7

10.4. Detection of clusters

The problem of the student flow segmentation arises often in an educational process control. Let 84 students studied the discipline course with volume 50 hours. The dean's office has got the information – the progress in terms of 100-points scale and the student attendance measured in hours. It is necessary to distribute the students in two multitude using the semester results.The program Prog_10_8 (listing 10.8) detected two clear-cut clusters of students (fig.91).

The segments have the centroids with coordinates:

1) b = 84.4 points, h = 43.5 hours;

2) b = 42.7 points, h = 22.9 hours.

The poor progress is correlated with lower attendance. It is well known fact in the educational practice.

```
Program Prog_10_8;
Uses GraphABC, ABCobjects;

Procedure Coord_BH;
  const x0=100;
  var j, k, xc, yc: integer;  var s: string;
  begin
   SetWindowSize(700,500);
   SetPenStyle(psClear);      SetPenColor(clBlack);
```

```pascal
    SetPenWidth(2);           Line(140, 20, 140, 420);
    Line(140, 420, 660, 420); SetPenWidth(1);
    Line(140, 20, 660, 20);   Line(660, 20, 660, 420);
  for k:=1 to 26 do
    begin
      Line(140+20*k, 420, 140+20*k, 428);
    end;
  for k:=1 to 13 do
  begin
    xc:=4*(k-1); s:=IntToStr(xc);
    var t:= new TextABC(96+40*k,428,12,s);
  end;
  for k:=1 to 20 do
    begin
      Line(140, 420-20*k, 148, 420-20*k);
    end;
  for k:=2 to 10 do
  begin
    yc:=10*(k-1);  s:=IntToStr(yc);
    var t:= new TextABC(116,450-40*k,12,s);
  end;
  var t1:=new TextABC(300,450,12, 'Attendance,  houre');
  var t2:=new TextABC(140,0,12,    'Pogress, points');
end;

Procedure Coord_DH;
  const x0=100;   var j, k, xc, yc: integer;   var s: string;
  begin
    SetWindowSize(700,500);   SetPenStyle(psClear);
    SetPenColor(clBlack);     SetPenWidth(2);
    Line(140, 20, 140, 420);  Line(140, 420, 660, 420);
    SetPenWidth(1);           Line(140, 20, 660, 20);
    Line(660, 20, 660, 420);
  for k:=1 to 26 do
    begin
      Line(140+20*k, 420, 140+20*k, 428);
    end;
  for k:=1 to 13 do
  begin
    xc:=4*(k-1);  s:=IntToStr(xc);
    var t:= new TextABC(96+40*k,428,12,s);
  end;
```

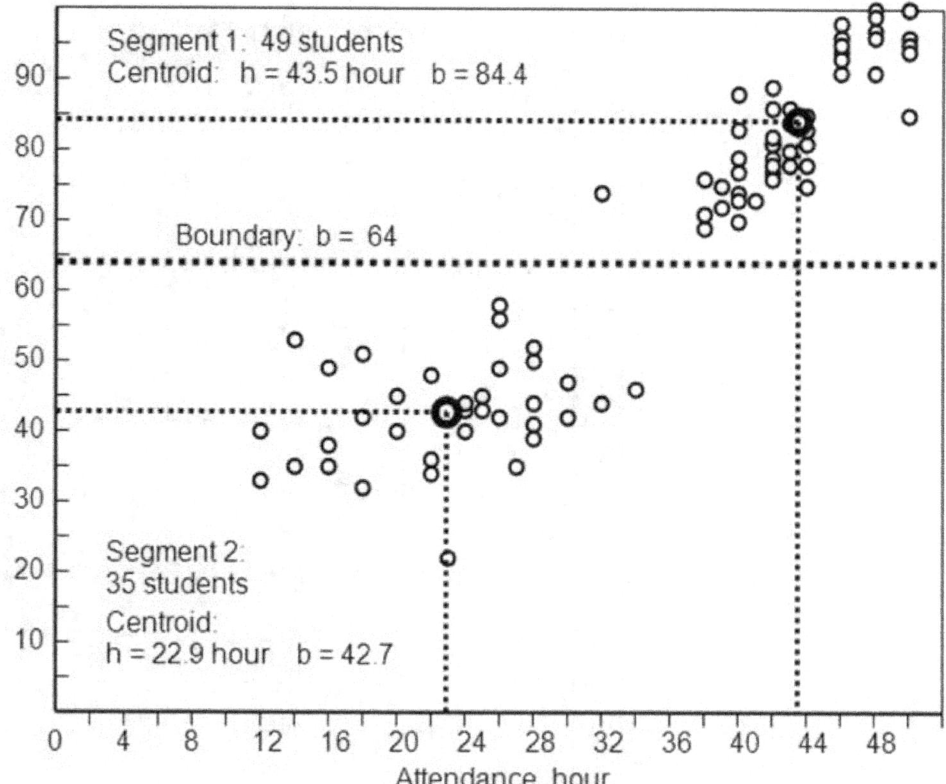

Figure 91 Segmentation of the student flow

```
for k:=1 to 20 do
   begin
      Line(140, 420-20*k, 148, 420-20*k);
   end;
for k:=2 to 10 do
  begin
     yc:=10*(k-1);  s:=IntToStr(yc);
     var t:= new TextABC(116,450-40*k,12,s);
  end;
  var t1:=new TextABC(340,450,12,  'Attendance,  hour');
end;


Procedure GenerClust;
const x0=140; y0=420; mx=10; my=4;
var j, k, n, sx, sy, x1, y1, x2, y2, xa, ya: integer;
var b, h, u, v, w, z: array[1..90] of integer;
var ax, ay, su, sv, sw, sz, em, av, au, aw, az as real;
var dx, dy, cx, cy: real;
```

```pascal
begin
  h[1]:=12;   h[2]:=14;   h[3]:=16;   h[4]:=16;
  h[5]:=18;   h[6]:=18;
  h[7]:=20;   h[8]:=22;   h[9]:=22;   h[10]:=24;
  h[11]:=24;  h[12]:=26;
  h[13]:=26;  h[14]:=24;  h[15]:=26;  h[16]:=27;
  h[17]:=28;  h[18]:=28;
  h[19]:=28;  h[20]:=28;  h[21]:=30;  h[22]:=38;
  h[23]:=38;  h[24]:=39;
  h[25]:=40;  h[26]:=40;  h[27]:=42;  h[28]:=40;
  h[29]:=39;  h[30]:=42;
  h[31]:=40;  h[32]:=40;  h[33]:=40;  h[34]:=42;
  h[35]:=42;  h[36]:=43;
  h[37]:=42;  h[38]:=43;  h[39]:=40;  h[40]:=42;
  h[41]:=42;  h[42]:=43;
  h[43]:=42;  h[44]:=41;  h[45]:=42;  h[46]:=43;
  h[47]:=44;  h[48]:=44;
  h[49]:=44;  h[50]:=44;  h[51]:=44;  h[52]:=46;
  h[53]:=46;  h[54]:=46;
  h[55]:=46;  h[56]:=46;  h[57]:=46;  h[58]:=48;
  h[59]:=48;  h[60]:=48;
  h[61]:=48;  h[62]:=50;  h[63]:=50;  h[64]:=50;
  h[65]:=50;  h[66]:=50;
  h[67]:=12;  h[68]:=14;  h[69]:=16;  h[70]:=18;
  h[71]:=20;  h[72]:=22;
  h[73]:=25;  h[74]:=23;  h[75]:=28;  h[76]:=32;
  h[77]:=34;  h[78]:=30;
  h[79]:=25;  h[80]:=26;  h[81]:=32;  h[82]:=38;
  h[83]:=48;  h[84]:=50;

  b[1]:=40;   b[2]:=53;   b[3]:=49;   b[4]:=35;
  b[5]:=51;   b[6]:=32;
  b[7]:=45;   b[8]:=36;   b[9]:=34;   b[10]:=43;
  b[11]:=40;  b[12]:=56;
  b[13]:=49;  b[14]:=44;  b[15]:=42;  b[16]:=35;
  b[17]:=41;  b[18]:=50;
  b[19]:=44;  b[20]:=52;  b[21]:=47;  b[22]:=69;
  b[23]:=71;  b[24]:=75;
  b[25]:=88;  b[26]:=83;  b[27]:=79;  b[28]:=74;
  b[29]:=72;  b[30]:=77;
  b[31]:=79;  b[32]:=70;  b[33]:=73;  b[34]:=76;
  b[35]:=78;  b[36]:=80;
```

```pascal
  b[37]:=81;  b[38]:=84;   b[39]:=77;  b[40]:=89;
  b[41]:=81;   b[42]:=78;
  b[43]:=82;  b[44]:=73;   b[45]:=86;  b[46]:=86;
  b[47]:=83;   b[48]:=78;
  b[49]:=85;  b[50]:=81;   b[51]:=75;  b[52]:=91;
  b[53]:=94;   b[54]:=93;
  b[55]:=98;  b[56]:=96;   b[57]:=95;  b[58]:=97;
  b[59]:=96;   b[60]:=100;
  b[61]:=99;  b[62]:=100; b[63]:=96;  b[64]:=95;
  b[65]:=100; b[66]:=94;
  b[67]:=33;  b[68]:=35;   b[69]:=38;  b[70]:=42;
  b[71]:=40;   b[72]:=48;
  b[73]:=45;  b[74]:=22;   b[75]:=39;  b[76]:=44;
  b[77]:=46;   b[78]:=42;
  b[79]:=43;  b[80]:=58;   b[81]:=74;  b[82]:=76;
  b[83]:=91;   b[84]:=85;

  for k:=1 to 84 do                 // Array display
     begin
       x1:=x0+Round(h[k]*mx); y1:=y0-Round(b[k]*my);
       SetPenStyle(psSolid);     SetPenWidth(2);
       SetPenColor(clBlack);     Circle(x1, y1, 4);
     end;
  // ===================Arithmetical average
   sx:=0; sy:=0;
   for k:=1 to 84 do
    begin
      sx:=sx+h[k]; sy:=sy+b[k];
    end;
   ax:=sx/84; ay:=sy/84;
   writeln('h = ',ax:3:2);        writeln('b = ',ay:3:2);
   x1:=x0+Round(ax*mx);  y1:=y0-Round(ay*my);
   SetPenColor(clBlack);     ay:=ay-3;
   SetPenWidth(4);
   var t9:=new TextABC(210,140,12, 'Boundary:  b =  64');
   y1:=y0-Round(ay*my);
   SetPenStyle(psDot);       Line(x0,y1,660,y1);
  //-------------------------------- Segmentation
   j:=0; n:=0;   su:=0; sv:=0; sw:=0; sz:=0;
   For k:=1 to 84 do
    begin
     if b[k]<ay then
```

```pascal
      begin
       j:=j+1;           u[j]:=h[k];      v[j]:=b[k];
        su:=su+u[j];    sv:=sv+v[j];
      end;
    if b[k]>ay then
      begin
       n:=n+1;           w[n]:=h[k];   z[n]:=b[k];
        sw:=sw+w[n];    sz:=sz+z[n];
      end;
   end;
  au:=su/j; av:=sv/j;          writeln(j);
  writeln(au:4:2);             writeln(av:4:2);
  x1:=x0+Round(au*mx);  y1:=y0-Round(av*my);
  SetPenStyle(psSolid);        SetPenColor(clBlack);
  SetPenWidth(5);              Circle(x1, y1, 7);
  SetPenColor(clBlack);        SetPenWidth(3);
  SetPenStyle(psDot);          Line(x0, y1, x1, y1);
  Line(x1, y1, x1, y0);        aw:=sw/n; az:=sz/n;
  writeln(n);                  writeln(aw:4:2);
  writeln(az:4:2);             x2:=x0+Round(aw*mx);
  y2:=y0-Round(az*my);  SetPenStyle(psSolid);
  SetPenColor(clBlack);        SetPenWidth(5);
  Circle(x2, y2, 6);           SetPenColor(clBlack);
  SetPenWidth(3);              SetPenStyle(psDot);
  Line(x0, y2, x2, y2);        Line(x2, y2, x2, y0);
end;
//===================== Main =====================
begin
 Coord_DH;   GenerClust;
 var t3:=new TextABC(170,30,12, 'Segment 1:  49 students');
 var t4:=new TextABC(170,48,12,
        'Centroid:   h = 43.5 hour    b = 84.4');
 var t5:=new TextABC(170,320,12, 'Segment 2:');
 var t6:=new TextABC(170,338,12, '35 students');
 var t7:=new TextABC(170,360,12, 'Centroid:');
 var t8:=new TextABC(170,378,12, 'h = 22.9 hour b = 42.7');
end.
```

Listing 10.8

Chapter 11. METHODS OF CRYPTOGRAPHY

11.1. Subject of cryptography

One may often hear at the 21st century the words "information defence". Now everybody has in the habit of using payments without cash transfer, payment cards and other financial documents intended for the confidential application. And as ever, the criminal attempts occur to get "non-sanctioned access" to the documents, in other words – the modern stealing.

The information security problem has arisen long before Christ. With appearance of a written language it became possible to send messages to the remoted addressee using material carriers – thin earhen slabs, small boards, birch barks, leather, papyrus before a paper invention.

The ruler edicts were being announced "at sight", all the rest messages were intended for a single recipient. They became secret being carefully hided from a stranger sight. In case of interception by an opponent, the message text was transformed in outward appearance in so manner it would not be read by a person which is not be let into the secret of the back transform. Alike text transformation is called "cryptography" (in Greece: cryptos – secret).

The numerical (digital) methods became the main technology of a cryptography. Due to the latin word "cifra", the text transformation is called ciphering and decoding is called deciphering. The illegal attempt to open ciphered text got the name "breaking in". The educated scientists consider alike action as a cryptoanalysis.

Two main activity spheres demanded always the skill to hide the message content – a diplomacy and a war action control. Now all participants of the economic process are to use elementary means of the information security since the personal computers and the digital communication channels became the everyday occurence. Security problems are actual especially in the bank sphere and in all juridical processes where agreements are to be signed.

The ciphering is not a single method to conceal some information. One may hide the very fact of a message transmission. There is a steganography, i.e. the science which deals with methods of concealing an object by means of another object. For instance, a message may be hidden in another text, in a picture or in a phonogram. A message may be written with invisible ink between lines of an open text. The reading technology is known to a recipient only. An open accessible object is to draw away one's attention from a truthful function of the container.

We'll consider in this chapter the cryptography methods only being seen in history development from naïve (as it seems to be now) to modern ciphering methods highly resistant to cryptoanalysis. The monograph does not pretend on deep research in cryptography and cryptoanalysis. They are not its problems. This chapter is to enlarge knowledge of the Pascal ABC instruments for text processing. The reader is to learn some little known (or forgotten) algebra parts however which are needed for a successful programming.

11.2. Ceasar's algorithm

One of the first ciphering methods went from deepness of ages. An algorithm of mono-alphabetical substitutions occurred. The ciphering principle is to change alphabet symbols of the message with other symbols of the same alphabet using the way being known to the communication participants only.

The history gave us a substitution method by means of the symbol cyclical shift at N alphabet positions. The algorithm is ascribed to the emperor July Ceasar which send his messages from the battle fields to the Senat.

We'll consider the cyphering using the english alphabet. Let the initial set contains 26 small letters written as a line in descending order from "a" to "z". The new alphabet for the secret message is got by means of the symbol shift at "sh" positions anticlockwise, for instance sh = 4. The cyclic shift means that the initial letters are not lost but are tranferred at the line end in result of the shift.

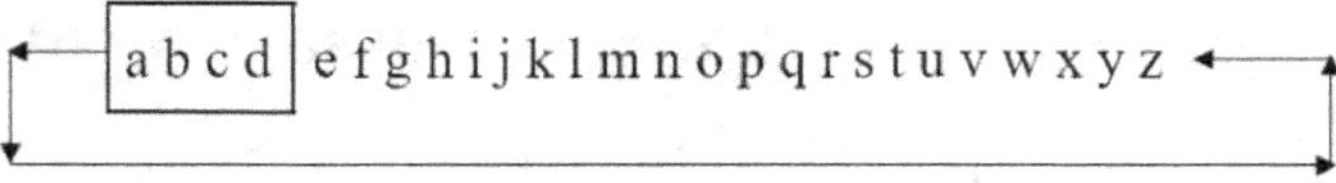

The derivative alphabet gets the next form:

$$e\ f\ g\ h\ i\ j\ k\ l\ m\ n\ o\ p\ q\ r\ s\ t\ u\ v\ w\ x\ y\ z\ \boxed{a\ b\ c\ d}$$

In this example the symbol "a" is substituted by the symbol "e", "b" by "f", "c" by "g", "d" by "h" and so futher.

Example 11.1. Do ciphering the latin utterance:

Radices litterarum amarae sunt - fructus dulces

(The roots of science are bitter, but the fruits are sweet)

Provide a possibility to change the secret shift from 5 to 14.

Solution.

The program Prog_11_1 (listing 11.1) ciphers the message. Key value "sh" and the message "mes" are tranferred into the subroutine "Coder". The simbol sequence (fig.92) depends on the key value.

Message = radiceslitterarumamaraesuntfructusdulces

Shift = 7

Send = yhkpjlzspaalyhyaththyhlzauamyajaazkasjz

Figure 92. Example of the ciphering by Ceasar's algorithm

Remark.

The omissions between message words are usually ignored. The gaps would be used by opponent to restore a message structure. One ought to give no additional information for a potential opponent.

```
Program Prog_11_1;
Uses ABCobjects;
var s: array[1..50] of string;
var m, sn: array[1..50] of integer;

Procedure Coder(sh: integer; mes: string);
begin
  var j, len, msg: integer;
  var snd: string;    len:=Length(mes);
  for j:=1 to len do
    begin
      case mes[j] of
        'a': m[j]:=1;   'h': m[j]:=8;   'o': m[j]:=15;   'v': m[j]:=22;
```

```pascal
    'b': m[j]:=2;     'i': m[j]:=9;      'p': m[j]:=16;    'w': m[j]:=23;
    'c': m[j]:=3;     'j': m[j]:=10;     'q': m[j]:=17;    'x': m[j]:=24;
    'd': m[j]:=4;     'k': m[j]:=11;     'r': m[j]:=18;    'y': m[j]:=25;
    'e': m[j]:=5;     'l': m[j]:=12;     's': m[j]:=19;    'z': m[j]:=26;
    'f': m[j]:=6;     'm': m[j]:=13;     't': m[j]:=20;
    'g': m[j]:=7;     'n': m[j]:=14;     'u': m[j]:=21;
  end;
  if (m[j]>sh) and (m[j]<27-sh) then sn[j]:=m[j]+sh;
  if m[j]<=sh then sn[j]:=sh+m[j];
  if m[j]>=27-sh then sn[j]:=8-sh;
  case sn[j] of
    1: s[j]:='a';     8: s[j]:='h';    15: s[j]:='o';    22: s[j]:='v';
    2: s[j]:='b';     9: s[j]:='i';    16: s[j]:='p';    23: s[j]:='w';
    3: s[j]:='c';    10: s[j]:='j';    17: s[j]:='q';    24: s[j]:=x';
    4: s[j]:='d';    11: s[j]:='k';    18: s[j]:='r';    25: s[j]:='y';
    5: s[j]:='e';    12: s[j]:='l';    19: s[j]:='s';    26: s[j]:='z';
    6: s[j]:='f';    13: s[j]:='m';    20: s[j]:='t';
    7: s[j]:='g';    14: s[j]:='n';    21: s[j]:='u';
  end;
  snd:=snd+s[j];
 end;
 var t3:=new TextABC(130,110,12, 'Shift = '+IntTostr(sh));
 var t4:=new TextABC(125,140,12,  'Send = '+snd);
end;
//========= Main =========
var mes: string;  var sh: integer;
begin
 mes:='radiceslitterarumamaraesuntfructusdulces';
 sh:=7; Coder(sh,mes);
 var t2:=new TextABC(100,80,12,'Message = '+mes);
end.
```

Listing 11.1

The ciphering is done with the next actions.

1. The message symbols are codered by the ordinal numbers of the latin alphabet.

2. The numbers y of substitute letters are calculated on the base of the symbol numbers in the initial alphabet using the value sh:

$$y = x + sh, \text{ if } x > sh \text{ and } x < 27 - sh;$$

$$y = x + sh, \text{ if } x \leq sh;$$

$$y = 8 - sh, \text{ if } x \geq 27 - sh;$$

3. The resulting numbers are transformed into the new alphabet letters and are joined into the letter sequence which is send through the communication channel.

Example 11.2.

Do the deciphering of the letter sequence received from the communication channel:

Received = 'yhkpjlzspaalyhybthyhlzbuamybjabzkbsjlz'

Remark.

Since the deciphering is made in law, the next things are known to the recipient:

- the open message text is written using the latin alphabet with small letters;

- the ciphering is made using the Ceasar's algorithm;

- the secret key is sh = 7.

These data are used in the program Prog_11_2 (listing 11.2) to decode the message (fig.93).

```
Program Prog_11_2;
Uses ABCobjects;
var s: array[1..50] of string;
var r: array[1..50] of integer;

Procedure Decoder(sh: integer; rcv: string);
begin
var j, len: integer;
var sn: array[1..50] of integer;
var msg: string;

len:=Length(rcv);  msg:=' ';
 for j:=1 to len do
  begin
   case rcv[j] of
```

```pascal
      'a': r[j]:=1;      'h': r[j]:=8;      'o': r[j]:=15;      'v': r[j]:=22;
      'b': r[j]:=2;      'i': r[j]:=9;      'p': r[j]:=16;      'w': r[j]:=23;
      'c': r[j]:=3;      'j': r[j]:=10;     'q': r[j]:=17;      'x': r[j]:=24;
      'd': r[j]:=4;      'k': r[j]:=11;     'r': r[j]:=18;      'y': r[j]:=25;
      'e': r[j]:=5;      'l': r[j]:=12;     's': r[j]:=19;      'z': r[j]:=26;
      'f': r[j]:=6;      'm': r[j]:=13;     't': r[j]:=20;
      'g': r[j]:=7;      'n': r[j]:=14;     'u': r[j]:=21;
    end;
    if r[j]>sh then sn[j]:=r[j]-sh;
    if r[j]<=sh then sn[j]:=r[j]+26-sh;
    case sn[j] of
      1: s[j]:='a';      8: s[j]:='h';    15: s[j]:='o';    22: s[j]:='v';
      2: s[j]:='b';      9: s[j]:='i';    16: s[j]:='p';    23: s[j]:='w';
      3: s[j]:='c';     10: s[j]:='j';    17: s[j]:='q';    24: s[j]:='x';
      4: s[j]:='d';     11: s[j]:='k';    18: s[j]:='r';    25: s[j]:='y';
      5: s[j]:='e';     12: s[j]:='l';    19: s[j]:='s';    26: s[j]:='z';
      6: s[j]:='f';     13: s[j]:='m';    20: s[j]:='t';
      7: s[j]:='g';     14: s[j]:='n';    21: s[j]:='u';
    end;
    msg:=msg+s[j];
  end;
var t3:=new TextABC(130,110,12,'Shift = '+IntTostr(sh));
var t4:=new TextABC(100,140,12,'Decoded = '+msg);
end;
```

Received = yhkpjlzspaalyhybththyhlzbuamybjabzkbsjlz

Shift = 7

Decoded = radiceslitterarumamaraesuntfructusdulces

Figure 93. Deciphering of the received letter sequence

```pascal
//=================Main ===================
var rcv: string; var sh: integer;
begin
  rcv:='yhkpjlzspaalyhybththyhlzbuamybjabzkbsjlz';
  sh:=7;
  Decoder(sh,rcv);
  var t2:=new TextABC(96,80,12,'Received = '+rcv);
end.
```

Listing 11.2

11.3. Polibius square

The ciphering algorithm is ascribed to Polibius, the ancient Greek historian (near 200 – 120 years B.C.). The same principle of substitution is suggested, but with changing of letters by decimal integer digits which are the numbers of lines and columnes in quare table where the alphabet symbols are placed. It is more convenient to use the english alphabet instead of Greek one. So, we'll take the square size 5 x 5 and place two symbols "i" and "j" together in a cell. Every cell is coded by the two-digit number from 11 to 55.

Table 4. Polibius square

	1	2	3	4	5
1	a	b	c	d	e
2	f	g	h	i, j	k
3	l	m	n	o	p
4	q	r	s	t	u
5	v	w	x	y	z

Example 11.3. Do ciphering the italian proverb:

Cento misure e un taglio solo
(100 times measure and cut off once only)

The ciphering makes the program Prog_11_3 (listing 11.3).

```
Program Prog_11_3;
Uses ABCobjects;
const M=30;

Procedure Coder(s: string);
begin
  var k, n: integer;   var send: string;
  var mes: array[1..M] of integer;   n:=Length(s);
  for k:=1 to n do
    begin
```

```pascal
  case s[k] of
   'a': mes[k]:=11;   'h': mes[k]:=23;   'p': mes[k]:=35;
   'v': mes[k]:=51;
   'b': mes[k]:=12;   'i': mes[k]:=24;   'q': mes[k]:=41;
  'w': mes[k]:=52;
   'c': mes[k]:=13;   'k' mes[k]:=25;    'r': mes[k]:=42;
   'x': mes[k]:=53;
   'd': mes[k]:=14;    'l': mes[k]:=31;   's': mes[k]:=43;
   y': mes[k]:=54;
   'e': mes[k]:=15;  m': mes[k]:=32;    't': mes[k]:=44;
  "z": mes[k]:=55;
   'f': mes[k]:=21;  'n': mes[k]:=33;   'u': mes[k]:=45;
   ' ': mes[k]:=66;
   'g': mes[k]:=22;   'o': mes[k]:=34;   'v': mes[k]:=51;
   end;
   send:=send+mes[k];
   end;
   writeln('   ',send);
end;
//============== Main ================
var s: string;
begin
  s:='cento misure e un taglio solo';   Coder(s);
end.
```

Listing 11.3

The program is alike two previous ones. The most important instruction block is the selector "case s[k] of" intended to change the proverb symbols by pairs of digit. The pairs are used to form the digit sequence at input of a communication channel:

```
1315334434663224434542156615664533 66441
                         122312434664 3343134
```

Remarks.

1. The coded message is folded in two parts since the digit sequence length is more than an admitted length of the text line in this book.

2. The pairs "66" occurred in the message though they are absent in the Polibius square. The gaps between words of the proverb are coded by the pair in the current example exclusively. It is made for an educational purpose only. The gaps are ignored in all rest examples.

The Polibius square is attractive due to the transition from letters to digits in a message. A numerical sequence seems to be more sofisticated to a person who does not know the ciphering algorithm and wants to break message. Hence, he is to waste more time to find secret key. It is the main the cipher oficcer can reckon upon. The time factor dominates in real situation when the information interchange takes place.

Let us see how the legal deciphering can be automated.

Example 11.4.

The numerical sequence has been received. It is known the mes-sage was ciphered using the Polibius square for the english alphabet. Read the message. Solution is given by the program Prog_11_4 (listing 11.4).

```
Program Prog_11_4;
Uses ABCobjects;
Procedure Decoder(s: string);
begin
  var k, n, j, x: integer;
  var y, z, dd, smb, mes: string;
  var send: array[1..29] of string;
  n:=Length(s);   j:=0; dd:=' ';
  for k:=1 to n do
    begin
```

```pascal
      x:= (k mod 2);
      if x=1 then  y:=s[k];
      if x=0 then
        begin
          z:=s[k]; j:=j+1;  dd:=y+z;    send[j]:=dd;
        end;
       dd:=' ';
     end;
     mes:=' ';
     for j:=1 to 29 do
       begin
         case send[j] of
           '11': smb:='a';   '12': smb:='b';    '13': smb:='c';
           '14': smb:='d';   '15': smb:='e';    '21': smb:='f';
           '22': smb:='g';   '23': smb:='h';    '24': smb:='i';
           '25': smb:='k';   '31': smb:='l';    '32': smb:='m';
           '33': smb:='n';   '34': smb:='o';    '35': smb:='p';
           '41': smb:='q';   '42': smb:='r';    '43': smb:='s';
           '44': smb:='t';   '45': smb:='u';    '51': smb:='v';
           '52': smb:='w';  '53': smb:='x';     '54': smb:='y';
           '55': smb:='z';   '66': smb:=' ';
         end;
         mes:=mes+smb;
        end;
      writeln(mes);
end;
//================== Main =========
var s: string;
begin
      s:='131533443466322443454215661 5
          664533664411223124346643343134';
      Decoder(s);
end.
```

Listing 11.4

One can see the result of deciphering:

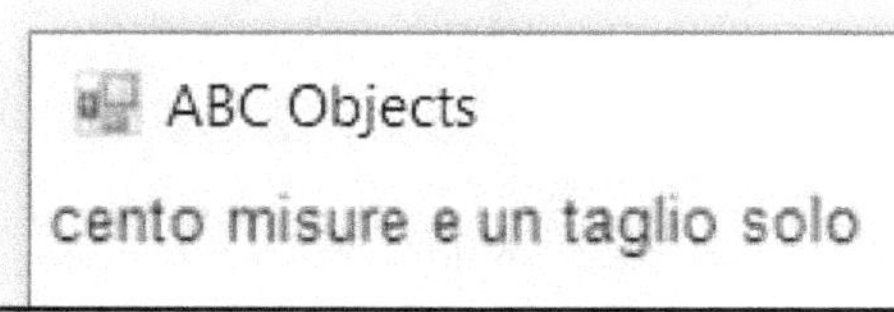

Select the digit pairs "even-odd" and then change the detected pairs by text letters from the square table "5 x 5". It is all that is needed.

11.4. Enciphering by the inverse symbols

The language Pascal ABC suggests two data types for processing text information and necessary functions for transforming data types. Let us remember that every symbol of text information is coded by a number from 0 to 255 when one-byte code is used in the ASCII system. Encipher tables are accessible for everybody in all informatic manuals and in the global net.

The library function Char(k) returns the symbol in accordance with the value "k". For instance, the function returnes the small letter "a" if k = 97. The library function Ord(b) returnes the number 98 since it is a code of the letter "b".

We'll substitute the message symbols with the code inversions instead of the codes to create difficulties for a potential breaker. A code inversion is an addition of the code to the number 256. For instance, the numerical value cod = 109 is a code of the symbol "m". The inversion is the difference sc = 156 − cod = 47. In this case the inverse symbol is

Char(sc) = 's'

Example 11.5. Encode the message with the inversion symbols:

a_barking_dog_seldom_bites

Solution.

The program Prog_11_5 (listing 11.5) transforms the message into the sequence of two-digit numbers. The gaps are substituted by symbols "_" with cod = 61. The output sequence is:

```
59615859424951465361564553614155485645476158514055411
```

Remark.

The program uses the inverse symbol array mes[k] which was modelled previously in the application Microsoft Excel.

```pascal
Program Prog_11_5;
const M=30;
Procedure Coder(s: string);
 begin
   var k, n: integer;
   var send: string;
   var mes: array[1..M] of integer;
   n:=Length(s);
   for k:=1 to n do
     begin
       case s[k] of
           'a': mes[k]:=59;    'j': mes[k]:=50;    's': mes[k]:=41;
           'b': mes[k]:=58;    'k': mes[k]:=49;    't': mes[k]:=40;
           'c': mes[k]:=57;    'l': mes[k]:=48;    'u': mes[k]:=39;
           'd': mes[k]:=56;    'm': mes[k]:=47;    'v': mes[k]:=38;
           'e': mes[k]:=55;    'n': mes[k]:=46;    'w': mes[k]:=37;
           'f': mes[k]:=54;    'o': mes[k]:=45;    'x': mes[k]:=36;
           'g': mes[k]:=53;    'p': mes[k]:=44;    'y': mes[k]:=35;
           'h': mes[k]:=52;    'q': mes[k]:=43;    'z': mes[k]:=34;
           'i': mes[k]:=51;    'r': mes[k]:=42;    '_': mes[k]:=61;
       end;
       send:=send+mes[k];
     end;
   writeln(' ',send);
 end;
//============= Main =========
var s: string;
begin
   s:='a_barking_dog_seldom_bites';
```

    Coder(s);
end.

Listing 11.5

Example 11.6. Decipher the received sequence:

59615859424951465361564553614155485645

47615851405541

It is known the apriory information:

1. A message is written in English with small letters.

2. The gaps are substituted by symbols "_".

3. The digit pairs are the inverse codes of the small letters in the ASCII system.

Solution.

The program Prog_11_6 (listing 11.6) reads and deciphers the received sequence:


```
Program Prog_11_6;
Uses ABCobjects;

Procedure Decoder(s: string);
begin
  var j, k, m, n, x: integer;
  var mes: string;
  var s1, s2:  array[1..30] of char;
, var smb: char;
  var d1, d2, sn,sd:  array[1..30] of integer;
  n:=Length(s);   j:=0; m:=0;
  for k:=1 to n do
    begin
      x:= (k mod 2);
      if x=1 then
        begin
```

```pascal
        j:=j+1; s1[j]:=s[k];
      end;
    if x=0 then
      begin
        m:=m+1;2[m]:=s[k];
      end;
  end;
  for k:=1 to 26 do
    begin
      case s1[k] of
        '0': d1[k]:=0;    '5': d1[k]:=5;
        '1': d1[k]:=1;    '6': d1[k]:=6;
        '2': d1[k]:=2;    '7': d1[k]:=7;
        '3': d1[k]:=3;    '8': d1[k]:=8;
        '4': d1[k]:=4;    '9': d1[k]:=9;
      end;
      case s2[k] of
        '0': d2[k]:=0;    '5': d2[k]:=5;
        '1': d2[k]:=1;    '6': d2[k]:=6;
        '2': d2[k]:=2;    '7': d2[k]:=7;
        '3': d2[k]:=3;    '8': d2[k]:=8;
        '4': d2[k]:=4;    '9': d2[k]:=9;
      end;
      sn[k]:=d1[k]*10+d2[k];    sd[k]:=156-sn[k];
    end;
  mes:=' ';
  for k:=1 to 26 do
    begin
      smb:=char(sd[k]); mes:=mes+smb;
    end;
  writeln(mes);
end;
//=============== Main ===============
var s: string;
begin
  s:='59615859424951465361564553361415548
      564547615851405541';
  Decoder(s);

end.
```

Listing 11.6

Remark.

Despite of the received sequence is written using decimal digits, it remains the text line. Hence, the digits are to be dealt with as graphical symbols which have not numerical values.

The program Prog_11_6 does the next actions:

1. It selects even and odd symbols s1[k], s2[k].

2. It transforms the symbol arrays s1[k], s2[k] into the number arrays d1[k], d2[k] using the operator "case d[k] of".

3. It fulfils assembling of two-digit codes where d1 is a ten digit and d2 is a unit digit.

4. It restores the values "x" of the initial symbols codes by means of the substration of inverse code from the number 156.

5. It transforms restored codes "x" into letters "smb": using the function Char(x).

6. It joints symbols smb into the output text line using the recursive expression message = message + smb.

11.5. Vigenere table

The transition from a mono-alphabetical substitution of symbols to a multi-alphabetical one was the significant step ahead in the cryptography. Abbot Johannes Trithemius suggested a square table containing N equal alphabets lines every being shifted at one step cyclicly in comparison with the previous line. He described the new principle in the book "Polygraphia" dated by 1528.

However, the cryptography history ascribes this idea to the french diplomat Blaise de Vigenere (1523-1596). He ciphered professionally being acquainted with the cryptography methods of the previous ages. Vigenere published the book "Traite des chiffres ou secretes manieres d'escrire" in the 1586. Probably, it has over-shadowed all books on the cryptography known at that time.

Now we'll study the cipher method that is called "Chiffre Vigenere" in modern manuals. This method is based on the known algorithm of mono-alphabetical substitution, but ciphering has been brought to perfection due to the next innovations:

1. The idea of Johannes Ttrithemius to use the square matrix with the derivative alphabets.

2. An application of the secret word ("parole" in french). Being repeated many times, it becames a gamma. The length of the gamma is equal to the message length.

The gammed multi-alphabetical ciphering provided the highest security of a message during the next 400 years up to the invention of the powerful programmed computers in the 20th century.

Let us consider the algorithm intended to cipher the message in English. We'll suppose that small letters are used. The alphabet is written 26 times in the square table (fig.94).

Due to the alphabet shift, the symbols are repeated along the rising diagonals. Numbers of lines serve as the ordinal numbers of the gamma symbols. One may write the gamma as the sequence of random symbols, but it is better to use some sensible combination of symbols that is convenient to remember.

Since there are not very long words in all european languages, we'll form the long gamma by means of repeating some arbitrary word so many times to get the lengh that is equal the message length. If the last repetition is out of the length, the excessive symbols are cut.

Example 11.7.

Encipher the message using 23 letters :

Strike while the iron is hot

If the ciphering would be made without a computer, one ought to do the next actions in fig.94:

G	M		1	2	3	4	5	6	7		20	21	22	23	24	25	26
u	s	1	a	b	c	d	e	f	g		t	u	v	w	x	y	z
s	t	2	b	c	d	e	f	g	h		u	v	w	x	y	z	a
e	r	3	c	d	e	f	g	h	i		v	w	x	y	z	a	b
c	i	4	d	e	f	g	h	i	j		w	x	y	z	a	b	c
h	k	5	e	f	g	h	i	j	k		x	y	z	a	b	c	d
a	e	6	f	g	h	i	j	k	l		y	z	a	b	c	d	e
n	w	7	g	h	i	j	k	l	m		z	a	b	c	d	e	f
c	h	8	h	i	j	k	l	m	n		a	b	c	d	e	f	g
e	i	9	i	j	k	l	m	n	o		b	c	d	e	f	g	h
u	l	10	j	k	l	m	n	o	p		c	d	e	f	g	h	i
s	e	11	k	l	m	n	o	p	q		d	e	f	g	h	i	j
e	t	12	l	m	n	o	p	q	r		e	f	g	h	i	j	k
c	h	13	m	n	o	p	q	r	s		f	g	h	i	j	k	l
h	e	14	n	o	p	q	r	s	t		g	h	i	j	k	l	m
a	i	15	o	p	q	r	s	t	u		h	i	j	k	l	m	n
n	r	16	p	q	r	s	t	u	v		i	j	k	l	m	n	o
c	o	17	q	r	s	t	u	v	w		j	k	l	m	n	o	p
e	n	18	r	s	t	u	v	w	x		k	l	m	n	o	p	q
u	i	19	s	t	u	v	w	x	y		l	m	n	o	p	q	r
s	s	20	t	u	v	w	x	y	z		m	n	o	p	q	r	s
e	h	21	u	v	w	x	y	z	a		n	o	p	q	r	s	t
c	o	22	v	w	x	y	z	a	b		o	p	q	r	s	t	u
h	t	23	w	x	y	z	a	b	c		p	q	r	s	t	u	v
		24	x	y	z	a	b	c	d		q	r	s	t	u	v	w
		25	y	z	a	b	c	d	e		r	s	t	u	v	w	x
		26	z	a	b	c	d	e	f		s	t	u	v	w	x	y

Figure 94. The Vigenere table

1. To write the ordinal number of the symbols above the Vigenere table.

2. To write the ordinal number of the alphabets lines in the column of the table.

3. To invent a parole and write the gamma in the column left of the alphabet numbers. For instance, the parole is a word "use chance" . The gamma is the parole in this case:

gamma = "usechanceusechanceusech"

4. To write the message in the column left of the gamma.

5. To read the message and the gamma synchronously. To select the column number for the current symbol of the message in the Vigenere table and the alphabet number for the current symbol of the gamma.

6. To select the symbol from the cell of the Vigenere table where cross selected column and line. The selected symbol is to substitute the current symbol of the message. When the end of the message will be reached, the cipher of the message will be got.

The program Prog_11_7 (listing 11.7) automates this algorithm and outputs the result on the display (fig.95).

```
message: = strikewhiletheironishot

gamma: = usechanceusechanceusech

send: = mlvkrejjmfwxjlieqrcklqa
```

Fig 95. Enciphering with the Vigenere table

```
Program Prog_11_7;
Uses ABCobjects;
var g, m: array[1..32] of integer;
var s: array[1..32,1..32] of string;

Procedure Coder(mes,gg: string);
begin
var j, k, n, len, gam, msg: integer;
var  snd: string;
s[1,1]:='a';   s[1,8]:='h';    s[1,15]:='o';   s[1,22]:='v';
```

```
s[1,2]:='b';   s[1,9]:='i';    s[1,16]:='p';   s[1,23]:='w';
s[1,3]:='c';   s[1,10]:='j';   s[1,17]:='q';   s[1,24]:='x';
s[1,4]:='d';   s[1,11]:='k';   s[1,18]:='r';   s[1,25]:='y';
s[1,5]:='e';   s[1,12]:='l';   s[1,19]:='s';   s[1,26]:='z';
s[1,6]:='f';   s[1,13]:='m';   s[1,20]:='t';
s[1,7]:='g';   s[1,14]:='n';   s[1,21]:='u';

s[2,1]:='b';   s[2,8]:='i';    s[2,15]:='p';   s[2,22]:='w';
s[2,2]:='c';   s[2,9]:='j';    s[2,16]:='q';   s[2,23]:='x';
s[2,3]:='d';   s[2,10]:='k';   s[2,17]:='r';   s[2,24]:='y';
s[2,4]:='e';   s[2,11]:='l';   s[2,18]:='s';   s[2,25]:='z';
s[2,5]:='f';   s[2,12]:='m';   s[2,19]:='t';   s[2,26]:='a';
s[2,6]:='g';   s[2,13]:='n';   s[2,20]:='u';
s[2,7]:='h';   s[2,14]:='o';   s[2,21]:='v';
```

Similar 22 array blocks s[3, n] … s[24, n]
are to be here (n = 1, 2, …, 26)

```
s[25,1]:='y';   s[25,8]:='f';    s[25,15]:='m';  s[25,22]:='t';
s[25,2]:='z';   s[25,9]:='g';    s[25,16]:='n' ; s[25,23]:='u';
s[25,3]:='a';   s[25,10]:='h';   s[25,17]:='o';  s[25,24]:='v';
s[25,4]:='b';   s[25,11]:='i';   s[25,18]:='p';  s[25,25]:='w';
s[25,5]:='c';   s[25,12]:='j';   s[25,19]:='q';  s[25,26]:='x';
s[25,6]:='d';   s[25,13]:='k';   s[25,20]:='r';
s[25,7]:='e';   s[25,14]:='l';   s[25,21]:='s';
s[26,1]:='z';   s[26,8]:='g';    s[26,15]:='n';  s[26,22]:='u';
s[26,2]:='a';   s[26,9]:='h';    s[26,16]:='o';  s[26,23]:='v';
s[26,3]:='b';   s[26,10]:='i';   s[26,17]:='p';  s[26,24]:='w';
s[26,4]:='c';   s[26,11]:='j';   s[26,18]:='q';  s[26,25]:='x';
s[26,5]:='d';   s[26,12]:='k';   s[26,19]:='r';  s[26,26]:='y';
s[26,6]:='e';   s[26,13]:='l';   s[26,20]:='s';
s[26,7]:='f';   s[26,14]:='m';   s[26,21]:='t';

len:=Length(gg);
for j:=1 to len do
  begin
    case gg[j] of
      'a': g[j]:=1;   'h': g[j]:=8;    'o': g[j]:=15;  'v': g[j]:=22;
      'b': g[j]:=2;   'i': g[j]:=9;    'p': g[j]:=16;  'w': g[j]:=23;
      'c': g[j]:=3;   'j': g[j]:=10;   'q': g[j]:=17;  'x': g[j]:=24;
      'd': g[j]:=4;   'k': g[j]:=11;   'r': g[j]:=18;  'y': g[j]:=25;
      'e': g[j]:=5;   'l': g[j]:=12;   's': g[j]:=19;  'z': g[j]:=26;
      'f': g[j]:=6;   'm': g[j]:=13;   't': g[j]:=20;
```

```pascal
    'g': g[j]:=7;   'n': g[j]:=14;  'u': g[j]:=21;
  end;
  case mes[j] of
   'a': m[j]:=1;    'h': m[j]:=8;    'o': m[j]:=15;  'v':  m[j]:=22;
   'b': m[j]:=2;    'i': m[j]:=9;    'p': m[j]:=16;  'w': m[j]:=23;
   'c': m[j]:=3;    'j': m[j]:=10;  'q': m[j]:=17;   'x': m[j]:=24;
   'd': m[j]:=4;    'k': m[j]:=11;  'r': m[j]:=18;   'y': m[j]:=25;
   'e': m[j]:=5;    'l': m[j]:=12;  's': m[j]:=19;   'z': m[j]:=26;
   'f': m[j]:=6;    'm': m[j]:=13;  't': m[j]:=20;
   'g': m[j]:=7;    'n': m[j]:=14;  'u': m[j]:=21;
  end;
  gam:=g[j];   msg:=m[j];   snd:=s[gam,msg];
  writeln('   '+snd);
 end;
 var t1:=new TextABC(196,120,14,
        'message: = strikewhiletheironishot');
 var t2:=new TextABC(210,150,14,
       'gamma: = usechanceusechanceusech');
 var t3:=new TextABC(232,180,14,
        'send: = mlvkrejjmfwxjlieqrcklqa');
end;
//=========== Main ===========
 var mes, gg: string;
 begin
  mes:='strikewhiletheironishot';
  gg:='usechanceusechanceusech';
  Coder(mes, gg);
 end.
```

Listing 11.7

The ciphering table is written as the array s[gam, msg] where "gam" is the symbol number in the gamma, "msg": is the number of the message symbol. The program measures the message length "len", numbers every symbol of the gamma and the message in cycle from 1 to "len". Then the program compiles the output cipher line using symbol snd[gam, msg] from the cell of the Vigenere table with coordinates "gam" and "msg".

Example 11.8

Now it is time to solve the reciprocal problems – to decipher above mentioned sequence. There is an information to develop the procedure for deciphering:

- the text is written in Engish;

- a cipher officer used the Vigenere table;

- the gamma is known.

We can use some blocks from the program Prog_11_7:

- two-dimential arrays from the Vigenere table;

- one-dimential array of the english alphabet.

Solution.

The program Prog_11_8 (listing 11.8) receives the sequence cipher from the fig.94 and returns the information (fig.96) using the parole "use chance":

```
Program Prog_11_8;
Uses ABCobjects;
var g: array[1..26] of integer;
var s: array[1..26,1..26] of string;

 Procedure Decoder(rcv,gg: string);
 begin
    var j, k, n, len, gam, rc: integer;
    var smb,mes: string;
  s[1,1]:='a';    s[1,8]:='h';    s[1,15]:='o';   s[1,22]:='v';
  s[1,2]:='b';    s[1,9]:='i';    s[1,16]:='p';   s[1,23]:='w';
  s[1,3]:='c';    s[1,10]:='j';   s[1,17]:='q';   s[1,24]:='x';
  s[1,4]:='d';    s[1,11]:='k';   s[1,18]:='r';   s[1,25]:='y';
  s[1,5]:='e';    s[1,12]:='l';   s[1,19]:='s';   s[1,26]:='z';
  s[1,6]:='f';    s[1,13]:='m';   s[1,20]:='t';
  s[1,7]:='g';    s[1,14]:='n';   s[1,21]:='u';
```

Similar 24 array blocks s[2, n] … s[25, n]
are to be here (n = 1, 2, …, 26)

ABC Objects

```
21      m      s
19      l      t
5       v      r
3       k      i
8       r      k
1       e      e
14      j      w
3       j      h
5       m      i
21      f      l
19      w      e
5       x      t
3       j      h
8       l      e
1       i      i
14      e      r
3       q      o
5       r      n
21      c      i
19      k      s
5       l      h
3       q      o
8       a      t
```

send: = mlvkrejjmfwxjlieqrcklqa

gamma: = usechanceusechanceusech

message: = strikewhiletheironishot

Figure 96. Deciphering of the send sequence

```
s[26,1]:='z';   s[26,8]:='g';    s[26,15]:='n';  s[26,22]:='u';
s[26,2]:='a';   s[26,9]:='h';    s[26,16]:='o';  s[26,23]:='v';
s[26,3]:='b';   s[26,10]:='i';   s[26,17]:='p';  s[26,24]:='w';
s[26,4]:='c';   s[26,11]:='j';   s[26,18]:='q';  s[26,25]:='x';
s[26,5]:='d';   s[26,12]:='k';   s[26,19]:='r';  s[26,26]:='y';
s[26,6]:='e';   s[26,13]:='l';   s[26,20]:='s';
s[26,7]:='f';   s[26,14]:='m';   s[26,21]:='t';
len:=Length(rcv);
for j:=1 to len do
  begin
    case gg[j] of
      'a': g[j]:=1;   'h': g[j]:=8;    'o': g[j]:=15;  'v': g[j]:=22;
      'b': g[j]:=2;   'i': g[j]:=9;    'p': g[j]:=16;  'w': g[j]:=23;
```

```pascal
      'c': g[j]:=3;    'j': g[j]:=10;  'q': g[j]:=17;  'x': g[j]:=24;
      'd': g[j]:=4;   'k': g[j]:=11;   'r': g[j]:=18;  'y': g[j]:=25;
      'e': g[j]:=5;    'l': g[j]:=12;  's': g[j]:=19;  'z': g[j]:=26;
      'f': g[j]:=6;  'm': g[j]:=13;    't': g[j]:=20;
      'g': g[j]:=7;  'n': g[j]:=14;  'u': g[j]:=21;
    end;
  gam:=g[j];  smb:=rcv[j];
  for k:=1 to 26 do
    begin
      if s[gam,k]=smb then
        begin
          case k of
      1: mes:='a';     8: mes:='h';  15: mes:='o';  22: mes:='v';
      2: mes:='b';     9: mes:='i';  16: mes:='p';  23: mes:='w';
      3: mes:='c';  10: mes:='j';  17: mes:='q';  24: mes:='x';
      4: mes:='d';  11: mes:='k';  18: mes:='r';  25: mes:='y';
      5: mes:='e';  12: mes:='l';  19: mes:='s';  26: mes:='z';
      6: mes:='f';  13: mes:='m';  20: mes:='t';
      7: mes:='g';  14: mes:='n';  21: mes:='u';
          end;
        end;
      end;
  writeln('  ',gam,'        ',smb,'        ',mes);
 end;
var t1:=new TextABC(200,120,12,
       'send: = mlvkrejjmfwxjlieqrcklqa');
var t2:=new TextABC(200,150,11,
       'gamma: = usechanceusechanceusech');

var t3:=new TextABC(200,180,12,
       'message: = strikewhiletheironishot');
end;
//========== Main ===========
 var rcv, gg: string;
 begin
  rcv:='mlvkrejjmfwxjlieqrcklqa';
  gg:='usechanceusechanceusech';
  Decoder(rcv, gg);
 end.
```

Listing 11.8

Remark.

It is recommended to check carefully an integrity of the message, the gamma and the received sequences. They must have an equal quantity of the symbols.

11.6. Additive Viary codes

Marquise de Viary, an officer of the french army, suggested the ciphering method in 1888. The method inherits the main principles of the Ceasar's and Vigenere's algorithms, i.e. the derivative shifted alphabets and the gamming.

New elements of Viary ciphering algorithm are:

1. Symbols of the open message P_k and the gamma G_k are coded by integer numbers, for instance by the ordinal numbers of the alphabet letters.

2. The ciphering is made by means of the operation

$$C_k = (P_k + G_k) \bmod N;$$

where N – quantity symbols in the alphabet.

Remark.

If numbers "x" and "y" are operands, the result is the integer rest of the division operation (x + y)/N.

Let us see the ciphering process.

Example 11.9.

The aphorism of George Bernard Shaw (1856-1950) is known:

I am an atheist and I thank God for it

Remark.

We'll cipher the message using the single alphabet. It is useful in educational purpose that the main innovation of the Viary algorithm would not be overshadowed.

Solution.

1. To do the preparations for ciphering, using secret word "aphorism"):

> **message:**   iamanatheistandithankgodforit
>
> **gamma:**   aphorismaphorismaphorismaphor

2. To number 26 letters of the english alphabet with numbers from 0 to 31.

3. To substitute the gamma symbols by the numerical codes.

4. To fulfill the summation of the message and gamma using mod = 26. If the resulp is less than 10, one ought to add the digit 0 left of one-digit code. The sequence must contain the even quantity of digits.

5. To join the symbol codes in a line of the ciphered sequence.

This algorithm is the base of the program Prog_11_9 (listing 11.9). We'll see on the screen the results (fig.97).

Figure 97. Additive ciphering

```pascal
Program Prog_11_9;
Uses ABCobjects;
var ss: array[1..26] of string;

Procedure Coder(s, g: string);
 begin
  var j, k,  len: integer;
  var m, p, mp, sp: array[1..30] of integer;
  var cod, send: string;   len:=Length(s);  cod:=' ';
  for j:=1 to len do
    begin
     case s[j] of
      'a': m[j]:=1;   'h': m[j]:=8;       'o': m[j]:=15;    'v': m[j]:=22;
      'b': m[j]:=2;   'i': m[j]:=9;       'p': m[j]:=16;    'w': m[j]:=23;
      'c': m[j]:=3;   'j': m[j]:=10;      'q': m[j]:=17;    'x': m[j]:=24;
      'd': m[j]:=4;   'k': m[j]:=11;      'r': m[j]:=18;    'y': m[j]:=25;
      'e': m[j]:=5;   'l': m[j]:=12;      's': m[j]:=19;    'z': m[j]:=26;
      'f': m[j]:=6;   'm': m[j]:=13;      't': m[j]:=20;
      'g': m[j]:=7;   'n': m[j]:=14;      'u': m[j]:=21;
     end;
     case g[j] of
      'a': p[j]:=1;   'h': p[j]:=8;       'o': p[j]:=15;   'v': p[j]:=22;
      'b': p[j]:=2;   'i': p[j]:=9;       'p': p[j]:=16;   'w': p[j]:=23;
      'c': p[j]:=3;   'j': p[j]:=10;      'q': p[j]:=17;   'x': p[j]:=24;
      'd': p[j]:=4;   'k': p[j]:=11;      'r': p[j]:=18;   'y': p[j]:=25;
      'e': p[j]:=5;   'l': p[j]:=12;      's': p[j]:=19;   'z': p[j]:=26;
      'f': p[j]:=6;   'm': p[j]:=13;      't': p[j]:=20;
      'g': p[j]:=7;   'n': p[j]:=14;      'u': p[j]:=21;
     end;
     mp[j]:=m[j]+p[j];
     if mp[j]< 26 then sp[j]:=mp[j]
     else sp[j]:=mp[j]-26 ;
     if sp[j]<10 then cod:=cod+0+sp[j]
     else cod:=cod+sp[j];
    end;
   writeln('     ',s);  writeln('     ',g);    write('    ',cod);
 end;
//========= Main =========
var s, g: string;
begin
 s:='iamanatheistandithankgodforit';
```

```
  g:='aphorismaphprismaphorismaphor';
  Coder(s,g);
end.
```

Listing 11.9

Example 11.10. Develop the program that deciphers the sequence:

$$101721160610132106250110192323222 1240$$
$$90303160781707005002412$$

It is known previously:

- the message is written in English with small letters;

- the sequence is a result of an additional ciphering;

- the gamma is made with parole "aphorism".

Soluion.

1. Calculate the quantity of symbols in the received sequence.

2. Select pairs of symbols.

3. Transform the pairs in two-digits decimal numbers and save them in the array s[j].

4. Transform the gamma symbols into the number array g[j] being the ordinal numbers of the russian alphabet letters.

5. Calculate the symbol codes dc[j] of the initial message.

6. Transform codes dc[j] into the symbols of the ruissuian alphabet.

The program Prog_11_10 (listing 11.10) realizes this algorithm and prints the result on the screen (fig.98).

```
Program Prog_11_10;
Uses ABCobjects;
var s, g: array[1..60] of integer;
var ms: array[1..60] of string;
```

Figure 98. Restoration of the message

```
Procedure Decoder(rc,gg: string);
begin
  var j, k, lens, leng, x, code, dcd: integer;
  var y, z, dd, smb, mes: string;
  var dc, msg: array[1..70] of integer;
  var snd: array[1..70] of string;
  lens:=Length(rc);
  for k:=1 to lens do
    begin
     x:= (k mod 2);   if x=1 then  y:=rc[k];
     if x=0 then
       begin
        z:=rc[k]; j:=j+1; dd:=y+z;  Val(dd,s[j],code);
       end;
   end;
  leng:=Length(gg);
  for j:=1 to leng do
    begin
     case gg[j] of
      'a': g[j]:=1;    'h': g[j]:=8;     'o': g[j]:=15;     'v': g[j]:=22;
      'b': g[j]:=2;    'i': g[j]:=9;     'p': g[j]:=16;     'w': g[j]:=23;
      'c': g[j]:=3;    'j': g[j]:=10;    'q': g[j]:=17;     'x': g[j]:=24;
      'd': g[j]:=4;    'k': g[j]:=11;    'r': g[j]:=18;     'y': g[j]:=25;
      'e': g[j]:=5;    'l': g[j]:=12;    's': g[j]:=19;     'z': g[j]:=26;
      'f': g[j]:=6;    'm': g[j]:=13;    't': g[j]:=20;
      'g': g[j]:=7;    'n': g[j]:=14;    'u': g[j]:=21;
     end;
    dc[j]:=s[j]+26-g[j];
    if dc[j]<26 then dcd:=dc[j] else dcd:=dc[j]-26;
      case dcd of
        1: ms[j]:='a';  8: ms[j]:='h';  15: ms[j]:='o'; 22: ms[j]:='v';
        2: ms[j]:='b';  9: ms[j]:='i';  16: ms[j]:='p'; 23: ms[j]:='w';
```

```pascal
      3: ms[j]:='c';  10: ms[j]:='j';   17: ms[j]:='q';   24: ms[j]:='x';
      4: ms[j]:='d';  11: ms[j]:='k';   18: ms[j]:='r';   25: ms[j]:='y';
      5: ms[j]:='e';  12: ms[j]:='l';   19: ms[j]:='s';   26: ms[j]:='z';
      6: ms[j]:='f';  13: ms[j]:='m';  20: ms[j]:='t';
      7: ms[j]:='g';  14: ms[j]:='n';  21: ms[j]:='u';
    end;
  end;
  writeln(' ',rc);   writeln(' ',gg);   writeln(' ',ms);
end;
//=============== Main ===============
var rc, gg: string;
begin
    rc:='10172116061013210625010919232 3
         2221240903031608170705002412';
    gg:='aphorismaphorismaphorismaphor';
    Decoder(rc,gg);
end.
```

Listing 11.10

11.7. Enciphering by the matrix algebra methods

The vector productions are often used in the matrix algebra. The text message may be transformed into the digit vector with components being the codes of an alphabet symbols. Hence, the enciphering may be done by means of the multiplication of the message number vector by a secret square number matrix [14].

There are stages of an enciphering algorithm:

1. Change an alphabet symbols with their ordinal numbers from 1 to N.

2. Write the coded text with the letter numbers.

3. Distribute the number sequence into the triads. If the quantity of the initial text symbols is not divisible by 3, then the

text line ought to be supplemented with one or two special symbols, for instance, the symbol "_".

    4. Create the square key matrix which satisfies two conditions:

- it is symmetrical relatinely the main diagonal;
- matrix determinant is equal to 1.

The matrix example is:

$$A = \begin{pmatrix} a_{11} & a_{12} & a_{13} \\ a_{21} & a_{22} & a_{23} \\ a_{31} & a_{32} & a_{33} \end{pmatrix} = \begin{pmatrix} 14 & 8 & 3 \\ 8 & 5 & 2 \\ 3 & 2 & 1 \end{pmatrix}$$

    5. Confirm that determinant det of the matrix A is equal to 1:

$$\det = a_{11}{}^*(a_{12}{}^*a_{33} - a_{23}{}^*a_{32}) - a_{12}{}^*(a_{21}{}^*a_{33} - a_{31}{}^*a_{23}) +$$
$$+ a_{13}{}^*(a_{21}{}^*a_{32} - a_{22}{}^*a_{31})$$

    6. Calculate the triad elements of the ciphered message:

$$A * \begin{pmatrix} s_{1j} \\ s_{2j} \\ s_{3j} \end{pmatrix} = \begin{pmatrix} 14*s_{1j} & 8*s_{2j} & 3*s_{3j} \\ 8*s_{1j} & 5*s_{2j} & 2*s_{3j} \\ 3*s_{1j} & 2*s_{2j} & 1*s_{3j} \end{pmatrix}$$

where j = 1, 2, 3, …, N – triad number in the message;

        s_{ij} – symbol ordinal number in the alphabet.

Remark.

The length of the ciphered message increases 3 times.

Example 11.11.

Develop the program which ciphers an italian proverb:

La poverita non fa vergogna

(The poverty is not shameful)

Solution.

The program Prog_11_11 (listing 11.11) forms 9 triads $s1$, $s2$, $s3$ and joins them into the line at the input of the communication channel (fig.99).

Remark.

The result is printed in two parts since the length the output line is more than the page width.

```pascal
Program Prog_11_11;
Uses ABCobjects;
var td, vp: array[0..35,0..35] of integer;
Procedure Coder(s: string);
 begin
  var j, k, n, da, e1, e2, e3: integer;
  var send1, send2, s1, s2, s3: string;
  var d: array[0..35] of integer;
  var a: array[0..22,0..22] of integer;
  var b: array[0..35,0..35] of integer;
  n:=Length(s);   j:=0;
 for k:=1 to n do
   begin
   case s[k] of
    'a': d[k]:=1;     'h': d[k]:=8;   'o': d[k]:=15;   'v': d[k]:=22;
    'b': d[k]:=2;     'i': d[k]:=9;   'p': d[k]:=16;   'w': d[k]:=23;
    'c': d[k]:=3;     'j': d[k]:=10;  'q': d[k]:=17;   'x': d[k]:=24;
    'd': d[k]:=4;     'k': d[k]:=11;  'r': d[k]:=18;   'y': d[k]:=25;
    'e': d[k]:=5;     'l': d[k]:=12;  's': d[k]:=19;   'z': d[k]:=26;
    'f': d[k]:=6;     'm': d[k]:=13;  't': d[k]:=20;   '-': d[k]:=27;
    'g': d[k]:=7;     'n': d[k]:=14;  'u': d[k]:=21;   '_': d[k]:=28;
   end;
```

```pascal
//----Triads before coding
      if k mod 3 = 0 then
        begin
          j:=j+1;  td[j,3]:=d[k];
            td[j,2]:=d[k-1]; td[j,1]:=d[k-2];
        end;
    end;
//-------Key matrix
    a[1,1]:=14; a[1,2]:=8;  a[1,3]:=3;
    a[2,1]:=8;   a[2,2]:=5;  a[2,3]:=2;
    a[3,1]:=3;   a[3,2]:=2;  a[3,3]:=1;
//-----Triads after coding
    send1:=' ';  send2:=' ';
    writeln('  k   s1    s2     s3');
    writeln('------------------------------');
    for k:=1 to 9 do
      begin
        e1:=a[1,1]*td[k,1];        e2:=a[1,2]*td[k,2];
        e3:=a[1,3]*td[k,3];        vp[k,1]:=e1+e2+e3;
        e1:=a[2,1]*td[k,1];        e2:=a[2,2]*td[k,2];
        e3:=a[2,3]*td[k,3];        vp[k,2]:=e1+e2+e3;
        e1:=a[3,1]*td[k,1];        e2:=a[3,2]*td[k,2];
        e3:=a[3,3]*td[k,3];        vp[k,3]:=e1+e2+e3;
        s1:=IntToStr(vp[k,1]);  s2:=IntToStr(vp[k,2]);
        if vp[k,3]>99 then s3:=IntToStr(vp[k,3])
            else s3 :='0'+IntToStr(vp[k,3]);
        writeln('  ',k,'  ',s1,'  ',s2,'  ',s3);
        if k<6 then send1:=send1+s1+s2+s3
                else send2:=send2+s1+s2+s3;
      end;
    writeln('------------------------------');
    writeln('  send_1:= ',send1);
  writeln('  send_2:= ',send2);
  end;
//============== Main ==================
var s: string;
begin
  s:='la_poverita_non_fa_vergogna';  Coder(s);
end.
```

Listing 11.11

```
🖳 ABC Objects

k   s1   s2   s3
---------------------------------
1  260  157  066
2  410  247  100
3  241  148  060
4  372  221  090
5  358  215  086
6  443  256  097
7  583  344  133
8  353  209  083
9  213  128  050
---------------------------------
send_1:=  260157066410247100241148060372221090358215086
send_2:=  443256097583344133353209083213128050
```

Figure 99. Matrix enciphering

Now we'll consider the deciphering algorithm.

1. Transform the received ciphered line into the sequence of decimal digits.

2. Distribute number sequence into triads.

3. Calculate elements "b" of the accessory matrix :"B" using elements "a" of the key matrix "A":

$$b_{11} = a_{22} * a_{33} - a_{23} * a_{32}; \quad b_{21} = -a_{21} * a_{33} + a_{23} * a_{31};$$

$$b_{12} = -a_{12} * a_{33} + a_{13} * a_{32}; \quad b_{22} = a_{11} * a_{33} - a_{13} * a_{31};$$

$$b_{13} = a_{12} * a_{23} - a_{13} * a_{22}; \quad b_{23} = -a_{11} * a_{23} + a_{13} * a_{21};$$

$$b_{31} = a_{21} * a_{32} - a_{22} * a_{31};$$

$$b_{32} = -a_{11} * a_{32} + a_{12} * a_{31};$$

$$b_{33} = a_{11} * a_{22} - a_{12} * a_{21}.$$

4. Divide every element of the accessory matrix by the determinant to get the inverse matrix:

$$A^{-1} = \begin{pmatrix} 1 & -2 & 1 \\ -2 & 5 & -4 \\ 1 & -4 & 6 \end{pmatrix}$$

5. Do deciphering of the received sequence by means of vector multiplication:

$$msg = A^{-1} * \begin{pmatrix} c_{1j} \\ c_{2j} \\ c_{3j} \end{pmatrix} = \begin{pmatrix} 1*c_{1j} & -2*c_{2j} & 1*c_{3j} \\ -2*c_{1j} & 5*c_{2j} & -4*c_{3j} \\ 1*c_{1j} & -4*c_{2j} & 6*c_{3j} \end{pmatrix}$$

where

$j = 1, 2, 3, \ldots, N$ – triad number of the initial message;

c_{ij} – 3-digit number which substitutes the message symbol;

$i = 1, 2, 3$ – the ordinal number of the digit in the triad.

Example 11.12.

Do deciphering two parts of the received sequence.

```
send_1:= 2601570664102471002411480603722210903582 15086
send_2:= 44325609758334413335320908321312805 0
```

Preliminary information is accessible:

1. The initial message is written in Italian.

2. The alphabet symbols have been substituted by the ordinal numbers.

3. The ciphered sequence has been got by means of matrix multiplication.

4. The secret matix is:

$$A = \begin{pmatrix} 14 & 8 & 3 \\ 8 & 5 & 2 \\ 3 & 2 & 1 \end{pmatrix}$$

Remark.

Let suppose the reverse matrix has been calculated before program developing. The deciphering procedure is shown in listing 11.12 and fig.100.

```pascal
Program Prog_11_12;
Uses ABCobjects;
var c: array[0..99] of integer;
var d1, d2, d3: array[1..90] of integer;
var vp: array[0..99,0..99] of integer; var b: string;
Procedure Transform(s: string);
begin
  var j, k, m, n, x, code, z1, z2, z3: integer;
  var s1,s2,s3: string;    var smb: char;
  var c1,c2,c3:  array[1..90] of integer;
  var rm: array[1..3,1..3] of integer;
//-------------------Reverse matrix--------------
  rm[1,1]:=1;    rm[1,2]:=-2;    rm[1,3]:=1;
  rm[2,1]:=-2;   rm[2,2]:=5;     rm[2,3]:=-4;
  rm[3,1]:=1;    rm[3,2]:=-4;    rm[3,3]:=6;
//-------------------Number triads--------------
  n:=Length(s);
  for k:=1 to n do
   begin
     x:= (k mod 3);
    if x=0 then
     begin
      m:=k div 3; j:=((m-1) mod 3)+1;
      n:=((m-1) div 3)+1;     s3:=s[k-2];
      Val(s3,d3[k],code);     s2:=s[k-1];
      Val(s2,d2[k],code);     s1:=s[k];
      Val(s1,d1[k],code);
      c[m]:=d3[k]*100+d2[k]*10+d1[k];
    end;
  end;
end;
```

m	c(m)	decod	received
1	260	12	l
2	157	1	a
3	66	28	_
4	410	16	p
5	247	15	o
6	100	22	v
7	241	5	e
8	148	18	r
9	60	9	i
10	372	20	t
11	221	1	a
12	90	28	_
13	358	14	n
14	215	15	o
15	86	14	n
16	443	28	_
17	256	6	f
18	97	1	a
19	583	28	_
20	344	22	v
21	133	5	e
22	353	18	r
23	209	7	g
24	83	15	o
25	213	7	g
26	128	14	n
27	50	1	a

Figure 100. Matrix deciphering of the message

```
Function Symb(num: integer):string;
begin
  case num of
    1: b:='a';      8: b:='h';     15: b:='o';     22: b:='v';
    2: b:='b';      9: b:='i';     16: b:='p';     23: b:='w';
    3: b:='c';     10: b:='j';     17: b:='q';     24: b:='x';
    4: b:='d';     11: b:='k';     18: b:='r';     25: b:='y';
    5: b:='e';     12: b:='l';     19: b:='s';     26: b:='z';
    6: b:='f';     13: b:='m';     20: b:='t';     27: b:='-';
    7: b:='g';     14: b:='n';     21: b:='u';     28: b:='_';
  end;
end;
Procedure Decoder;
begin
```

```pascal
  var j, k, m, n, z1, z2, z3, num: integer;
  var rcv: string;
  writeln(' m    c(m)  decod  received');
  writeln('------------------------------------------------' );
  m:=1;  k:=-2;  rcv:=' ';
  while m<26 do
    begin
      k:=k+3;   z1:=1*c[m]-2*c[m+1]+1*c[m+2];
      z2:=-2*c[m]+5*c[m+1]-4*c[m+2];
      z3:=1*c[m]-4*c[m+1]+6*c[m+2];  num:=z1;  Symb(num);
      writeln(' ',k,'      ',c[m],'        ',z1,'            ',b);
      rcv:=rcv+b; num:=z2; Symb(num);
      writeln(' ',k+1,'      ',c[m+1],'        ',z2,'            ',b);
      rcv:=rcv+b;  num:=z3;  Symb(num);
      writeln(' ',k+2,'      ',c[m+2],'        ',z3,'            ',b);
      rcv:=rcv+b; m:=m+3;
    end;
    var t1:= new TextABC(215,60,12,'Restored message:');
    var t2:= new TextABC(212,85,12,rcv);
end;
//================= Main =====================
var s: string;    var num: integer;
begin
  s:='26015706641024710024114806037222109035821 50
     8644325609758334413335320908321312 8050';
  Transform(s);
  Decoder;
end.
```

Listing 11.12

Remark.

Three columns of the table in fig.100 contains:

1) c(m)   – triads of the received sequence;

2) decod   – deciphered numbers;

3) received   – restored message.

11.8. Brute force attack

We would remind you that cryptoanalysis is an investigation of the enciphered message in purpose of illegal access to its content. Alike "analysis" is called a computer attack or a break in criminal termines. The aim of the cryptoanalysis is to open algorithm of ciphering or to fit a parole, secret key.

A preliminary information is necessary for a successful attack, for instance:

- analysis of the subject (an intercepted message);

- the space of a knowledge;

 - the languge of the initial message;

- the length and the sructure of the message et cetera.

Example 11.13.

Emulate the break attempt. There are two problems:

1. Encipher the latin athorism, using the Ceasar's algorithm with secret alphabet shift value sh = 7:

        Si vis pacem – para bellum

(If you want peace, be prepared for war)

2. Decipher the coded text line by the brute force attack.

Solution.

The program Prog_11_13 (listing 11.13) solves the first problem (fig.101).

```
Program Prog_11_13;
Uses ABCobjects;
var s: array[1..50] of string;
var m, sn: array[1..50] of integer;

Procedure Coder(sh: integer; mes: string);
begin
  var j, len, msg: integer;
  var snd: string;
```

```pascal
    len:=Length(mes);
      case mes[j] of
        'a': m[j]:=1;    'h': m[j]:=8;      'o': m[j]:=15;    'v': m[j]:=22;
        'b': m[j]:=2;    'i': m[j]:=9;      'p': m[j]:=16;    'w': m[j]:=23;
        'c': m[j]:=3;    'j': m[j]:=10;     'q': m[j]:=17;    'x': m[j]:=24;
        'd': m[j]:=4;    'k': m[j]:=11;     'r': m[j]:=18;    'y': m[j]:=25;
        'e': m[j]:=5;    'l': m[j]:=12;     's': m[j]:=19;    'z': m[j]:=26;
        'f': m[j]:=6;    'm': m[j]:=13;     't': m[j]:=20;
        'g': m[j]:=7;    'n': m[j]:=14;     'u': m[j]:=21;
      end;
      if (m[j]>sh) and (m[j]<27-sh) then sn[j]:=m[j]+sh;
      if m[j]<=sh then sn[j]:=sh+m[j];
      if m[j]>=27-sh then sn[j]:=8-sh;  if m[j]=21 then sn[j]:=2;
      case sn[j] of
        1: s[j]:='a';     8: s[j]:='h';    15: s[j]:='o';    22: s[j]:='v';
        2: s[j]:='b';     9: s[j]:='i';    16: s[j]:='p';    23: s[j]:='w';
        3: s[j]:='c';    10: s[j]:='j';    17: s[j]:='q';    24: s[j]:='x';
        4: s[j]:='d';    11: s[j]:='k';    18: s[j]:='r';    25: s[j]:='y';
        5: s[j]:='e';    12: s[j]:='l';    19: s[j]:='s';    26: s[j]:='z';
        6: s[j]:='f';    13: s[j]:='m';    20: s[j]:='t';
        7: s[j]:='g';    14: s[j]:='n';    21: s[j]:='u';
      end;
      snd:=snd+s[j];
    end;
    var t3:=new TextABC(200,110,12, 'Shift = '+IntTostr(sh));
    var t4:=new TextABC(195,140,12, 'Send = '+snd);
  end;
  //================== Main ====================
var mes: string;
var sh: integer;
begin
  mes:='sivispacemparabellum';
  sh:=7;  Coder(sh, mes);
  var t0:=new TextABC(240,20,12,  'Algorithm of Ceasar');
  var t1:=new TextABC(270,50,12,  'Enciphering');
  var t2:=new TextABC(170,80,12,  'Message = '+mes);
end.
```

Listing 11.13

The program Prog_11_14 (listing 11.14) solves the second problem. We'll suppose that a breaker does not know the secret shift of the initial alphabet (fig.101), so he is to fit value "sh" step by step.

```
Program Prog_11_14;
Uses ABCobjects;
var s: array[1..50] of string;
var r: array[1..50] of integer;

Procedure Decoder(sh: integer; rcv: string);
begin
  var j, len: integer;
  var msg: string;
  var sn: array[1..50] of integer;
  len:= Length(rcv);    msg:=' ';
  for j:=1 to len do
    begin
  case rcv[j] of
    'a': r[j]:=1;      'h': r[j]:=8;       'o': r[j]:=15;     'v': r[j]:=22;
    'b': r[j]:=2;      'i': r[j]:=9;       'p': r[j]:=16;    'w': r[j]:=23;
    'c': r[j]:=3;      'j': r[j]:=10;    'q': r[j]:=17;    'x': r[j]:=24;
    'd': r[j]:=4;      'k': r[j]:=11;    'r': r[j]:=18;     'y': r[j]:=25;
    'e': r[j]:=5;      'l': r[j]:=12;    's': r[j]:=19;    'z': r[j]:=26;
    'f': r[j]:=6;    'm': r[j]:=13;    't': r[j]:=20;
    'g': r[j]:=7;    'n': r[j]:=14;    'u': r[j]:=21;
  end;
  if r[j]>sh then sn[j]:=r[j]-sh;
  if r[j]<=sh then sn[j]:=r[j]+26-sh;
  case sn[j] of
    1: s[j]:='a';    8: s[j]:='h';    15: s[j]:='o';    22: s[j]:='v';
    2: s[j]:='b';    9: s[j]:='i';    16: s[j]:='p';    23: s[j]:='w';
    3: s[j]:='c';  10: s[j]:='j';    17: s[j]:='q';    24: s[j]:='x';
    4: s[j]:='d';  11: s[j]:='k';    18: s[j]:='r';    25: s[j]:='y';
    5: s[j]:='e';  12: s[j]:='l';    19: s[j]:='s';    26: s[j]:='z';
    6: s[j]:='f';  13: s[j]:='m';  20: s[j]:='t' ;
    7: s[j]:='g';  14: s[j]:='n';  21: s[j]:='u';
    24: s[j]:='x';
  end;
    msg:=msg+s[j];
```

```
    end;
  if sh<10 then writeln('      ',sh,'   ',msg)
             else writeln('    ',sh,'   ',msg);
end;
//================ Main ==============
var rcv: string; var sh: integer;
begin
 rcv:='zpcpzwhjltwhyhilssbt';
 writeln('    sh          Message');
 writeln(' ------------------------------------------------');
 for sh:=1 to 14 do
   begin
     Decoder(sh,rcv);
   end;
end.
```

Listing 11.14

Algorithm of Ceasar

Enciphering

Message = sivispacemparabellum

Shift = 7

Send = zpapzwhjltwhyhilssbt

Figure 101. The ciphered message

The program Prog_11_14 deciphered the coded message 14 times with different values of parameter sh = 1, 2, ..., 14 and printed results on the screen (fig.102).

The intelligent answer is found in the 7th line, so the coded message has been broken successively.

11.9. Statistics of european alphabets

Every language has a specific repetition of symbols as well repetition of symbol combinations (with 2 or 3 symbols). Reiteration frequence is an important information for cryptoanalists.

Statistical properties of european languges are well studied and published. However, one ought to investigate the language statistics without assistance when he is studying the numerical analysis and the Pascal ABC. Let us see how it may be realized.

```
sh        Message
---------------------------------------------
 1   yooyvgiksvgxghkrrs
 2   xnnxufhjrufwfgjqqzr
 3   wmzmwtegiqtevefippyq
 4   vlylvsdfhpsdudehooxp
 5   ukxkurcegorctcdgnnwo
 6   tjwjtqbdfnqbsbcfmmvn
 7   sivispacemparabellum
 8   rhuhroabdloaqaadkktl
 9   qgtgqnaacknapaacjjsk
10   pfsfpmaabjmaoaabirj
11   oereolaaailanaaahhqi
12   ndqdnkaaahkamaaaggph
13   mcpcmjaaagjalaaaffog
14   lbobliaaafiakaaaeenf
```

Figure 102. Variants of deciphering

Example 11.14.

Investigate the symbol reiteration in the business english language using a text fragment.

Solution.

The program Prog_11_15 (listing 11.15) is to estimate the frequency of every alphabet symbol and to show results:

- symbol repetition in % to the suggested letter array;

- frequence hystogram (fig.103).

```pascal
Program Prog_11_15;
Uses GraphABC, ABCobjects;
const M=50;

Procedure ClearArray;
var j, k, N: integer;   var f: array[1..M] of integer;

begin
  for j:=1 to 26 do
    begin
      f[j]:=0;
    end;
end;

Procedure CoordSys;
  const x0=100;
  var j, k, vc, yc: integer;     var s: string;
  begin
    SetWindowSize(900,500); SetPenStyle(psClear);
    SetPenColor(clBlack);       SetPenWidth(3);
    Line(140, 50, 140, 440);   Line(140, 440, 800, 440);
    SetPenWidth(1);             Line(800, 50, 800, 440);
    for k:=1 to 26 do
      begin
        Line(140+25*k, 440, 140+25*k, 448);
      end;
    for k:=1 to 13 do
      begin
        Line(140, 440-30*k, 800, 440-30*k);
      end;
for k:=1 to 26 do
    begin
      xc:=2*k;
      Case k of
          1: s:='a'      8: s:='h';    15: s:='o';     22: s:='v';
          2: s:='b';     9: s:='i';    16: s:='p';     23: s:='w';
          3: s:=c;      10: s:='j';    17: s:='q';     24: s:='x';
          4: s:='d';    11: s:='k';    18: s:='r';     25: s:='y';
          5: s:='e';    12: s:='l';    19: s:='s';     26: s:='z'
```

```pascal
    6: s:='f';      13: s:='m';    20: s:='t';
    7: s:='g';      14: s:='n';    21: s:='u';
  end;
  var t:= new TextABC(137+25*k,452,12,s);
  end;
 for k:=1 to 13  do
  begin
   yc:=1*(k-1); s:=IntToStr(yc);
   var t:= new TextABC(115,430-30*(k-1),12,s);
  end;
end;

Procedure Stat(s:string);
begin
  var k, j, N, sum: integer;
  var f: array[1..M] of integer;
  var p: array[1..26] of real;
  var a: array[1..26] of string;
  var xp, yp: integer;
  var h1, h2: string;
  ClearArray;     N:=Length(s);
 for k:=1 to N do
  begin
   case s[k] of
'a': f[1]:=f[1]+1;  'h': f[8]:=f[8]+1;     'o': f[15]:=f[15]+1;
 'v': f[22]:=f[22]+1;
'b': f[2]:=f[2]+1;   'i': f[9]:=f[9]+1;      'p': f[16]:=f[16]+1;
 'w': f[23]:=f[23]+1;
'c': f[2]:=f[3]+1;    'j': f[10]:=f[10]+1;  'q': f[17]:=f[17]+1;
 'x': f[24]:=f[24]+1;
'd': f[4]:=f[4]+1;   k': f[11]:=f[11]+1;   'r': f[18]:=f[18]+1;
 'y': f[25]:=f[25]+1;
'e': f[5]:=f[5]+1;   'l': f[12]:=f[12]+1;  's': f[19]:=f[19]+1;
 'z': f[26]:=f[26]+1;
 'f': f[6]:=f[6]+1; 'm': f[13]:=f[13]+1;   't': f[20]:=f[20]+1;
'g': f[7]:=f[7]+1; 'n': f[14]:=f[14]+1;  'u': f[21]:=f[21]+1;
    end;
   end;
  sum:=0;
 for k:=1 to 26 do
  begin
   sum:=sum+f[k];
```

```pascal
      end;
   for k:=1 to 26 do
     begin
      p[k]:= f[k]*100/sum;
      Case k of
        1: a[k]:='a';    2: a[k]:='b';    3: a[k]:='c';    4: a[k]:='d';
        5: a[k]:='e';    6: a[k]:='f';    7: a[k]:='g';    8: a[k]:='h';
        9: a[k]:='i';   10: a[k]:='j';   11: a[k]:='k';   12: a[k]:='l';
       13: a[k]:='m';  14: a[k]:='n';   15: a[k]:='o';   16: a[k]:='p';
       17: a[k]:='q';  18: a[k]:='r';   19: a[k]:='s';   20: a[k]:='t';
       21: a[k]:='u';  22: a[k]:='v';   23: a[k]:='w';   24: a[k]:='x';
       25: a[k]:='y';  26: a[k]:='z';
      end;
      writeln(' ',a[k],'    ',p[k]:3:1);
     end;
   for k:=1 to 26 do
    begin
     xp:=140+25*k;     yp:=440-Round(p[k]*30);
     SetPenWidth(6);  SetPenColor(clBlack);
     Line(xp, 440, xp, yp);
    end;
   h1:='Business  english  statistics';
   var t1:= new TextABC(200,20,14,h1);
   h2:='Array: 1156  symbols';
   var t2:= new TextABC(500,20,12,h2);
end;
//=============== Main =====================
var s:string;
   begin
      CoordSys;
      s:='dear sirs, we thank you for your comments and pro-
posals on our draft contract for the construction';
      s:=s+' of  the assembly plant. we are glad to tell you that
having studied your amendments, ewe accept';
      s:=s+' them on the whole and are ready to introduce them
into our contract. at the same time we would like';
      s:=s+' to re-state that in accordance with our contract for
the construction of the car assembly plant on';
      s:=s +' a "turn-key" basis,the contractor assumes full re-
sponsibility for the organization and execution';
```

s:=s+' of all civil works. the civil works shall be carried out by both russian specialists and local firms';

s:=s+' engaged as subcontractors. the works to be executed by your local firms shall be supervised by our';

s:=s+' competent specialists, the cost of their services is included in the contract price. we have already';

s:=s+' competent specialists, the cost of their services is included in the contract price. we have already';

s:=s+' made enquires about the firms and believe that they have sufficient experience, competence and';

s:=s+' faclities to execute the works we intended to entrust to them. we ask you to consider the present';

s:=s+' letter and if the contract is of interest to you let us know by email when when you are ready';

s:=s+' to sign it. We look forward to your reply. yours faithfully, jeffrey jacobson';

    Stat(s);
end.

Listing 11.15

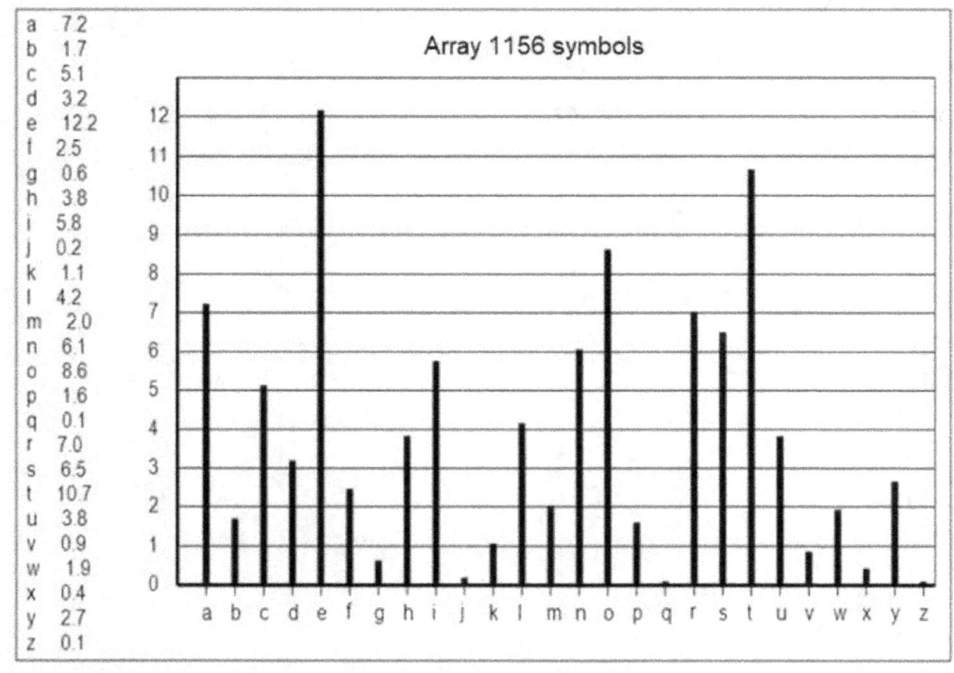

Figure 103. The symbol repetition in the business English

Conclusion.

The program has detected the most frequently symbols used in the business english – a, e, i, n, o, r, s, t. Rarely used symbol are j, q, x, z.

This information helps to find the truth letters in a coded symbol sequence even if a substitution would be done. Besides, one may find that an unknown text is written in English despite of ciphering.

Example 11.15.

Investigate the symbol reiteration in I talian language using a paper fragment.

Solution.

The program Prog_11_16 (listing 11.16) is to estimate the frequency of every alphabet symbol and to show results:

- symbol repetition in % to the suggested letter array;

- frequence hystogram (fig.104).

```pascal
Program Prog_11_16;
Uses GraphABC, ABCobjects;
const M=50;

Procedure ClearArray;
var j, k, N: integer;
var f: array[1..M] of integer;
begin
  for j:=1 to 26 do
    begin
      f[j]:=0;
    end;
end;

Procedure CoordSys;
  const x0=100;
  var j, k, xc, yc: integer;
  var s: string;
  begin
```

```pascal
SetWindowSize(820,500);  SetPenStyle(psClear);
SetPenColor(RGB(1,1,1));  SetPenWidth(3);
Line(140, 20, 140, 440);        Line(140, 440, 800, 440);
SetPenWidth(1);                 Line(800, 20, 800, 440);
  for k:=1 to 26 do
    begin
      Line(140+25*k, 440, 140+25*k, 448);
    end;
  for k:=1 to 15 do
    begin
      Line(140, 440-30*k, 800, 440-30*k);
    end;
  for k:=1 to 26 do
  begin
    xc:=2*k;
    Case k of
       1: s:='a';      8: s:='h';    15: s:='o';     22: s:='v';
       2: s:='b';      9: s:='i';    16: s:='p';     23: s:='w';
       3: s:='c';     10: s:='j';    17: s:='q';     24: s:='x';
       4: s:='d';     11: s:='k';    18: s:='r';     25: s:='y';
       5: s:='e';     12: s:='l';    19: s:='s';     26: s:='z';
       6: s:='f';     13: s:='m';    20: s:='t';
       7: s:='g';     14: s:='n';    21: s:='u';
    end;
    var t:= new TextABC(137+25*k,452,12,s);
  end;
  for k:=1 to 15  do
  begin
    yc:=1*(k-1);
    s:=IntToStr(yc);
    var t:= new TextABC(115,430-30*(k-1),12,s);
  end;
end;

Procedure Stat(s:string);
begin
  var k, j, N, sum: integer;
  var f: array[1..M] of integer;
  var p: array[1..26] of real;
  var a: array[1..26] of string;
  var xp, yp: integer;
  var h1, h2: string;
```

```pascal
ClearArray;
N:=Length(s);
for k:=1 to N do
  begin
    case s[k] of
'a': f[1]:=f[1]+1;    'h': f[8]:=f[8]+1;      'o': f[15]:=f[15]+1;
 v': f[22]:=f[22]+1;
'b': f[2]:=f[2]+1;     'i': f[9]:=f[9]+1;      'p': f[16]:=f[16]+1;
'w': f[23]:=f[23]+1;
'c': f[3]:=f[3]+1;     'j': f[10]:=f[10]+1; 'q': f[17]:=f[17]+1;
 'x': f[24]:=f[24]+1;
'd': f[4]:=f[4]+1;     'k': f[11]:=f[11]+1; 'r': f[18]:=f[18]+1;
 'y': f[25]:=f[25]+1;
'e': f[5]:=f[5]+1;     'l: f[12]:=f[12]+1;  's': f[19]:=f[19]+1;
  'z': f[26]:=f[26]+1;
'f':  f[6]:=f[6]+1;   'm': f[13]:=f[13]+1;  't': f[20]:=f[20]+1;
'g': f[7]:=f[7]+1;    'n': f[14]:=f[14]+1;  'u': f[21]:=f[21]+1;
    end;
  end;
 sum:=0;
 for k:=1 to 26 do
  begin
    sum:=sum+f[k];
  end;
for k:=1 to 26 do
  begin
   p[k]:= f[k]*100/sum
   Case k of
    1: a[k]:='a';       8: a[k]:='h';     15: a[k]:='o';    22: a[k]:='v';
    2: a[k]:='b';       9: a[k]:='i';     16: a[k]:='p';    23: a[k]:='w';
    3: a[k]:='c';     10: a[k]:='j';     17: a[k]:='q';    24: a[k]:='x';
    4: a[k]:='d';     11: a[k]:='k';     18: a[k]:='r';    25: a[k]:='y';
    5: a[k]:='e';     12: a[k]:='l';     19: a[k]:='s';    26: a[k]:='z';
    6: a[k]:='f';     13: a[k]:='m';     20: a[k]:='t';
    7: a[k]:='g';     14: a[k]:='n';     21: a[k]:='u';
   end;
   writeln('  ',a[k],'      ',p[k]:4:2);
  end;
 for k:=1 to 26 do
  begin
   xp:=140+25*k;     yp:=440-Round(p[k]*30);
```

```pascal
    SetPenWidth(6);   SetPenColor(clBlack);
    Line(xp, 440, xp, yp);
  end;
  h1:='Statistic of italian language';
  var t1:= new TextABC(200,0,12,h1);
  h2:='Array: 1661  symbols';
  var t2:= new TextABC(500,0,12,h2);
end;
//============== Main ========================
var s:string;
begin
  CoordSys;
  s:='arte, paesaggio, clima, cucina, "dolce vita", la creativita del
"made in itali", la qualita della';
  s:=s+' nostra manifattura. l"italia vanta risorse che fanno im-
pallidire gran parte dei concorrenti.';
  s:=s+' e che ci offrono oggi il trampolino ideale per spiccare
un salto in avanti nell"economia globale.';
  s:=s+' durante gli anni cinquanta e sessnta i nostri prodotti
sedussero i consumatori europei e americani,';
  s:=s+' l"italia divento la meta turistica pui ambita, il boom
diede lavoro a millioni di persone. con uno';
  s:=s+' scatto d"orgoglio, possiamo ripetere quell"esperienza
su scala piu ampia. il "made in italy" ha ancora';
  s:=s+' un potenziale di mercato enorme, il patrimonio naturale
e culturale del nostro territorio ci';
  s:=s+' consentirebbe di diventare il "giardino del planeta". e,
piu ancora che nel secondo dopoguerra,';
  s:=s+' da questo nuovo miracolo economico potrebbe scaturi-
re un fiume di posti di lavoro. solo un sogno?';
  s:=s+' per ora ovviamente si, ma non irrealizzabile. a ben ve-
dere, anzi, qualche segnale di movimento puo';
  s:=s+' gia essere colto. secondo confindustria, la produzione
di beni "belli e ben fatti (i due tratti';
  s:=s+' dell "made in italy") e in netta ripresa. parliamo di ali-
mentare, ar  redamento, abbigliamento o';
  s:=s+' tessile-casa, calzature, occhialeria. oreficeria-
gioielleria. Questi settori occupano oggi mezzo';
  s:=s+' millione di addetti, ma i margini di espansione sono
molto ampi nei prossimi cinque anni, nel paesi';
```

```
s:=s+' emergenti (cina, brasile, india, russia, turchia, emirati)
ci saranno duecento millioni di "nuovi';
    s:=s+' richi in piu rispetto a oggi, tutti potenziali consumatori di
prodotti italiani. si stimano svariati';
    s:=s+' milliardi aggiutivi di eaportazioni, con rilevanti ricadute
occupazionali anche nel mezzogiorno.';
  Stat(s);
end.
```

Listing 11.16

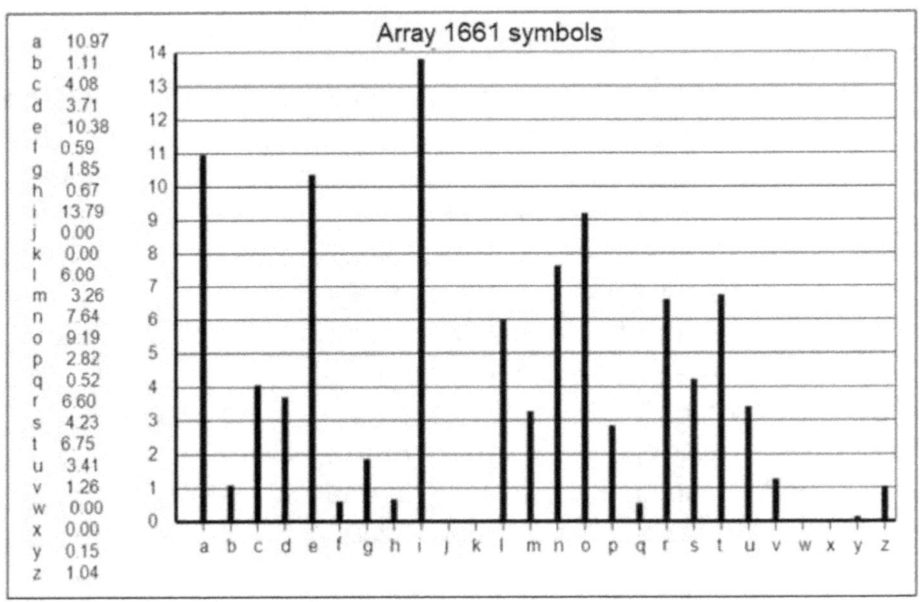

Figure 104. Symbol repetition in the italian newspaper text

Conclusion.

The most frequent symbols in the investigated text are a, e, i, l, n, o, r, t. The letters j, k, q, w, x, y came from Teutonic languages. They are met rarely in Italian language. An Italian text is recognized easily due to the statistical properties.

11.10. Mod operations

One ought to know some functions and operations in the higher algebra to understand the modern cryptograhy methods. The most important of them are:

- great common divisor;

- high degree of an integer number;

- coprime numbers;

- mod operations (add and inverse);

- unidirectional functions;

- hash-functions;

Great common divisor.

Definition.

Let "a" and "b" are positive integer numbers. The great common divisor for them is the maximum number "c" that divides both "a" and "b" without rest:

$$c = gcd(a, b)$$

There is the Euclid algorithm to find a "gcd". At first, we'll denote variables

q – quotent; r – rest.

$$a = b^*q_1 + r_1, \qquad 0 < r_1 < b;$$
$$b = r_1^*q_2 + r_2, \qquad 0 < r_2 < r_1;$$
$$r_1 = r_2^*q_3 + r_3, \qquad 0 < r_3 < r_2;$$
$$\dots \qquad\qquad \cdot \cdot$$
$$r_{k-2} = r_{k-1}^*q_k + r_k, \quad 0 < r_k < r_k;$$
$$r_{k-1} = r_k^*q_{k+1}.$$

One must fulfil the series of division operations where the new dividend gets value of the previous divisors, the new divisors is being given the value of the previous rest:

The last rest coincides with $gcd(a, b)$, if it is not equal to 0.

Example 11.16. Find gcd(7038, 5797).

Solution.

The program Prog_11_17 (listing 11.19) finds gcd = 17 (fig.105).

```pascal
Program Prog_11_17;
Uses ABCobjects;

Procedure GCD(a,b: integer);
var k, prod, gcd: integer;
var q, r, dvd, dvs: array[0..20] of integer;
var s1, s2, s3: string;
begin
writeln('   k     dvd      dvs      q        r');
writeln('------------------------------------------------');
  dvd[0]:=a;   dvs[0]:=b;
  q[0]:=dvd[0] div dvs[0];
  r[0]:=dvd[0] mod dvs[0];
  k:=0;
  while r[k]>0 do
   begin
     k:=k+1;  dvd[k]:=dvs[k-1];
     dvs[k]:=r[k-1];
     q[k]:=dvd[k] div dvs[k];
     prod:=dvs[k]*q[k];
     r[k]:=dvd[k]-prod;
     if r[k]=0 then
       begin
         gcd:=r[k-1];   break;
       end;
  writeln('   ',k,'     ',dvd[k],'   ',dvs[k],'    ',q[k],'    ',r[k]);
  end;
  s1:=IntToStr(a);
  var t1:=new TextABC(70,150,12,'a = '+s1);
  s2:=IntToStr(b);
  var t2:=new TextABC(70,170,12,'b = '+s2);
  s3:=IntToStr(gcd);
  var t3:=new TextABC(54,190,12,'gcd = '+s3);
end;
```

```
//======Main=====
var a, b: integer;
begin
  a:=7038;  b:=5797;
  GCD(a,b);
end.
```

Listing 11.17

Denotes are in fig.105:

k   - cycle step;

dvd   - dividend;

dvs   - divisor;

q   - quotient;

r   - rest.

```
 ABC Objects

  k     dvd     dvs     q       r
  -----------------------------------------           a = 7038
  1     5797    1241    4       833                    b = 5797
  2     1241    833     1       408
  3     833     408     2       17                     gcd = 17
```

Figure 105. Great common divisor. Variant 1

Let us run the program with $a = 3230$ and $b = 1216$ (fig.106).

```
 ABC Objects

  k     dvd     dvs     q       r
  -----------------------------------------           a = 3230
  1     1216    798     1       418                    b = 1216
  2     798     418     1       380
  3     418     380     1       38                     gcd = 38
```

Figure 106. Great common divisor. Variant 2

High degree of integer number

The mod degree operation is often used in cryptography:

$$w = x^d \bmod N,$$

where "x" and "d" are integer numbers and $d \gg 1$.

The result of the direct operation x^d is overflow of the computer digit spacing. One may avoid overflow if the operation is to be done at three stages:

1. Transform the exponent d into binary system:

$$d = d_0 * 2^k + \ldots + d_{k-1} * 2 + d_k,$$

where $d_0 = 1$; d_k – binary digits.

2. Set x0 = x, and then calculate

$$x_j = (x_{j-1})^2 * x^{d(j)} \bmod N, \quad j = 1, 2, \ldots, k$$

3. Set $w = x_k$.

Example 11.18.

Calculate $w = 22^{123} \bmod 53$.

Solution.

The program Prog_11_18 (listing 11.18):

- transforms the exponent d (fig.107, left);
- calculates w for $x = 22$ and $N = 53$ (fig.107, right).

```
Program Prog_11_18;
Uses ABCobjects;
var z: integer;
var a, b: array[0..10] of integer;

Procedure BinCod(d: integer);
var j, k: integer;
begin
  a[0]:=d div 2; b[0]:=d-2*a[0];
   writeln('------------------'); writeln('   j    d = ',d);
   writeln('------------------'); writeln('   0','        ',b[0]);
```

```pascal
      j:=0;
    while (a[j]>0) or (b[j]>0) do
       begin
          j:=j+1; a[j]:=a[j-1] div 2;     b[j]:=a[j-1]-2*a[j];
          if (a[j]>0) or (b[j]>0) then
             writeln('   ',j,'         ',b[j]);
       end;
       z:=j-1;
  end;

  Procedure Power(x,d,n: integer);
  var j, k, g, h, cod: integer;
  var s: array[0..9] of integer;


    begin
      writeln('-------------------------------');
      writeln('  x = ',x);  writeln('  d = ',d);
      writeln('  cod = (x ^ d)  mod  ',n);
      s[0]:=x;
      for j:=1 to z do
        begin
          if b[j]=1 then g:=x else g:=1;
          h:=s[j-1]*s[j-1];    s[j]:=g*h mod n;
        end;
      cod:=s[j];
      writeln('  cod = ',cod);
    end;
```

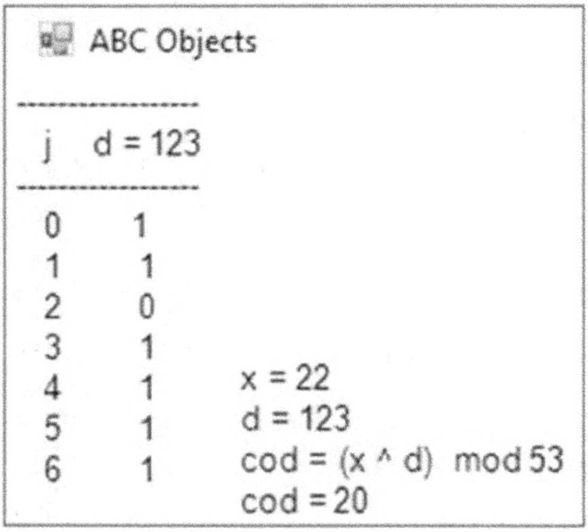

<table>
<tr><td colspan="2">🖥 ABC Objects</td></tr>
<tr><td colspan="2">--------------------</td></tr>
<tr><td>j</td><td>d = 123</td></tr>
<tr><td colspan="2">--------------------</td></tr>
<tr><td>0</td><td>1</td></tr>
<tr><td>1</td><td>1</td></tr>
<tr><td>2</td><td>0</td></tr>
<tr><td>3</td><td>1</td></tr>
<tr><td>4</td><td>1</td></tr>
<tr><td>5</td><td>1</td></tr>
<tr><td>6</td><td>1</td></tr>
</table>

Figure 107. Mod degree

```
//=== Main ======
var d, n, x: integer;
begin
  d:=123;
  BinCod(d);
  x:=22;   n:=53;
  Power(x, d, n);
end.
```

Listing 11.18

Mod inverse

The condition is often used in the asymmetrical ciphering:

$$c*d \bmod N = 1, \quad (*)$$

where: "c" and "d" – coprime numbers; c < N, d < N.

The task usualy arises: the number "s" is given; it necessaty to find number "c", which satisfies the condition (*).

Definition.

The number "c" is called an inversion of the number "d" with "mod N" if it satisfies the condition (*). It is written as an expression

$$c = d^{-1} \bmod N$$

Example 11.19.

The numbers are given: N = 77, d = 62. Find inversion denoted "c".

Solution.

Program Prog_11_19 (listing 11.19) finds inversion "c" (fig.108)

Program Prog_11_19;
Uses ABCobjects;

Procedure Invers(d,n: integer);

```
invers  condit
--------------------
  33      44
  34      29
  35      14
  36      76
  37      61
  38      46
  39      31
  40      16
  41       1
```

Number d = 62

Module N = 77

Inversion c = 41

Figure 108. Mod inversion

```pascal
var k, c, r: integer;
var s1, s2, s3: string;
begin
  writeln(' invers  condit');  writeln('--------------------');
  r:=2; k:=0;
  while r>0 do
    begin
      k:=k+1;  r:=d*k mod n;
      if r=1 then
        begin
          c:=k;  break;
        end;
      if k>32 then writeln('   ',k,'        ',r);
    end;
  s1:=IntToStr(c);  s2:=IntToStr(d);  s3:=IntToStr(n);
  var t1:=new TextABC(110,70,11,'Number d = '+s2);
  var t2:=new TextABC(110,90,11,'Module N = '+s3);
  var t3:=new TextABC(110,110,11,'Inversion c = '+s1);
  c:=d*k mod n;  writeln('   ',k,'       ',c);
end;
//=========Main=========
var d,n: integer;
begin
  d:=62;  n:=77;  Invers(d,n);
end.
```

Listing 11.19

11.11. Ciphers with asymmetrical keys

The general property of all above mentioned ciphering methods is both correspondents are being to have equal keys. Such keys are called symmetrical.

Centures-old experience and good sense promt that any cunning lock may be breaked by a skill man. It is a question of time only. Hence, one ought to change lock and key sometimes. However, how can it be done in cryptography? One may transfer the secret key tete-a-tete once only when the correspondens meet to arrange matters about future distant information interchange. Then they will be separated with great distances, and the next meeting will be in significant time or may be even never. How may the new key be transmitted to another end of the communication channel?

A vicious circle arises – the very key becomes an object of transmission. One must have the supplement secret channel to transfer the new parole. Such channel developing is a very difficult technical problem. However, mathematicians have found that one may manage without any supplement channel. They proved that one ought to use two asymmetrical keys, one key being open (non-secret). An open key may be transmitted upon the undefended channel if the key is derived from the secret key which is out of the opponent access.

The algebraical idea of asymmetrical keys is based on using unidirectional function $y = f(x)$, such that a special function exists:

$$x = f^{-1}(y).$$

Property of the special function is that one may easy to calculate value $y(x)$, but reciprocal task to calculate $x(y)$ proves to be extremely difficult. The solution is possible theoretically but cryptoanalysis will take too many time that may be excessive even for a modern computer.

So, centures-old dream of cipher specialists has come true. The witty solution found three mathematicians – Rivest, Shamir, Adleman (Massachusetts institute of tecnology, USA).

The base of the algorithm is the function:

$$y = a^x \bmod N,$$

where a – number to be transmitted upon a undefended channel;

N – coprime number, which can be divided by N and 1 only;

x – integer number (parole) from the agregate {1, 2, …, N}.

The enciphering reliability of the parole is secured by the complicacy of discrete logarythm calculating:

$$x = \log_a y \bmod N.$$

Further we'll use name "algorythm RSA".

Algorythm of key selection:

1. Select random coprime numbers P, Q. Values P, Q and P*Q are to be secret.

2. Calculate value of the Euler function eu = (P – 1)*(Q – 1).

3. Select the first key1 that satisfies conditions:

$$1 < key1 \le eu; \quad \gcd(key1, eu) = 1.$$

4. Select the second key2 using condition:

$$(key1*key2) \bmod eu = 1$$

Remarks.

1. Keys key1, key2 are equal in rights: any of them can be secret.

2. An abonent must get one of the number pairs:

(key1, N) or (key2, N).

Enciphering and deciphering algorythms

Let us consider that it is necessary to transmit secretly a parole to our abonent through the undefended channel where interchange is planned using symmetrical keys.

1. Select open key "key1".

2. Substitite every symbol $s(k)$ of the parole with its alphabet ordinal number ps(k), the number being enciphered using formula:

$$cod(k) = [ps(k)]^{opk} \bmod N$$

3. The sequence $cod(k) = [ps(k)]^{opk} \bmod N$ is supplemented with the codes of the open key and multiplied by coprime number N; then it is transmitted upon the open communication channel.

4. At another end of the channel the received sequence cod(k) is transformed into the sequence of the alphabet ordinal numbers using formula:

$$s\_num(k) = [cod(k)]^{s(k)} \bmod N$$

5. Restore the parole symbols using numbers s_num(k).

Example 11.19.

Do enciphering parole "enterprise" intended for transmission through undefended channel using the algorithm RSA.

Remark.

We'll start using small values of the coprime numbers. They make easy the procedure of program checking.

Solution.

1. Select coprime numbers P = 5, Q = 11; N = P*Q = 55.

2. Calculate eu = (P-1)*(Q-1) = 40.

3. Select random value of the open key key1 = opk = 27.

4. Develop the procedure fragment which determs the parole length (length = 10) and creates array pd[k]. The sequence is added with code opk and number N.

5. Encipher every parole element by RSA algorithm and prepare the sequence for transmitting upon the undefended channel (fig.109).

```
P = 5   Q = 11
N = 55
eu = 40
Coded word = enterprise
opk  = 27
----------------------------------------
wrd        ps          cod
----------------------------------------
  e         05           25
  n         14           09
  t         20           15
  e         05           25
  r         18           17
  p         16           36
  r         18           17
  i         09           04
  s         19           24
  e         05           25
----------------------------------------
  aux opk  = 17
     aux N  = 4
  Send = 25091525173617042425 1704
```

Figure 109. Encipherig parole by RSA algorithm

The message contains 12 pairs of decimal digits. Every one-digit number is added with the 0 on the left. The first 10 digits present parole, the 11th pair is the key, the last pair is number N.

We'll hide code "opk" in complement to number 44; "N" is complement to 59. Our abonent is not to remember all these secrets since he has the deciphering program in an electronical file which adaptes automatically to all changes of parole at the sending end of the channel.

Remark.

You are shown here an educational example only. Of course, there are possible hundreds other variants of the enciphering algorithms.

The program Prog_11_20 (listing 11.20) codes the parole.

```
Program Prog_11_20;
Uses ABCobjects;
var n, eu, opk, sk, z: integer;
var a, b, ps: array[0..15] of integer;

Procedure StartData(p, q, opk: integer; wrd: string);
begin
  writeln('  P = ',p,'  Q = ',q);  n:=p*q;
  writeln('  N = ',n); eu:=(p-1)*(q-1);  writeln('  eu = ',eu);
  writeln('  Coded word = ',wrd);       writeln('  opk  = ',opk);
end;

Procedure BinCod(opk: integer);
var j,k,d: integer;
begin
  d:=opk; a[0]:=d div 2; b[0]:=d-2*a[0];    j:=0;
  while (a[j]>0) or (b[j]>0) do
    begin
      j:=j+1;  a[j]:=a[j-1] div 2;  b[j]:=a[j-1]-2*a[j];
    end;
    z:=j-1;
end;
```

```pascal
Procedure Coder(wrd: string; p,q: integer);
var j, k, key1, len, g, h, ak, an: integer;
var pwr, wh, cod: array[0..15] of integer;
var s: array[0..15] of integer; var s1, s2, s3, s4, snd: string;
begin
  an:=59-n; ak:=44-opk; len:=Length(wrd);
  writeln('----------------------------------');
  writeln('  wrd        ps        cod');
  writeln('----------------------------------');
  for k:=1 to len do
    begin
    case wrd[k] of
     'a': ps[k]:=1;       'j': ps[k]:=10;       's': ps[k]:=19;
     'b': ps[k]:=2;       'k': ps[k]:=11;       't': ps[k]:=20;
     'c': ps[k]:=3;       'l': ps[k]:=12;       'u': ps[k]:=21;
     'd': ps[k]:=4;       'm': ps[k]:=13;       'v': ps[k]:=22;
     'e': ps[k]:=5;       'n': ps[k]:=14;       'w': ps[k]:=23;
     'f': ps[k]:=6;       'o': ps[k]:=15;       'x': ps[k]:=24;
     'g': ps[k]:=7;       'p': ps[k]:=16;       'y': ps[k]:=25;
     'h': ps[k]:=8;       'q': ps[k]:=17;       'z': ps[k]:=26;
     'i': ps[k]:=9;       'r': ps[k]:=18;       '_': ps[k]:=27;
    end;
    BinCod(opk);
    if z=1 then cod[k]:= ps[k]*ps[k]*ps[k] mod n;
    if z>1 then
      begin
        s[0]:=ps[k];
        for j:=1 to z do
          begin
            if b[j]=1 then g:=ps[k] else g:=1;
            h:=s[j-1]*s[j-1];  s[j]:=g*h mod n;
          end;
        cod[k]:=s[j];
      end;
    if ps[k]<10 then s1:='0'+IntToStr(ps[k])
       else s1:=IntToStr(ps[k]);
    if cod[k]<10 then s2:='0'+IntToStr(cod[k])
       else s2:=IntToStr(cod[k]);
    writeln('   ',wrd[k],'          ',s1,'            ',s2);
    snd:=snd+s2;
  end;
```

```
  writeln('--------------------------------------');
  if ak< 10 then s3:='0'+IntToStr(ak) else s3:=IntToStr(ak);
  if an<10  then s4:='0'+IntToStr(an) else s4:=IntToStr(an);
  writeln('  aux opk  = ',ak);   writeln('      aux N  = ',an);
  snd:=snd+s3+s4;                 writeln('  Send = ',snd);
end;
//============= Main ===============
var p, q: integer; var wrd: string;
begin
  p:=5; q:=11; opk:=27;   wrd:='enterprise';
  StartData(p,q,opk,wrd);
  BinCod(opk);    Coder(wrd,p,q);
end.
```

Listing 11.20

Example 11.20.

Encipher the line "rc": that is an enciphered word. Restore the parole.

rc = 250915251736170424251704

Algorithm.

1. Distribute the line at symbol pairs.

2. Transform pairs into digits and write them into array s[k].

3. Decipher open key "opk" and number N.

4. Find value of Euler function eu = (P-1)*(Q-1).

5. Calculate the secret key "sk" from equation:

$$(opk*sk) \bmod eu = 1$$

6. Find degree of every number s[k] mod N from the first 10 digits.

7. Transform the degree results into letters of the parole.

Solution is found by the program Prog_11_21 (listing 11.21 and fig.110).

```
Program Prog_11_21;
Uses ABCobjects;
var a, b, s: array[0..60] of integer;
var eu, n, opk, sk, u: integer;
```

```pascal
Procedure Decod_Opk_N(rc: string);
var j, k, ak, an, len, x, code: integer;
var tk, tn, y, z, dd, word: string;
begin
  len:=Length(rc); for k:=1 to len do
  for k:=1 to len do
    begin
      x:= (k mod 2); if x=1 then  y:=rc[k];
      if x=0 then
        begin
          z:=rc[k]; j:=j+1; dd:=y+z; Val(dd,s[j],code);
        end;
        if (k mod 2)=0 then
          begin
            s[k]:=s[j];
          end;
    end;
    tk:= rc[len-3]+rc[len-2]; tn:= rc[len-1]+rc[len];
    writeln('    ak = ',tk,'    an = ',tn);
    Val(tk,ak, code);
    Val(tn,an, code);
    opk:=44-ak;   n:=59-an;
    writeln('  opk = ',opk,'     N = ',n);
end;

Procedure SecretKey(p,q: integer);
var c, r, x, eu, sk: integer;
begin
  eu:=(p-1)*(q-1); r:=2; x:=0;
  while r>1 do
    begin
      x:=x+1; r:=opk*x mod eu;
      if r=1 then
        begin
          sk:=x;   break;
        end;
    end;
  writeln('    sk = ',sk,'     eu = ',eu);
end;
```

```pascal
Procedure BinCod(sk: integer);
var j, k: integer;
begin
  a[0]:=sk div 2;  b[0]:=sk-2*a[0];
  writeln('-------------------------------');
  writeln('    j      a      sk = ',sk);
  writeln('-------------------------------');
  writeln('   0','        ',a[0],'          ',b[0]);   j:=0;
  while (a[j]>0) or (b[j]>0) do
    begin
      j:=j+1;    a[j]:=a[j-1] div 2;    b[j]:=a[j-1]-2*a[j];
      if (a[j]>0) or (b[j]>0)
          then writeln('   ',j,'         ',a[j],'          ',b[j]);
    end;
    u:=j-1;
end;

Procedure Decoder(rc: string;sk,n: integer);
var j, k, g, h, len, m, p, q, x, code, an, ak: integer;
var ps, cod, sn: array[0..60] of integer;
var w: array[0..60] of string;
var y, z, dd, word: string;
begin
  writeln('  ---------------------------------');
  writeln('  cod[k]     sn[k]       smb');
  writeln('  ---------------------------------');
  len:=Length(rc);  m:=Round(len/2)-2;
  for k:=1 to m do
    begin
    if u=1 then sn[k]:=s[k]*s[k]*s[k] mod n;
      case sn[k] of
        1: w[k]:='a';     8: w[k]:='h';   15: w[k]:='o';   22: w[k]:='v';
        2: w[k]:='b';     9: w[k]:='i';   16: w[k]:='p';   23: w[k]:='w';
        3: w[k]:='c';    10: w[k]:='j';   17: w[k]:='q';   24: w[k]:='x';
        4: w[k]:='d';    11: w[k]:='k';   18: w[k]:='r';   25: w[k]:='y';
        5: w[k]:='e';    12: w[k]:='l';   19: w[k]:='s';   26: w[k]:='z';
        6: w[k]:='f';    13: w[k]:='m';   20: w[k]:='t';
        7: w[k]:='g';    14: w[k]:='n';   21: w[k]:='u';
      end;
    if u>1 then
      begin
       sn[0]:=s[k];
```

```pascal
     for j:=1 to u do
       begin
         if b[j]=1 then g:=s[k] else g:=1;
         h:=sn[j-1]*sn[j-1];      sn[j]:=g*h mod n;
       end;
   for k:=1 to m do
     begin
       if u=1 then sn[k]:=s[k]*s[k]*s[k] mod n;
        case sn[k] of
          1: w[k]:='a';      8: w[k]:='h';   15: w[k]:='o';   22: w[k]:='v';
          2: w[k]:='b';      9: w[k]:='i';   16: w[k]:='p';   23: w[k]:='w';
          3: w[k]:='c';     10: w[k]:='j';   17: w[k]:='q';   24: w[k]:='x';
          4: w[k]:='d';     11: w[k]:='k';   18: w[k]:='r';   25: w[k]:='y';
          5: w[k]:='e';     12: w[k]:='l';   19: w[k]:='s';   26: w[k]:='z';
          6: w[k]:='f';     13: w[k]:='m';   20: w[k]:='t';
          7: w[k]:='g';     14: w[k]:='n';   21: w[k]:='u';
        end;
      if u>1 then
        begin
         sn[0]:=s[k];
         begin
           if b[j]=1 then g:=s[k] else g:=1;
           h:=sn[j-1]*sn[j-1];    sn[j]:=g*h mod n;
         end;
         case sn[k] of
           1: w[k]:='a';      8: w[k]:='h';   15:  w[k]:='o';   22:  w[k]:='v';
           2: w[k]:='b';      9: w[k]:='i';   16:  w[k]:='p';   23:  w[k]:='w';
           3: w[k]:='c';     10: w[k]:='j';   17:  w[k]:='q';   24:  w[k]:='x';
           4: w[k]:='d';     11: w[k]:='k';   18:  w[k]:='r';   25:  w[k]:='y';
           5: w[k]:='e';     12: w[k]:='l';   19:  w[k]:='s';   26:  w[k]:='z';
           6: w[k]:='f';     13: w[k]:='m';   20:  w[k]:='t';
           7: w[k]:='g';     14: w[k]:='n';   21:  w[k]:='u';
         end;
       end;
     writeln('     ',s[k],'        ',sn[k],'        ',w[k]);
     word:=word+w[k];
     end;
   writeln(' -------------------------------------------------------------');
   writeln('  Received = ', rc);
   writeln('  Decoded word  =  ',word);
end;
```

```
//============= Main =============
var num, p, q: integer;
var rc: string;
begin
  rc :='2509152517361704242517 04';
  Decod_Opk_N(rc);    p:=5; q:=11;
  SecretKey(p,q);       sk:=3; BinCod(sk);
  Decoder(rc,sk,n);
end.
```

Listing 11.21

Figure 110. Deciphering of the parole

Remark.

The security of the RSA ciphering depends on values P and Q. The large values are preferable. Repeat the example 11.21 using increased values if you have time.

11.12. Digital signature

Historically a signature confirmed always authenticity of a paper document. This property remained when documents became electronic, but other signature functions were added. Now electronical digit signature (EDS) provides other document properties:

- integrity, absense of changes and falsifications after signing;

- date of signing;

- impossibility to refuse from the document.

Modern spaces of EDS application:

- financial operations;

- state and administrative control;

- taxation;

- social services;

- commerce

and so on.

The EDS carries more information in comparison with the manuscript graphical signature. The EDS has form of relatively small file which is derivated from the document and further accompanies it. The main part of the EDS is a hash-function of the document. It is a special document pattern which allows distinct it from another document. In some sense hash-function is alike a finger-print. Hence, hash-function gives a means to confirm authenticity of the document.

The preparation of digitalized document for the secure trasmission through a communication channel is alike to operation of message enciphering that was considered above in this chapter.

Algorithm RSA was used in the first state EDS standards of USA and Russia (near 1990). Modern standards are based on the elliptical functions. They provide the best security with smaller volumes of the EDS file. The monograph is intended for those who understands algebraical principles of the RSA. We'll not concern special parts of algebra that demand profound knowledge of higher mathematic. We are not studying here higher algebra and theory of the information security. Our task is to enlarge skill in the Pascal ABC but in very handsome application of the language.

Mathematical principles

We are to solve two problems when deal with the EDS.

Problem 1.

How to find a hash-function which can be a pattern of the document (or its digest, its compressed reflection).

A hash-function is any function $y = h(x_1, x_2, \ldots, x_n)$ that sets an integer number "n" of fixed length in accordance with a vector $x_1, x_2, \ldots, x_n$ of arbitrary length. Many hash-functions satisfied the condition but not every may be called cryptographical being able to correspond supplement demands:

- $h(x)$ is to be computed quick enough and easily;

- restoration of message $x(y)$ is to be practically impossible even if value "y" is known;

- if "x" is known, it would be difficult to find another message $x' \neq x$, such that $h(x) = h(x')$;

- it would be difficult to find any pair of different messages x' and x, for which h(x) = h(x').

Problem 2.

Develop pair of mutual supplementary keys (secret key "sk" and open key "opk") and compute number "ds" which is fit as a digit signature to hash-function "hf".

$$ds = (hf)^{sk} \bmod N$$

Number pair (opk, N) and a message are to be transmitted through an undefended channel to an abonent, the message being ciphered using open key. The abonent is to calculate hash-function "w". If w = y, it means that the signature is true.

Example 11.21.

Transmit the enciphered message together with its hash-function hf and the signature ds using the open channel.

 Message = camomiles_are_torned

Remark.

You ought not to laugh at hash-function and the signature are not files but two-digit numbers only, since the Message is a very short sequence. They are quite enough to learn securty method and at the same time do not sophisticate the educational example.

Solution.

1. Program Prog_11_22 (listing 11.22) computes the hash-function (figure 111). The alphabet ordinal numbers of symbols are printed in the second column. The growing sum of these numbers is in the third column. Total sum is equal 244.

```pascal
Program Prog_11_22;
Uses ABCobjects;

var m, n, eu, sk, z, sgn, sum: integer;
var a, b, ps: array[0..60] of integer;
var send: string;

Procedure Hash(wrd: string);
var j, k, hf, key1, len, g, h, sum, x, xs: integer;
var pwr, wh, cod, s: array[0..15] of integer;
var s1, s2, s3, snd: string;
begin
  len:=Length(wrd); sum:=0; m:=231;
  writeln('   Symb   Code   GrowSum');
  writeln('-----------------------------------------------');
  for k:=1 to len do
    begin
      case wrd[k] of
        'a': ps[k]:=1;    'i': ps[k]:=9;     'q': ps[k]:=17;
        'y': ps[k]:=25;
        'b': ps[k]:=2;    'j': ps[k]:=10;    'r': ps[k]:=18;
        'z': ps[k]:=26;
        'c': ps[k]:=3;    'k': ps[k]:=11;    's': ps[k]:=19;
       '_': ps[k]:=27;
        'd': ps[k]:=4;    'l': ps[k]:=12;    't': ps[k]:=20;
        'e': ps[k]:=5;    'm': ps[k]:=13;    'u': ps[k]:=21;
        'f': ps[k]:=6;    'n': ps[k]:=14;    'v': ps[k]:=22;
        'g': ps[k]:=7;    'o': ps[k]:=15;    'w': ps[k]:=23;
        'h': ps[k]:=8;    'p': ps[k]:=16;    'x': ps[k]:=24;
      end;
    sum:=sum+ps[k];
    writeln('        ',wrd[k],'          ',ps[k],'              ',sum);
  end;
  writeln('-----------------------------------');
  hf:=sum mod m;
  writeln('   Sum = ',sum,'     M = ',m);
  writeln('   Hash = ',hf);
end;
//==========Main===========
 var s1, s2, s3, wrd: string;
 var hf: integer;
 begin
```

```
    wrd:='camomiles_are_torned';
    Hash(wrd);
end.
```

Listing 11.22

```
 ABC Objects

 Symb   Code   GrowSum
-----------------------------------
   c       3         3
   a       1         4
   m      13        17
   o      15        32
   m      13        45
   i       9        54
   l      12        66
   e       5        71
   s      19        90
          27       117
   a       1       118
   r      18       136
   e       5       141
          27       168
   t      20       188
   o      15       203
   r      18       221
   n      14       235
   e       5       240
   d       4       244
-----------------------------------
 Sum = 244      M = 231
 Hash = 13
```

Figure 111. Computing of the hash-function

The hash-function (i.e. message digest) w = 13. This number is a result of the operation hf = Sum mod N; N =231 is one of the EDS secrets.

2. Encipher the Message using RSA algorithm:

- take coprimes P = 5, Q = 11;

- calculate module M = P*Q = 55;

- calculate Euler function eu = (P - 1)*(Q - 1) = 40;

- select the open key opk = 3;

- compute the secret key sk = 27 (see the part 11.10).

3. Calculate the EDS:

$$ds = hf^{sk} \bmod N.$$

Two procedures BinCod and Power (listing 11.23 and fig.112) do this operation.

Program Prog_11_23;

Uses ABCobjects;
var z: integer;
var a, b: **array**[0..10] **of** integer;

Procedure BinCod(sk: integer);
var j, k: integer;
begin
  a[0]:=sk **div** 2; b[0]:=sk-2*a[0];
  writeln('------------------');
  writeln(' j   sk = ',sk);
  writeln('------------------');
  writeln(' 0','   ',b[0]);   j:=0;
  while (a[j]>0) **or** (b[j]>0) **do**
    begin
     j:=j+1;
     a[j]:=a[j-1] **div** 2;   b[j]:=a[j-1]-2*a[j];
     if (a[j]>0) **or** (b[j]>0) **then** writeln('   ',j,'   ',b[j]);
    end;
    z:=j-1;
end;

Procedure Power(hf, sk, n: integer);
var j, k, g, h, ds: integer;
var s: **array**[0..9] **of** integer;
  begin
    writeln('----------------------------------');
    writeln(' hf = ',hf);   writeln(' sk = ',sk);
    writeln(' ds = (hf ^ sk) mod ',n);
    s[0]:=hf;

```
    for j:=1 to z do
      begin
        if b[j]=1 then g:=hf else g:=1;
        h:=s[j-1]*s[j-1];    s[j]:=g*h mod n;
      end;
    ds:=s[j];
    writeln(' ds = ',ds);
  end;
//===== Main ======
var sk, n, hf: integer;
begin
  sk:=27;           BinCod(sk);
  hf:=13;  n:=55;   Power(hf,sk,n);
end.
```

Listing 11.23

```
-------------------------
 j   sk = 27
-------------------------
 0       1
 1       1
 2       0
 3       1
 4       1
-------------------------
 hf = 13
 sk = 27
 ds = (hf ^ sk)  mod 55
 ds = 7
```

Figure 112. Calculating of the digit signature (ds)

The procedure Coder (listing 11.24) does this work and shows the result (fig.113).

```
Program Prog_11_24;
Uses ABCobjects;
var z: integer;
var a, b: array[0..10] of integer;
var ps: array[0..30] of integer;
```

```pascal
Procedure Coder(wrd: string; hf,n,ds: integer);
var j, k, len, g, h, opk, sum, m, p, q: integer;
var cod, s, ps: array[0..60] of integer;
var snd, sc, s1,s 2, s3, s4,s5,s6: string;

begin
  len:=Length(wrd); m:=231;
  for k:=1 to len do
    begin
    case wrd[k] of
      'a': ps[k]:=1; 'i':  ps[k]:=9;     ' q': ps[k]:=17; 'y':  ps[k]:=25;
      'b': ps[k]:=2; 'j':  ps[k]:=10;    'r': ps[k]:=18; 'z':  ps[k]:=26;
      'c': ps[k]:=3; 'k':  ps[k]:=11;    's': ps[k]:=19; '_':  ps[k]:=27;
      'd': ps[k]:=4; 'l':  ps[k]:=12;    't': ps[k]:=20;
      'e': ps[k]:=5; 'm': ps[k]:=13;    'u': ps[k]:=21;
      'f': ps[k]:=6; 'n':  ps[k]:=14;    'v': ps[k]:=22;
      'g': ps[k]:=7; 'o':  ps[k]:=15;   'w': ps[k]:=23;
      'h': ps[k]:=8; 'p':  ps[k]:=16;   'x':  ps[k]:=24;
     end;
    sum:=sum+ps[k];
   end;
  hf:=sum mod m;
  writeln(' symb  s_num     cod');
  writeln('-----------------------------------');
  for k:=1 to len do
   begin
    s[0]:=ps[k]; opk:=3;  BinCod(opk);
    for j:=1 to z do
      begin
       if b[j]=1 then g:=ps[k] else g:=1;
       h:=s[j-1]*s[j-1];   s[j]:=g*h mod 55;
      end;
     cod[k]:=s[j];
     writeln('       ',wrd[k],'          ',ps[k],'              ',
     cod[k]);          sc:=IntToStr(cod[k]);
     if cod[k]<10 then snd:=snd+'0'+sc else snd:=snd+sc;
    end;
    writeln(' Send = ',snd);   s1:=IntToStr(opk);
    if opk>9 then snd:=snd+s1 else snd:=snd+'0'+s1;
    writeln(' Send = ',snd);
    s2:=IntToStr(hf);
```

```pascal
    if hf>9 then snd:=snd+s2 else snd:=snd+'0'+s2;
    writeln(' Send = ',snd);
    p:=5; s3:=IntToStr(p);
    if p>9 then snd:=snd+s3 else snd:=snd+'0'+s3;
    writeln(' Send = ',snd);
    q:=11; s4:=IntToStr(q);
    if q>9 then snd:=snd+s4 else snd:=snd+'0'+s4;
    writeln(' Send = ',snd);       s5:=IntToStr(n);
    if n>9 then snd:=snd+s5 else snd:=snd+'0'+s5;
    writeln(' Send = ',snd);       s6:=IntToStr(ds);
    if ds>9 then snd:=snd+s6 else snd:=snd+'0'+s6;
    writeln(' Send = ',snd);
end;
//================= Main ====================
var hf, len, sk, n, ds: integer;      var wrd: string;
begin
  wrd:='camomiles_are_torned';
  hf:=13; sk:=27; n:=55;  BinCod(sk);  Power(hf,sk,n);
  ds:=7;
  Coder(wrd,hf,n,ds);
end.
```

Listing 11.24

4. Do enciphering of the message. Print on the screen the message symbols (symb), their alphabet numbers (s_num) and the ciphered codes (cod).

5. Add the hash-function "hf", the digit signature "ds" and number N to to the message cipher (fig.113).

At last one must be shure the received document is authentical. One ought to do the next actions for this purpose:

1. Transform the text line into the sequence of two-digit decimal numbers.

2. Distribute the components of the message:

- a ciphered text line "rc";

- an open key "opk";

- data for computing of secret key "sk" (usually random numbers P and Q or their derivatives);

- a digit signature "ds".

3. Calculate hash-function of received document:

$$w = (ds)^{opk} \bmod N, \text{ where } N = P*Q$$

4. Compare received and computed values of hash-function and do conclusion (fig.114).

```
ABC Objects

symb  s_num   cod

  c      3      27
  a      1       1
  m     13      52
  o     15      20
  m     13      52
  i      9      14
  l     12      23
  e      5      15
  s     19      39
        27      48
  a      1       1
  r     18       2
  e      5      15
        27      48    Send = 27015220521423153948010215482520024915 09
  t     20      25    Send = 27015220521423153948010215482520024915 0903
  o     15      20    Send = 27015220521423153948010215482520024915090313
  r     18       2    Send = 270152205214231539480102154825200249150903 1305
  n     14      49    Send = 270152205214231539480102154825200249150903 130511
  e      5      15    Send = 2701522052142315394801021548252002491509031305 1155
  d      4       9    Send = 2701522052142315394801021548252002491509031305 115507
```

Figure 113. Cipher, hash, signature

Conclusion.

Coincidence w and hr (both values are equal 13) confirms the authenticity and the integrity of the received electronical document.

The program Prog_11_25 (listing 11.25) has realized the checking algorithm.

```
Program Prog_11_25;
Uses ABCobjects;
var m, n, hf, p, q, opk, sk, ds, z: integer;
var a, b, s: array[0..66] of integer;
var send:  string;

Procedure Decoder(rc: string);
var j, k, len, x, code: integer;
var sn: array[1..66] of integer;
var y, z, dd, s1, s2, s3, s4, s5, s6:  string;
begin
  writeln(' ');
  writeln('   rc = ',rc);
  len:=Length(rc);
  j:=0;
```

ABC Objects
rc = 27015220521423153948010215482520024915090313051107
j   opk = 3
0   1
1   1
ds = 7
opk = 3
w = (ds^opk) mod N
w = 13
opk = 3   P = 5   N = 55
hf = 13   Q = 11   ds = 7

Figure 114. Verification of the message

```
  for k:=1 to len do
    begin
      x:=(k mod 2);
      if x=1 then y:=rc[k];
      if x=0 then
        begin
          z:=rc[k];  j:=j+1;
          dd:=y+z;
```

```pascal
          Val(dd,s[j],code);
        end;
      if (k mod 2)=0 then
          begin
             s[k]:=s[j]; j:=Round(k/2);
          end;
    end;
   j:=40;
   while j<50 do
     begin
       j:=j+2;
     end;
   opk:=s[42];  s1:=IntToStr(opk);
   var t1:=new TextABC(170,40,12,' opk = '+s1);
   hf:=s[44];  s2:=IntToStr(hf);
   var t2:=new TextABC(170,65,12,' hf = '+s2);
   p:=s[46];  s3:=IntToStr(p);
   var t3:=new TextABC(250,40,12,' P = '+s3);
   q:=s[24];  s4:=IntToStr(q);
   var t4:=new TextABC(250,65,12,' Q = '+s4);
   n:=p*q;    s5:=IntToStr(n);
   var t5:=new TextABC(320,40,12,' N = '+s5);
   ds:=s[25];    s6:=IntToStr(ds);
   var t6:=new TextABC(320,65,12,' ds = '+s6);
end;

Procedure BinCod(opk: integer);
var j, k: integer;
begin
  a[0]:=opk div 2;  b[0]:=opk-2*a[0];
  writeln('-------------------');  writeln(' j     opk = ',opk);
  writeln('-------------------');  writeln('  0','        ',b[0]);
  j:=0;
  while (a[j]>0) or (b[j]>0) do
    begin
      j:=j+1;
      a[j]:=a[j-1] div 2;   b[j]:=a[j-1]-2*a[j];
      if (a[j]>0) or (b[j]>0) then writeln('  ',j,'        ',b[j]);
    end;
    z:=j-1;
end;
```

```pascal
Procedure Power(ds,opk,n: integer);
var j, k, g, h, w: integer;
var s: array[0..9] of integer;

begin
  writeln('-------------------------------');
  writeln(' ds = ',ds);  writeln(' opk = ',opk);
  writeln(' w = (ds^opk) mod  N');   s[0]:=ds;
  for j:=1 to z do
    begin
      if b[j]=1 then g:=ds else g:=1;
      h:=s[j-1]*s[j-1]; s[j]:=g*h mod n;
    end;
  w:=s[j];  writeln(' w = ',w);
  writeln('-------------------------------');
end;
//============= Main =================
var rc, s1, s2, s3, s4, s5, s6: string;
begin
    rc:='2701522052142315394801021548252002
        4915090313051107';
    Decoder(rc);  BinCod(opk);
    Power(ds,opk,n);
end.
```

Listing 11.25

CONCLUSION

The author attempted to collect in the monograph the main things which a beginner investigator ought to know and to be good at solution of problems that he would meet with.

The investigations seem to be the most interesting part of any profession where a specialist has firstly met a problem that nobody has not still tried to solve, on his opinion. However, he is needy some instruments and methods but the mathematics, he has been taught before, can not give him such possibility. It is the very chance when he can get the methods of the numerical analysis and its program realization using the language Pascal ABC due to this book.

This book continues an edition series devoted to the numerical analysis that is started not so long ago in Russia, near 2001 according author's observation. Numerical analysis was taught usually to students of the mathematical faculties but the most promoted specalists was showing keen interest still in 1950s. Several books were edited in 1980s.

This mathematical direction developed intensively in USA as it may be seen from the capital editions [6, 7, 8, 9]. Shortly after russian books has appeared in Moscow University [5] and in technical insitutes [10, 11]. In some time the teachers of the informatic has accumulated necessary experience in applying of the numerical methods [12, 13].

The author has edited the lection course "Numeric analysis methods and modelling" in 2011 [3]. The course is based on the problem solutions using instruments of the application Microsoft Excel and language Visual Basic for Applications. The book has been remaked and edited in 2017 [2]. The language Pascal ABC is used to solve the same and many new technical and economical problems. The current monograph is a version of above mentioned book in author's translation.

SUMMARY

1. Nowadays index of a professional skill involves using computer for experimental data processing, current analysis of business state, trend detection, preparation of tactical and strategical decisions.

2. A computer may be an effecitive instrument only if the user has aquired practical skill of the mathematical analysis. Educational institutions as a rule learn to use known formulae for analysis and calculations. In practice situation often arises where the exact mathematical descriptions prove to be inaccessible if classical methods are used. In such case one ought to apply approximation methods with well-grounded decision error. It means that numerical analysis methods are to be used which were being developed during previous 300 years.

3. The second condition of computer effectiveness is user's practical skill in programming. One ought to have habit to develop a program even if in one high level language. The Pascal is one of the best educational programming languge. The version Pascal ABC is popular now, at least in Russia.

4. Pascal ABC quality is estimated highly not only in the educational process. This monograph confirms its great possibilities in decision of many practical problems. There are:
- drawing and analysis of functional dependences;
- solution of trannscendetal equations;
- numerical differentiation and integration;
- solution of non-homogeneous differential equations;
- interpolation of grid functions;
- extrapolation of numerical sequences;
- filtering of numerical flows;

- dynamical modelling in mechanics, automatic control systems, electrical circuits, economics and so on;

- information security of communication channels.

5. Solution of all problems are confirmed in the monograph with the code listings. The listings are useful as inquiry materials. There are many examples for beginners concerning problems: data type transforms, number rounding, moving object drawing, combination text and numeric information in pictures and so on.

NAME INDEX

Adleman Leonard Max (born1945), USA; a comupter scientist, one of the creators of the RSA encryption algorithm. He is also known for the creation of the field of DNA computer. Subchapter 11.11

Archimedes of Siracuses (287 – 212 B.C.), Greece; mathematician, physicist, engineer, inventor. He proved a range of geometrical theorems.
Subchapter 1.2

Bessel Friedrich Wilhelm (1784 – 1846), Germany; astronomer, mathe-matician, physicist, geodesist. He generalized a special type of functions, named Bessel functions (after his death).   Introduction

Cartesius Renatus (1596-1650), France; philosopher, mathematician, physicist, mechancc; founder of the analytical geometry.
Subchapters 1.2, 3.2

Cauchy Augustin Louis (1789 - 1857), France; mathematician and physicist, a pioneer in the analytical function theory, differential equations, number theory, geometry, mathematical physics.   Introduction, Subchapter 5.1

Chebyshev Pafnutiy Lvovich (1821 - 1894), Russia; mathematician and mechanic; valuable contributions in the integral calculus, number theory, function interpolation and approximation.   Introduction, Subchapter 3.1

Domar Evsey David (1914 - 1997), USA; economist, co-author of the Domar-Harrod model. Subchapter 10.2

Euclides of Alexandria (near 300 B.C.), Greece; hystorian and mathematician; the main work "The Elements"; the creator

of Euclidian geometry; the contributions in the stereometry, number theory. Subchapters 10.3, 11.10

Euler Leonard (1707 - 1783), Swiss; mathematician; founder of
the scientific schools in Germany and Russia; works in the
stereometry, analysis of infinitesimal, differential and variation calculus, mathematical physics, optics, sky mechanics. Introduction, Subchapters 5.1, 11.11

Fermat Pierre (1601 - 1665), France; one of the founders of analytical geometry, number theory, mathematical analysis,
probability theory.
Introduction, Subchapter 1.4

Gauss Carl Friedrich (1777 - 1855), Germany; contributed in the
number theory, algebra, statistics, analysis, differential
geometry, probability theory, matrix theory, geodesy, ge-
ophisics, mechanics, electrostatics, magnetic fields, optics, astronomy. Introduction, Subchapters 3.1, 7.4

Harrod Roy Forbs (1900 - 1978), USA; economist; author of "In-
ter-national Economics", former standard textbook; co-
author of Domar-Harrod model.   Subchapter 10.2

Kantorovich Leonid Vitalyevich (1912 - 1986), Russia; mathematician and economist; known for his theory and development of techniques for the optimal allocation of resources; founder of linear programming; winner of Nobel
Memorial Prize in 1975.   Introduction

Keynes Jhon Maynard (1883 - 1946), Great Britain; economist,
investigations in the macroecomicm, economical cycles
theory (recessions and depressions).   Subchapter 10.2

Kolmogorov Andrey Nikolaevich (1903 - 1987), Russia; mathematician; significant contributions to the mathematics of
probability theory, topology, intuistionic logic, turbulence,

classical mechanics, algorithmical information theory and computational complexity. Introduction

Kutta Martin Wilhelm (1867 - 1944), Germany; co-author of Runge-Kutta numerical method of differential equation decision. Subchapter 4.2

Lagrange Josef Louis (1736 - 1813), Italy, France; contributions in the mathematical anaysis, number theory, algebra, function interpolation, differential equations, variation calculus. Introduction, Subchapters 3.1, 4.1, 4.2

Laplace Pierre Simon (1749 - 1827), France; significant contributions in the mathematics, physics, astronomy, number theory, differential equations, mathematical physics, sky mechanics. Subchapter 5.3

Legendre Adrien-Marie (1752 - 1833), France; mathematician; developed concepts of polynomials and function transformations, named after Legendre; significant works in the number theory, elliptical integrals, geometry, geodesy. Subchapters 4.2, 7.4

Leibnitz Gotfrid Wilhelm (1646 - 1716), Germany; contributions in philosophy, hystory, mathematical logics, combinatorics, analytical mathematics, physics, mechanics. Introduction, Subchapter 3.1

Leontyev Vasiliy Vasilyevich (1905 - 1999), Russia, USA; economist; works in enomical mathematics, econometrics. Winner of Nobel Memorial Prize in 1973. Subchapter 10.2

Mendeleev Dmitriy Ivanovich (1834 - 1907), Russia; encylopaedist; author of the Periodic table of elements; significant works in chemistry, physics, metrology, economics, geology, oil-refining, aeronautics. Subchapter 7.1

Newton Isaac (1643 - 1727), Great Britain; mathemtician, astronomer, physicist, theologian; he has opened the world gravity law; contributions in functional analysis, decision of the transcendent equa-tions, interpolation, differential and integral calculus. Introduction, Subchapters, Subchapters 1.3, 3.1, 4.1

Pascal Blaise (1623 - 1662), France; mathematician, physicist, french classical writer, catholic theologian, inventor of mechanical calculators; mathematical works in the analysis, probability theory, projective geometry.   Introduction, Subchapters 1.2, 6.1, Summary

Poisson Simeon Denis (1781 - 184), France; mathematician, physicist, mechanic, engineer. Famous contributions in the probability theory, mathematical phycics, sky mechanics. Introduction, Subchapter 6.1

Polibius (200 – 120 B.C.), Greece; historian; author of General History in 40 volumes (period of Rome Republic rising). Subchapter 11.3

Pontryagin Lev Semyonovich (1908 - 1988), Russia; mathematician; works in the topology, continious group theory, differential equations, differential games.   Introduction

Rivest Ronald Linn (born 1947), USA; cryptographer, one of the author of enciphering algotrithm RSA. Subchapter 11.11

Runge Karl David Tolme (1856- 1927), Germany; mathematician and physicist; co-author of Runge-Kutta numerical method of diffe-rential equation decision. Subchapter 85

Shamir Adi (born 1952), Israel; cryptographer, one of the author of enciphering algotrithm RSA; co-inventor of the Feige-Fiat-Shamir identification scheme. Subchapter 11.11

Shannon Claud Eldwood (1916 - 2001), USA; founder of the information theory; famous works in the stohastic system theory, finite automates theory, automatic control theory. Introduction

Simpson Thomas (1710 - 1761), Great Britain; mathematician, works in the geometry, stereometry, trigonometry, probability theory, integral calculus. Introduction, Subchapter 3.1

Viary (marquise de Viary, the 19th centure), France; military officer, cryptographer, author of encrypting method (1888) using algebraical equations.   Subchapter 11.6

Vigenere Blaise (1523 - 1596), France; diplomat, translator, alchemist, cryptographer; author of the book "Traicte des Chiffres ou Secretes Manieres d'Escrire".   Subchapters 11.5, 11.6

Vitushkin Anatoliy Georgievich (1931 - 2004), Russia; mathematician; main works in the approximation theory, complex geometry, information theory; founder of moscow school of the complex analysis and approximation theory. Introduction

Wiener Norbert (1894 - 1964), USA; mathematician and philosopher; the originator of cybernetics, a formalization of the notion of feedback, with implications for engineering, systems control, computer science, biology, and the organization of society.   Introduction

Wirth Niklaus (born 1934), Swiss; computer scientist; developer of several programming languages, including Pascal; the pioneer in several topics in sofrware engineering; winner of several international prizes for developing a sequence of innovative computer languages.   Introduction

REFERENCES

1. Komlev N. Manual for self-tuition in Pascal ABC and Turbo.
   Moscow: "Solon-Press", 2013. – 257 p., in russian.
   ISBN 978-5-91359.112-8

2. Zhavoronkov L. Numerical analysis and modelling in Pascal ABC.
   Moscow: "Onto-Print", 2017. – 366 p., in russian.
   ISBN 978-5-906886-71-2

3. Zhavoronkov L. Numerical methods and digital models.
   Moscow: "MIBA", 2011. – 415 p., in russian.
   ISBN 978-5-94102-058-4

4. Zhavoronkov L., Lapenko V. Amplitude selector.
   Invention cerificate № 530444, 1976, USSR

5. Zaguskin V. Numerical methods of equation decision.
   Moscow: "State Publishing House. Phys.-Math. Literature",
   1960. – 216 p., in russian.

6. Jong M.T. Methods of discrete signal and system analysis. USA,
   McGraw-Hill Book Company, 1982, vol.1 – 160 p., vol.2 -
   309 p.

7. Kuo B.C. Digital control systems. USA, University of Illinois.
   New York: Holt, Reinehart and Winston,
   Moscow: "Macinery-building", 1986. – 448 p. , in russian

8. Mathews J.H., Fink K.D. Numerical methods using MATH-
   LAB. USA, Prentice Hall, 1999. Moscow: "Williams", 2001
   – 720 p., in russian
   ISBN 0-13-270042-5 (in eng)
   ISBN 5-8459-0162-6 (in rus)

9. Huang T.S. (editor), Eclund J.O., Nussbaumer G.J.and others.

Quick algorithms in picture digital processing. USA, 1980.
Moscow: "Radio and Communication", 1984. – 224 p., in russian

10. Lapchik M.P., Ragulina M.I., Henner E.K. Numerical methods.

Moscow: "Information Center Academy", 2007 – 384 p., in russian
ISBN 978-5-7695-4016-5

11. Kireev V.I., Panteleev A.V. Numerical methods. Decision examples.

Moscow: "Higher School", 2008. – 480 p., in russian
ISBN 5-06-004763-6

12. Koldaev V.D. Numerical methods and programming.

Moscow: "Forum", 2009, 2013. – 336 p., in russian
ISBN 978-5-8199-0333-9

13. Garber G.Z. Numericfl methods and programming in VBA Excel.

Moscow: "Print Com", 2009., - 432 p., in russian
ISBN 978-5-98585-015-4

14. Babash A.V, Baranova E.K., Melnikov G.N.Information security.

Moscow: "KNORUS", 2012. – 136 p., in russin
ISBN 978-5-406-01170-6

15. Barichev S.G., Goncharov V.V., Serov R.E.

Modern criptography principles
Moscow: "Hot Line – Telecom", 2011. – 175 p., in russian.
ISBN 978-5-9912-0182-7

APPLICATION

№№	File	Content	Sub-chapter
1	Prog_1_1	Parabola family	1.1
2	Prog_1_2	Archimedes spiral	1.2
3	Prog_1_3	Pascal snail	1.2
4	Prog_1_4	Epicycloid	1.2
5	Prog_1_5	Hypocycloid	1.2
6	Prog_1_6	Algebraic equation system	1.3
7	Prog_1_7	Algebraic equation	1.3
8	Prog_1_8	Transcendental equation	1.3
9	Prog_1_9	Extreme points 1	1.4
10	Prog_1_10	Extreme points 2	1.4
11	Prog_2_1	The first derivative	2.2
12	Prog_2_2	The 1st and the 2nd derivatives	2.2
13	Prog_2_3	Numerical differentiaton	2.2
14	Prog_3_1	Pulse square	3.1
15	Prog_3_2	Vase volume	3.2
16	Prog_3_3	Liquid level	3.2
17	Prog_4_1	Lagrange polynomial 1	4.2
18	Prog_4_2	Lagrange polynomial 2	4.2
19	Prog_5_1	1st-order differential equation	5.2
20	Prog_5_2	2nd-order differential equation	5.3
21	Prog_6_1	Equal-weight filter	6.3
22	Prog_6_2	Exponential filter	6.4
23	Prog_6_3	Median filter	6.5
24	Prog_7_1	Zero-order extrapolation	7.2
25	Prog_7_2	1st-order extrapolation	7.2
26	Prog_7_3	2nd-order extrapolation	7.2
27	Prog_7_4	Linear trend	7.4
28	Prog_7_5	Exponential trend	7.5
29	Prog_7_6	Parabolic trend	7.5
30	Prog_8_1	Pendulum oscillations	8.2
31	Prog_8_2	Pendulum swings	8.2
32	Prog_8_3	Steam-engine	8.3
33	Prog_8_4	Basketball trajectories	8.4
34	Prog_8_5	Automatic control system	8.5

№№	File	Content	Sub-chapter
35	Prog_8_6	Inertial link 1	8.6
36	Prog_8_7	Inertial link 2	8.6
37	Prog_8_8	Inertial link 3	8.6
38	Prog_8_9	Differential link 1	8.7
39	Prog_8_10	Differential link 2	8.7
40	Prog_8_11	Differential link 3	8.7
41	Prog_8_12	Resonance link 1	8.8
42	Prog_8_13	Resonance link 2	8.8
43	Prog_8_14	Amplitude selector	8.9
44	Prog_9_1	Direct pursuit	9.2
45	Prog_9_2	Proportional approach	9.3
46	Prog_9_3	Beam approach	9.4
47	Prog_10_1	Web model 1	10.1
48	Prog_10_2	Web model 2	10.1
49	Prog_10_3	Web model 3	10.1
50	Prog_10_4	Domar-Harrod model 1	10.2
51	Prog_10_5	Domar-Harrod model 2	10.2
52	Prog_10_6	Centroid 1	10.3
53	Prog_10_7	Centroid 2	10.3
54	Prog_10_8	Clusters	10.4
55	Prog_11_1	Ceasar's algorithm 1	11.2
56	Prog_11_2	Ceasar's algorithm 2	11.2
57	Prog_11_3	Polibius square 1	11.3
58	Prog_11_4	Polibius square 2	11.3
59	Prog_11_5	Inverse symbols 1	11.4
60	Prog_11_6	Inverse symbols 2	11.4
61	Prog_11_7	Vigenere table 1	11.5
62	Prog_11_8	Vigenere table 2	11.5
63	Prog_11_9	Viary code 1	11.6
64	Prog_11_10	Viary code 2	11.6
65	Prog_11_11	Matrix enciphering	11.7
66	Prog_11_12	Matrix deciphering	11.7
67	Prog_11_13	Force attack 1	11.8
68	Prog_11_14	Force attack 2	11.8
69	Prog_11_15	English language statistic	11.9
70	Prog_11_16	Italian language statistic	11.9

№№	№№	№№	Sub-chapter
71	Prog_11_17	Great common divisor	11.10
72	Prog_11_18	Mod degree	11.10
73	Prog_11_19	Mod inversion	11.10
74	Prog_11_20	RSA 1	11.11
75	Prog_11_21	RSA 2	11.11
76	Prog_11_22	Hash-function	11.12
77	Prog_11_23	Mod degree	11.12
78	Prog_11_24	Signature	11.12
79	Prog_11_25	Verification	11.12

Scientific edition

Leonid Zhavoronkov
Numerical analysis in Pascal ABC

Monograph

Moscow, 2019